AF862248

CONCISE ENCYCLOPEDIA OF SELF-PROPAGATING HIGH-TEMPERATURE SYNTHESIS

CONCISE ENCYCLOPEDIA OF SELF-PROPAGATING HIGH-TEMPERATURE SYNTHESIS

History, Theory, Technology, and Products

Edited by

INNA P. BOROVINSKAYA

ALEXANDER A. GROMOV

EVGENY A. LEVASHOV

YURI M. MAKSIMOV

ALEXANDER S. MUKASYAN

ALEXANDER S. ROGACHEV

ELSEVIER

Elsevier
Radarweg 29, PO Box 211, 1000 AE Amsterdam, Netherlands
The Boulevard, Langford Lane, Kidlington, Oxford OX5 1GB, United Kingdom
50 Hampshire Street, 5th Floor, Cambridge, MA 02139, United States

Library of Congress Cataloging-in-Publication Data
A catalog record for this book is available from the Library of Congress

British Library Cataloguing-in-Publication Data
A catalogue record for this book is available from the British Library

ISBN: 978-0-12-804173-4

Publisher: John Fedor
Acquisition Editor: Anita Koch
Editorial Project Manager: Amy Clark
Production Project Manager: Vijayaraj Purushothaman
Cover Designer: Alan Studholme

Typeset by SPi Global, India

This book is dedicated to the memory of SHS father-founder Alexander G. Merzhanov (1931–2013).

CONTENTS

CONTRIBUTORS

Roza G. Abdulkarimova

Iñigo Agote

Anahit G. Aleksanyan

Murat Alkan

Mikhail I. Alymov

Alexander P. Amosov

Singanahally T. Aruna

Vladimir B. Balashov

Florence Baras

Tatiana V. Barinova

Frédéric Bernard

Georgy V. Bichurov

Inna P. Borovinskaya

Vyacheslav N. Borshch

Boris Sh. Braverman

Giacomo Cao

Ludmila N. Chukhlomina

Seda K. Dolukhanyan

Sergey M. Fomenko

Zhengyi Fu

Ramil M. Gabbasov

Evgeny V. Golosov

Irena Gotman

Alexander A. Gromov

Manuel Gutiérrez

Elasar Y. Gutmanas

Tatyana I. Ignatieva

Hyong Il Won

Volya I. Itin

Mikhail B. Ivanov

Oksana V. Ivanova

Nikolay G. Kasatsky

Tamara A. Khabas

Suren L. Kharatyan

Boris B. Khina

Alexander I. Kirdyashkin

Evgeny G. Klimchuk

Yury R. Kolobov

Sergey V. Konovalihin

Igor' Yu. Konyashin

Dmitry Y. Kovalev

Dmitry Yu. Kovalev

Peter M. Krishenik

Georgiy I. Ksandopulo

Alexander E. Kudryashov

Mikhail M. Kulak

Victoria V. Kurbatkina

Miguel A. Lagos

Oleg V. Lapshin

Galina V. Lavrenchuk

Jong Hyeon Lee

Olga K. Lepakova

Evgeny A. Levashov

Jerzy Lis

Yury M. Maksimov

Zulkhair A. Mansurov

Khachatur V. Manukyan

Karen S. Martirosyan

Alexander G. Merzhanov

Roman V. Minin

Nina N. Mofa

Alexander S. Mukasyan

Yury S. Nayborodenko

Hayk Nersisyan, Sr.

Osamu Odawara

Giorgi Onashvilli

Roberto Orrù

Evgeny I. Patsera

Joshua M. Pauls

Yury S. Pogozhev

Vasiliy I. Ponomarev

Artem Yu. Potanin

Valentina K. Prokudina

Jan A. Puszynski

Nina I. Radishevskaya

Larisa G. Raskolenko

Miguel A. Rodríguez

Alexander S. Rogachev

Vitaly G. Salamatov

Vitaliy G. Salamatov

Vladimir N. Sanin

Ara Sargsyan

Yury B. Scheck

Anatoly A. Shiriev

Konstantin G. Shkadinsky

Dmitry V. Shtansky

Christopher E. Shuck

Konstantin L. Smirnov

Alexander M. Stolin

Alexander E. Sytschev

Laszlo Takacs

Viktor F. Tarasenko

Boris V. Trifonov

Ahmet Turan

Valery I. Uvarov

Vladimir I. Vereshchagin

Vladimir I. Vershinnikov

Timothy P. Weihs

Chang Whan Won

Galina G. Xanthopoulou

Onuralp Yücel

Vladimir I. Yukhvid

Vladimir V. Zakorzhevsky

Sergey A. Zelepugin

PREFACE

Combustion is an ancient, ubiquitous phenomenon that has a variety of applications and that persists in astonishing us by its continuous expansion into new fields of science and technology. Combustion was initially used only as a source for heating homes and cooking food. Currently, combustion is found in car and rocket engines, metallurgy and catalysis, waste management, and material synthesis. Safety issues, including those related to the destructive nature of uncontrolled fires, also require a real understanding of the combustion process.

This *Encyclopedia* includes articles prepared by more than 300 scientists and teams from 20 different countries that provide an overview of many aspects of the constructive application of the combustion phenomenon, for example, in the fabrication of advanced materials.

Regular studies in this direction began 50 years ago in 1967, when a group of researchers headed by Professor Alexander Merzhanov at the Academy of Science of USSR discovered a specific autowave, gasless combustion regime, which was later referred to as "solid flame."

It is worth noting that several isolated attempts were made earlier to produce solid materials by heterogeneous combustion. However, those efforts did not lead to a method of controllable synthesis of materials. In turn, the "solid flame" phenomenon not only stimulated the advance of fundamental combustion science but also resulted in the development of a novel, effective method for material synthesis, self-propagating high-temperature synthesis (SHS), and combustion synthesis.

Today SHS-based technologies provide valuable alterations to traditional methods of materials fabrication, such as powder metallurgy, conventional and force sintering, casting, extrusion, hot-isostatic pressure sintering, and others. Intensive and creative work by scientists from all over the world has established a multitude of unique diagnostics for the investigation of SHS processes. The peculiarities and mechanisms of combustion and structure formation were investigated and established for a variety of reactive systems, which allows synthesis of different compounds, including refractory carbides, borides, nitrides, and oxides. Two- and three-dimensional SHS models were developed that adequately describe experimentally observed effects and relationships. New materials have been synthesized, including composite and functionally graded materials, as well as 1D and 2D nanocrystals with unique properties. A new field of science that combines material science and combustion, that is, structural macrokinetics, has been established.

In this *Encyclopedia*, we collected brief articles on all the previously mentioned aspects of SHS process. This book may be of interest not only to scientists working in SHS field but also to researchers in multidisciplinary fields such as chemical engineering, metallurgy, material science, combustion, explosion, and chemistry of solids. This work is the first attempt of its kind in the SHS field and aims to capture the broad scope of this field rather than provide an in-depth representation of the state of the art, which has been accomplished in many recently published books and reviews.

We acknowledge Mr. J. Pauls, Mr. C. Shuck, Ms. O. Ivanova, and S. Turlo for their help in the preparation of this publication.

The authors will gratefully accept any comments and suggestions submitted to gromov@tpu.ru.

Sincerely yours,
Editors

Abbreviations

Alexander S. Mukasyan

CS	combustion synthesis
DTA	differential thermal analysis
EDS	energy dispersive analysis
ETE	electro thermal explosion
FC	filtration combustion
FESEM	field emission scanning electron microscopy
HEBM	high energy ball milling
HSMVR	high speed micro video recording
IALC	impregnated active layer combustion
IISC	impregnated inert support combustion
MA	mechanical activation
MASHS	mechanically activated self-propagating high-temperature synthesis
RMLF	reactive multilayer foil
SEM	scanning electron microscopy
SCS	solution combustion synthesis
SHS	self-propagating high-temperature synthesis
SRW	scintillation reaction wave
SPS	spark plasma sintering
STIM	synthetic instrumental material
TE	thermal explosion
TEM	transmission electron microscopy
TGA	thermal gravimetric analysis
TRXRD	time-resolved X-ray diffraction
VCS	volume combustion synthesis
XRD	X-ray diffraction

Concise Encyclopedia of Self-Propagating High-Temperature Synthesis
http://dx.doi.org/10.1016/B978-0-12-804173-4.00001-6

Abrasives

Valentina K. Prokudina

Titanium carbide, synthesized from elements in SHS reactors, was the first SHS abrasive material. This application was made possible because of the high abrasive ability of SHS TiC, which exceeds the abrasive ability of TiC obtained by other methods (carbo-thermic reduction and quasimelting) by 1.5–2 times.

Electron microscopy investigations of the surface of TiC powders, obtained by various routes, showed that powders of SHS TiC possessed a faceted shape with sharp cutting edges, while particles of reduced TiC were characterized by crumbly uncut surfaces without sharp edges [1].

On the basis of SHS TiC, grinding pastes KT and KTIOL were developed and produced commercially beginning in 1976. TiC powders of various particles sizes, binders, and surfactants served as components of these pastes. Industrial application of these pastes had a great economic effect because of the expelling of expensive diamond-based pastes.

Agglomerated TiC powders used in pastes consist of discrete but bonded grains. This structure gives the pastes some peculiar properties.

These pastes possess the unique ability to simultaneously grind and polish the machined surface of the processed material because of the unusual effect of self-destruction of particle agglomerates during processing. Additional tiny and sharp peaks, formed after the grain destruction, actively participate in the micro-cutting processes. Conventional abrasives such as chromia and alumina do not possess similar properties.

Table 1 Results of grinding brass specimens with KT paste

Paste's grain content	TiC powder concentration (w%)	Surface roughness (μm)	Grinding depth (μm)
160/125	70	–	200
100/80	60	–	140
63/50	50	–	80
60/40	50	0.25	47
28/20	40	0.16	32
14/10	35	0.10	20
7/5	30	0.06	12
3/2	25	0.04	7
1/0	20	0.03	3

Concise Encyclopedia of Self-Propagating High-Temperature Synthesis
http://dx.doi.org/10.1016/B978-0-12-804173-4.00002-8

The peculiar property mentioned is illustrated by the results produced when grinding brass gauge specimens. Table 1 shows the results [2].

When the synthesis was performed in an open pit instead of a reactor, more harsh grains were obtained (up to 1000 μm, while the largest grains obtained within reactors were less than 630 μm in diameter). All structural and mechanical properties remained unchanged [3].

SHS powders of compositional and magnetic-abrasive materials such as $TiB_2 + Al_2O_3$ and $B_4C + Al_2O_3$ also proved to be good prospects for abrasive applications. These materials are obtained within the SHS-reactor, using a mixture of oxides, metals, carbon, and iron-containing components as raw materials.

These abrasive materials have a twofold change in the microhardness of the individual particles because of the combination of a hard phase and a binder within each particle.

These materials may be characterized as materials with fixed abrasive ability. In Fig. 1, a comparison between abrasive ability of these abrasives with commercial synthetic abrasive material, white electrocorundum is shown.

Other technological branches of SHS, such as SHS-sintering, SHS-casting, and SHS-compaction, are also able to produce abrasives of simple and complex compounds with outstanding properties and a variety of applications. $Al_2O_3 + ZrO_2 + TiB_2$ oxide powders, obtained by SHS-sintering in an open pallet, are characterized by high uniformity of strength distribution and stability of mechanical properties throughout the entire operation.

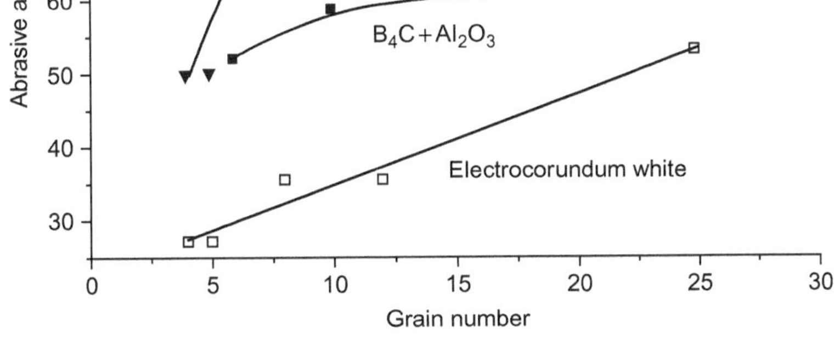

Fig. 1 Abrasive ability of SHS reactor composites in comparison with white electrocorundum.

Morphologically, this effect is manifested through smoother topography of the particles surface.

These properties of SHS oxide ceramics were applied most effectively in the final machining/processing of glass. The most conventional abrasive, obtained by the SHS-casting, is chromia-doped corundum (pink corundum). It is obtained as a coproduct of chromium carbide SHS-casting [4]. Pink corundum is used as a precursor for the preparation of abrasive circles, bars, and other grinding instruments.

Comparative studies on the durability of tools, made from SHS pink corundum and electrocorundum proved the preponderance of SHS abrasives.

In cases of abrasive circles (based on ceramic bindings) and bars (based on bakelite and vulcanite bindings) durability was 1.5–2 times higher; in the case of cutting circles, durability was 2.0–2.5 times higher. Additionally, the performance of machining increased 1.4–1.8 times, and the cost decreased by 20%.

SHS abrasives, obtained by SHS dynamic compaction with consequent milling, were marked by higher particle strength compared to other SHS abrasives.

Materials with identical composition, obtained by SHS dynamic compaction and SHS in a reactor, possess different mechanical properties—compacted material have relatively higher particle strength; this difference increases with the growth of particles.

As shown in Fig. 2, the highest particle strength (up to 343 MPa) was obtained in the case of SHS-compacted alloy STIM-3B (synthetic hard ally for tools—titanium carboboride with Ni binding) with the grain fraction 80 (1000/800 μm). This is still not

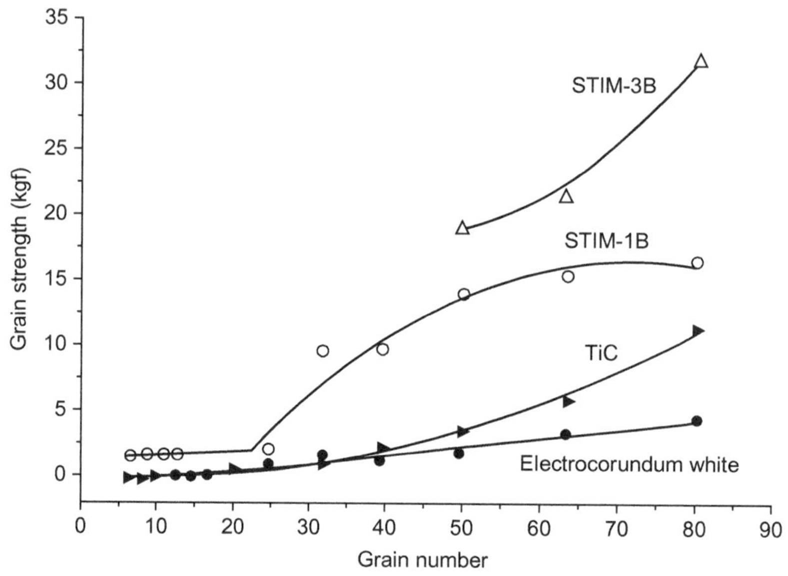

Fig. 2 Relation between grain strength and diameter in cases of SHS abrasives and white electrocorundum.

the apotheosis of properties of SHS-compacted abrasives. The materials known as SAPS (synthetic abrasive powder alloy with titanium carbide base) have a particle strength equal to 2157 MPa for 3500/3000 μm grains.

A series of abrasive materials were synthesized through SHS dynamic compaction.

These materials differ according to the composition of bindings (Fe-W, Fe-Cr, Ni-Cr-W), grain diameter (from 10 to 5000 μm) and properties, so they have the potential for use in a variety of abrasive tools (grains, circles, heads, and rods) and related applications from abrasive machining to final grinding (harsh grinding, dross, and spew removing) [5]. The presence of iron in alloys provides magnetic properties and makes magnetic separation of wastes after grinding of nickel and nonferrous alloys possible. Grains of these alloys possess isometric form and do not require further spheroidization. Because the circles are electroconductive, grains within the circles can be oriented in magnetic fields.

In comparison with foreign analogs, circles made by these alloys show an increase in grinding ratio by ~3 times. These materials attract additional interest because the raw materials for their production come from the waste products produced by grinding and cutting metals and hard alloys.

REFERENCES

[1] Merzhanov AG, Karyuk GG, Borovinskaya IP, Sharivker SYu, Moshkovskii EI, Prokudina VK, et al. Powder Metall (Poroshkovaya metallurgia) 1981;10:50–5 [in Russian].

[2] Amosov AP, Borovinskaya IP, Merzhanov AG. Powder technology of SHS-materials. Moscow: Mashinostroenie; 2007. p. 567 [Book, in Russian].

[3] Merzhanov AG, Borovinskaya IP, Prokudina VK, Pesotskaya NS, Nasonova MA. Nauka-Proizvodstvy 1998;8(10):5–12 [in Russian].

[4] Silyakov SL, Pesotskaya NS, Yukhvid VI. Int J SHS 1994;3(1):99–107.

[5] Samarin AN, Klyuev MM, Pityulin AN, et al. Aviacionnaya promyshlennost 1986;(3):22–4 [in Russian].

Acoustic Effects During SHS

Yury M. Maksimov, Ramil M. Gabbasov, Boris B. Khina, Evgeny A. Levashov

The effect of capillary titanium spreading was found on carbon grains during the synthesis of titanium carbide [1]. The combustion rate was weakly dependent on the temperature in the Ti-C system with coarse titanium particles ($d_{Ti} = 1100$ μm) [2], which indicated capillary combustion mode in this case. On the contrary, the combustion rate was strongly dependent on the initial temperature and the kinetic combustion mode for the fine titanium particles ($d_{Ti} = 100$ μm). Detailed studies of this phenomenon were published in [3]. The samples of Ti-C mixes with various densities, with a titanium particle size of 1100 μm and 100 μm, was burned under argon flow. High-frequency vibrations of the magnetostrictive transducer with a capacity of 4 kW and a frequency of 18 kHz were applied to the mixture. The initial powder Ti-C mixture was pressed into a steel crucible which was connected with the high-frequency vibrations emitter. The maximum increase in the combustion rate (factor of 2.7) of the mixtures was observed with the large titanium particles, which was explained by the effect of capillary spreading of liquid titanium in the ultrasonic field. The increase in the combustion rate was not observed for the mixtures with fine titanium particles. In this case, the combustion mode was close to the diffusive one (Fig. 1).

The combustion rate of the Ti-C-Ni and Ti-C-Ni-Mo systems, including the low-melting inert component, was studied versus the parameters of the ultrasonic field in [4].

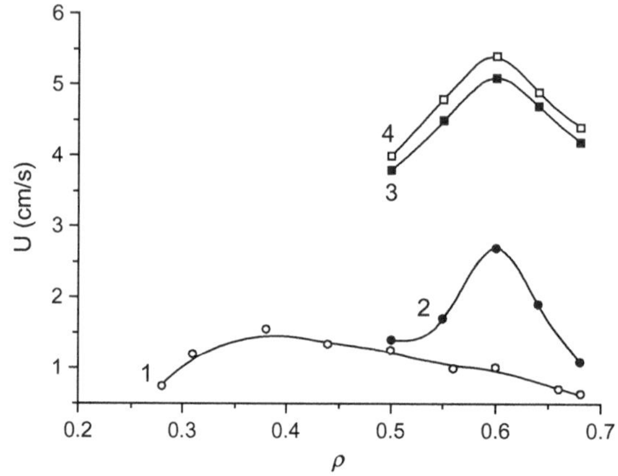

Fig. 1 Combustion rate of the Ti + 2B mixture versus initial density without ultrasound (1, 4) and with ultrasound (2, 3) action: 1, 2—$d_{Ti} \sim 1100$ μm; 3, 4—$d_{Ti} \sim 100$ μm.

Concise Encyclopedia of Self-Propagating High-Temperature Synthesis
http://dx.doi.org/10.1016/B978-0-12-804173-4.00003-X

Fine particles of titanium ($d_{Ti} < 50\ \mu m$), nickel ($d_{Ti} < 10\ \mu m$), and molybdenum ($d_{Ti} < 10\ \mu m$) were used in reaction. The paper presented a model in which spreading of the nickel melt over the surface of carbon grains led to blocking of the reaction surface (limiting the wetting angle of liquid Ni/C $\theta < 90$ degrees).

Adding a copper melt to the initial mixture with the limiting wetting angle $\theta < 90$ degrees enhanced the effect of the ultrasonic field on the combustion rate. Fine particles of reagents increased the homogeneity of combustion products, milled the grains of carbide, and enhanced physical and mechanical characteristics of the material. The lower the wettability of a high-melting reagent by the low-melting component, the stronger the effect of the ultrasonic field on the combustion rate [4]. In addition to these examples, the acoustic field had a favorable effect on SHS-compaction of hard alloys, and the properties of the STIM alloys improved the structure and properties of SHS welding, significantly influencing the superconducting properties during synthesis of superconductors [5,6]. Thus the acoustic vibrations influenced the kinetics of the SHS process, the structure, and the properties of reaction products.

The phenomena of acoustic emission during SHS were less studied. The acoustic vibrations during SHS were first reported in [7]. The experiments showed that the reaction systems (Ni-Al, Ti-B, Mo-B, et al.) generated the acoustic vibrations with the frequency from 5 Hz to 1.1 MHz with a pulse power up to 17 W during SHS. Combustion of different systems was characterized by the individual set of dynamic parameters for acoustic emission in the modes of low-frequency discrete pulses and high-frequency self-induced vibrations.

The spatial acoustic emission area was shown to be localized near the combustion wave. The possible reasons for the effect were the heat and chemical self-induced vibrations of the combustion wave, the process of formation and development of cracks in the sample, and the process of phase and chemical transformations of the medium. However, until recently this effect has been poorly studied and requires further research.

REFERENCES

[1] Shkiro VM, Borovinskaya IP. Capillary spreading of liquid metal during the combustion of titanium and carbon. Combust Explos Shock Waves 1976;12(6):945.

[2] Nekrasov EA, Maksimov YuM, Ziyatdinov MKh, Steinberg AS. Influence of capillary spreading on propagation of combustion wave in gas-free systems. Combust Explos Shock Waves 1978;14(5):26–32.

[3] Kirdyashkin AI, Maksimov YuM, Merzhanov AG. Effects of capillary flow on combustion in a gas-free system. Combust Explos Shock Waves 1981;17:591–6.

[4] Andreev VA, Levashov EA, Maltsev VM, Khavsky NN. Combustion of multicomponent systems in ultrasonic field. Combust Explos Shock Waves 1987;23(6):65–9.

[5] Levashov EA, Merzhanov AG, Khavsky NN. Prospects for the use of acoustic fields in SHS. J Eng Phys 1993;65(4):503–6.

[6] Klubovich VV, Kulak MM, Khina BB. Ultrasound in the process of self-propagating high-temperature synthesis. BNTU: Minsk; 2006. 208p.

[7] Kirdyashkin AI, Gabbassov RM, Maksimov YuM, Salamatov VG. Acoustic emission during self-propagating high-temperature synthesis. Combust Explos Shock Waves 2013;49(6):62–7.

Aerospace Applications

Alexander P. Amosov

The first studies of SHS application in space began in 1990 and were aimed at studying SHS regularities by conducting experiments under microgravity conditions [1]. To simulate microgravity, the researchers used drop tower methods with a duration of 2–10 s of microgravity, the parabolic flight of aircraft (18–30 s), and the flight of sounding rockets (up to 20 min). In studies of SHS system Ti-C-borax, B_2O_3-C-Al, samples with unusually high open porosity and greatly increased length compared to samples obtained under terrestrial conditions with gravity of 1 g were obtained. Investigations of SHS systems Ni-Al-B, Zr-Al-Fe_2O_3, Ga-P, In-P, and Ga-In-P showed a significant grain refinement of SHS products under microgravity conditions. In general, the results of the first studies also with Zn-S, Ti-B, Ni-Al, Ni-Al-Ti-B, Ti-C-Ni-Mo, Ti-C-Ni, and Ti-Fe systems demonstrated the possibility of implementing SHS in microgravity and that the rate of combustion, phase, and chemical composition of the SHS products are practically independent of the magnitude of gravity, but that this magnitude may affect the structure of SHS products if a liquid phase occurs in the process of combustion.

A new period in the study of the process and products of SHS in space began in 1997 under the leadership of A. G. Merzhanov by conducting experiments aboard the Russian MIR Space Station with almost unlimited duration of microgravity [1]. The latter was very important, as the duration of the structure formation of SHS products can substantially exceed the duration of the microgravity in the methods of dropping tower and parabolic flight. The use of powder mixtures with gasifying additives, of which the combustion forms an intermediate layer of molten reactants and the combustion proceeds with great gas emission (Ti + xC, Ti + 2B, Ti + xC + Ni + Al, Ni + Al, etc.), led to the synthesis of highly porous homogeneous ceramic-metal foam with a closed porosity, which cannot be obtained under terrestrial conditions. SHS of an aluminum particle suspension in vacuum, covered with a layer of nickel, led to the formation of a highly porous product with a skeletal structure. When using thermite compositions with aluminum (NiO + Al, NiO + Ni + Al, CrO_3 + Al, CrO_3 + TiO_2 + NiO + Al, etc.), which burn with melting of the reactants and formation of liquid products phase separation occurs in the synthesis products but that the structure, shape and surface of the burnt samples differ from those in terrestrial conditions. This line of research was continued till 2010 aboard the International Space Station [2]. The researchers studied the interaction of the molten products of thermite SHS systems (NiO-Al, NiO-Ni-Al, and NiO-Ni-Al-Ti) with the

Concise Encyclopedia of Self-Propagating High-Temperature Synthesis
http://dx.doi.org/10.1016/B978-0-12-804173-4.00004-1

substrates of titanium, molybdenum, tantalum, niobium, and in particular, the conditions of formation of protective coatings of intermetallic compounds with good adhesion properties on titanium substrate.

Parallel to these fundamental studies on the mechanism of SHS in space, other applied lines of investigations for application of SHS in space were developed, as a result of advantages such as the ability to conduct SHS processes because of the internal heat evolution without the use of permanent external sources of heating, in any atmosphere or in vacuum, using simple light-weight equipment [3–5]. First application is the use of SHS for obtaining materials and coatings, welding, and soldering when performing technological and repair works onboard a spacecraft. Second, the SHS process can be used in space as a source of thermal energy in devices for heating or cutting structural elements, as well as to generate thermal radiation. Third, the SHS process can be used to generate gases such as oxygen as well as for generating aerosols, including fire-extinguishing ones. Nanoenergetic gas generators from the thermite mixtures of nano-powders of fuel and oxidizer (Al-Bi_2O_3, Al-I_2O_5, etc.), which have high-energy density and generate pulses of high pressure because of a rapid gas release is also important application [4]. Such generators can be included in the composition of microelectromechanical systems (MEMS) with possible use as solid-fuel pulsed thrusters for digital systems of spacecraft motion correction [5].

It should be separately noted that numerous studies use SHS for in situ obtaining of materials from raw materials on the nearest planets such as Mars and on the Moon [6]. The surfaces of these planetary bodies are covered with a powdery layer of regolith composed of particles of complex oxides of different metals and silicon. It is desirable to obtain building structures (plates, bricks, beams, coatings) from mineral raw materials found on Earth with a minimum consumption of energy, for example, by adding aluminum powder and implementing thermite SHS reactions. Researchers added ilmenite and iron oxide powders to the mixture of regolith and aluminum, used the additive polytetrafluoroethylene to activate the thermite reaction. Other researchers did not use thermite reactions, but a highly exothermic reaction $Ti + 2B = TiB_2$ for sintering the regolith has been applied. In the end they used Mg and brought the regolith content up to 87–92 wt% and obtained a strong product through the technology of SHS-consolidation. The SHS process using regolith can be applied for production for not only structural but also functional materials: solar cells, semiconductors, catalysts, and others.

Some of the preceding space applications of SHS can be used in the field of aviation. Many opportunities are available for applying SHS under terrestrial conditions in order to obtain materials to manufacture aerospace equipment: thermal protection for reentry spacecraft [7]; high-temperature sintered ceramics and cast cermets for the manufacture of parts of gas turbine engines instead of nickel superalloys; lightweight, heat-resistant aluminum-ceramic composites and other materials—as well as the SHS technologies used in production.

REFERENCES

[1] Merzhanov AG, Rogachev AS, Rumanov EN, Sanin VN, Sytchev AE, Shcherbakov VA, et al. Influence of microgravity on self-propagating high-temperature synthesis of refractory inorganic compounds. Cosmic Res 2001;39(2):210–23.

[2] Shcherbakov VA, Sytschev AE, Sachkova NV. Interaction of SHS-produced melt with a Ti surface in microgravity conditions. Int J SHS 2010;19(2):141–9.

[3] Miyazaki E, Odawara O. SHS technology for in-situ resource utiluzation in space. Int J SHS 2003; 12(4):323–32.

[4] Moore JJ, Yi HC, Guigne JY. The application of self-propagating high-temperature (combustion) synthesis (SHS) for in-situ fabrication and repair (ISFR), and in-situ resource utilization (ISRU). Int J SHS 2005;14(2):131–49.

[5] Martirosyan KS. Nanoenergetic gas-generators: principles and applications. J Mater Chem 2011;21:9400–5.

[6] Xanthopoulou G. SHS for an in-situ resource utilization on the Moon and Mars. In: XII International symposium SHS (SHS 2013), 21–24 October, South Padre Island, TX, Book of Abstracts; 2013. p. 319–20.

[7] Vekinis GA, Xanthopoulou GG. Hybrid thermal protection system for spacecraft based on MgO SHS refractories. SHS, In: 10th International symposium SHS, Tsakhkadzor, 6–11 July, Armenia; 2009. p. 152–3. Book of Abstracts.

Alloys of Transition Metals

Anahit G. Aleksanyan, Seda K. Dolukhanyan

In the Laboratory of High Temperature Synthesis of A.B. Nalbandyan Institute of Chemical Physics of National Academy of Science, Republic of Armenia, a new method of synthesis of binary and multicomponent alloys has been developed, named "hydride cycle" (HC). HC is a novel, highly efficient method for the fabrication of refractory metal alloys. Current technologies of alloy production (mechanical alloying, arc melting, powder metallurgy, etc.) are associated with certain technological difficulties, whereas the HC method avoids most of them (particularly, it excludes melting). HC is a high-tech method based on the combination of SHS of transition metal hydrides and heat treatment of a mixture of hydrides, resulting in alloy formation [1–7].

The intent prompting the HC method is use of cheap SHS-synthesized powders of metal hydrides as starting materials. The hydrides of desired transition (III–V group) metals are mixed, compacted, and dehydrogenated-sintered at 900°C to 1150°C, which results in the formation of corresponding alloy. The advantage of HC is that the alloy formation takes place at relatively low temperature (900°C to 1150°C) with short-time exposition (1 to 2 h). It is important that the metals with very different melting points and different densities can be alloyed without melting. In general, the reactions of binary alloy formation can be expressed as

$$x\mathrm{Me}^1\mathrm{H}_2 + (1-x)\mathrm{Me}^2\mathrm{H}_2 + \text{heat} \rightarrow \mathrm{Me}^1{}_x\mathrm{Me}^2_{(1-x)} + \mathrm{H}_2\uparrow \quad (1)$$

$$x\mathrm{Me}^1\mathrm{H}_2 + (1-x)\mathrm{Me}^2 + \text{heat} \rightarrow \mathrm{Me}^1{}_x\mathrm{Me}^2_{(1-x)} + \mathrm{H}_2\uparrow \quad (2)$$

For direct synthesis of the alloys of given composition and crystalline structure, in HC the mechanism of formation of alloys in dependence on the ratio of initial components were studied in binary systems: TiH_2-ZrH_2, TiH_2-VHx, TiH_2-$NbHx$, TiH_2-Ni, TiH_2-Co, TiH_2-Al, Cu-Nb, Cu-Ti, and so on; and ternary systems: TiH_2 -ZrH_2-HfH_2, TiH_2-VHx-Mn, TH_2-$NbHx$-Al, TiH_2-ZrH_2-Ni, etc. It was found that during HC, the alloys with different crystalline structures may come from depending on the composition of the charge and the experimental conditions (initial density, heating rate, exposition time, etc.). As a result, in the identical system, hcp, bcc, ω-phase, O-phase, approximants,

Concise Encyclopedia of Self-Propagating High-Temperature Synthesis
http://dx.doi.org/10.1016/B978-0-12-804173-4.00005-3

intermetallides, and so on may be produced. Thus Ti_xZr_{1-x}, Ti_2Zr, Ti_xNb_{1-x}, Ti_xV_{1-x}, TiNi, TiCo, ZrNi, Zr_2Ni, α_2-Ti_3Al, γ-TiAl, $Ti_{0.125}Al_{0.75}Nb_{0.125}$, CuNb, $Ti_{0.37}V_{0.38}Mn_{0.25}$, $TiV_{0,8}Mn_{1.2}$, $TiV_{1.2}Mn_{0.8}$, and so on were synthesized.

The interaction of HC alloys obtained in TiH_2-ZrH_2, TiH_2-VHx, TiH_2-$NbHx$, TiH_2-ZrH_2-HfH_2, TiH_2-VHx-Mn, TH_2-$NbHx$-Al, TiH_2- ZrH_2-Ni, and other systems with hydrogen in the combustion mode also has been investigated. It was shown that the bulk alloy sample may interact with hydrogen in SHS mode and form high hydrogen-content hydrides (H/Me $\geq$2) [8–10].

REFERENCES

[1] Aleksanyan AG, Dolukhanyan SK, Mantashyan AA, Mailyan DG, Ter-Galstyan OP, Shekhtman VSh. New technique for producing the alloys based on transition metals. In: Baranovski B, et al., editors. Carbon nanomaterials in clean energy hydrogen systems. NATO science series, ICHMS'2007. Dordrecht: Springer; 2008. p. 783–94.
[2] Dolukhanyan SK, Aleksanyan AG, Ter-Galstyan OP, Shekhtman VSh, Sakharov MK, Abrosimova GE. Russ J Phys Chem B 2007;2(6):563–9.
[3] Aleksanyan AG, Dolukhanyan SK, Shekhtman VSh, Khasanov SS, Ter-Galstyan OP, Martirosyan MV. Int J Hydrogen Energy 2012;37:14234–9.
[4] Aleksanyan AG, Dolukhanyan SK, Shekhtman VSh, Huot J, Ter-Galstyan OP, Mnatsakanyan NL. J Alloys Compd 2011;509:786–9.
[5] Aleksanyan AG, Mayilyan DG, Dolukhanyan SK, Shekhtman VSh, Ter-Galstyan OP. Int J SHS 2010;19(1):34–9.
[6] Shekhtman VSh, Hakobyan HG, Aleksanyan AG, Dolukhanyan SK, Ter-Galstyan OP, Sakharov MK. Int J Hydrogen Energy 2011;36:1206–8.
[7] Dolukhanyan SK, Ter-Galstyan OP, Aleksanyan AG, Hakobyan HG, Mnatsakanyan NL, Shekhtman VSh. Khimicheskaya Phizika 2015;34(9):1–8.
[8] Hakobyan H, Aleksanyan A, Dolukhanyan S, Mnatsakanyan N, Shekhtman V. Solid compounds of transition elements. In: Bobet J-L, Chevalier B, Fruchart D, editors. Solid state phenomena, vol. 170. Switzerland: Trans Tech Publications; 2011. p. 354.
[9] Mayilyan D, Dolukhanyan S, Aleksanyan A. Int J SHS 2012;21:38–40.
[10] Dolukhanyan SK, Aleksanyan AG, TerGalstyan OP, Shekhtman VSh, Mnatsakanyan NL. Int J SHS 2014;23(2):78–82.

Aluminides

Iñigo Agote, Miguel A. Lagos

Combustion synthesis methods can be used to obtain different aluminides (TiAl, NiAl, FeAl, CoAl, etc.) starting from the elemental powders. Compared to conventional aerospace materials, aluminides of titanium, nickel, iron, cobalt, niobium, and so on, with various compositions offer attractive properties for potential structural applications. A density of 3.7–3.9 g/cm^3, which is less than half of that of nickel-based superalloys and more than 10% lower than titanium, plus a high elastic module of 175 GPa attract attention for aircraft engine, aerospace, and automotive applications.

Combustion synthesis of the aluminides typically begins when the system reaches the initiation temperature, and a SHS reaction begins at the metal interface, that is, between the aluminium and the iron, nickel, or titanium metals: Fig. 1A shows the reaction between Ti and Al. Note how the reaction begins approximately at the melting point of aluminium. The heat generated from the reaction melts the aluminium, allowing rapid reaction with the adjacent metal particle. In this technique, the aluminium reacts completely, forming an aluminide.

Because of its high specific mechanical properties and corrosion resistant behavior, TiAl is one of the most attractive aluminides. According to the stoichiometry, there are different TiAl aluminides: Ti_3Al, $TiAl_3$, and TiAl. However, the one that has attracted most interest is the equiatomic titanium aluminide (gamma TiAl). Gamma TiAl-based intermetallic alloys have been studied extensively as potential replacements for nickel-based superalloys [1–5] for different application fields, such as turbine blades

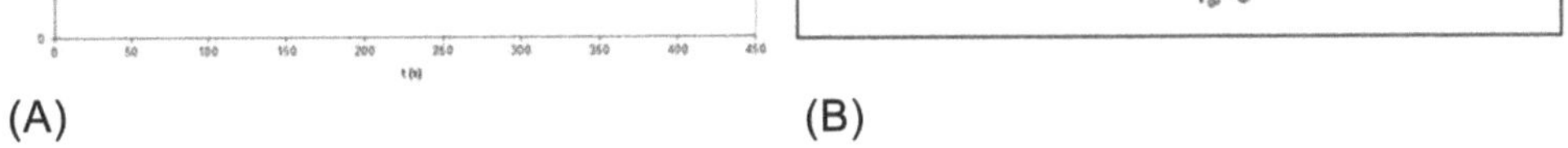

Fig. 1 (A) Thermogram of the Ti-Al system. (B) Adiabatic temperature (T_{ad}) variation with the reagents preheating (T_o) temperature.

Concise Encyclopedia of Self-Propagating High-Temperature Synthesis
http://dx.doi.org/10.1016/B978-0-12-804173-4.00006-5

for aircraft engines, stationary turbines, and space vehicles. These alloys have a relatively high yield strength at elevated temperatures (430 MPa at 750°C), good creep (elongation <0.2% after 50 h at 750°C and 80 MPa) characteristics and excellent oxidation/corrosion resistance. The combination of these properties, associated with their low density (3.9–4.1 g/cm^3), make them very attractive as high-temperature structural materials for aerospace, automotive, and other applications.

Because of the low adiabatic temperature of the Ti + Al reaction (T_{ad} = 1244°C), the system needs additional activation energy to support the synthesis process in self-sustaining mode. Fig. 1B shows the variation of the adiabatic temperature as a function of the preheating temperature. Note that when the initial reagents temperature is above 350°C, the process is favored toward being carried out in self-sustaining mode. A combustion synthesis process allows obtaining TiAl compounds in three main different stages: powder, porous, or dense. The synthesis and densification of TiAl-based compounds is, technologically, the most attractive route to obtain TiAl-based parts. Dense TiAl parts have been obtained by pseudo-HIP (in self-sustaining mode) and SPS (in thermal explosion mode) means. The properties of the parts are similar to other TiAl compounds obtained by conventional casting means: a fully lamellar microstructure can be obtained with good tensile properties and low porosity (less than 1%) [6–8]. Furthermore the use of SPS allows obtaining very fine microstructures with improved properties (see Fig. 2A) [9]. The optimization of the synthesis and densification process of TiAl-based material allows obtaining high-quality component aerospace applications such a blades (see Fig. 2B).

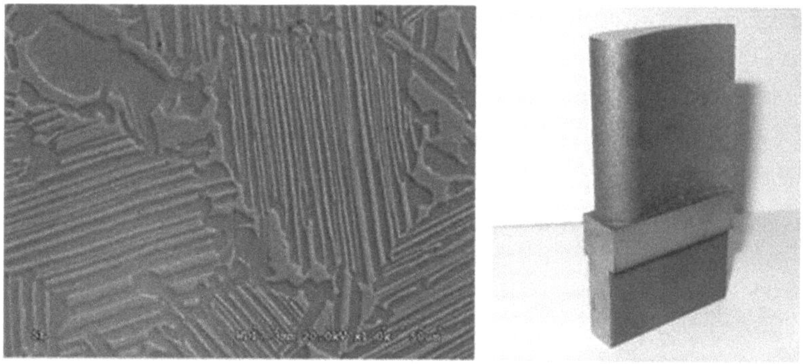

(A) (B)

Fig. 2 (A) SEM picture of TiAl. (B) TiAl blade obtained by SHS.

REFERENCES

[1] Schultz U, Leyens C, Fritscher K, Peters M. Aerosp Sci Technol 2003;7(1):73–80.
[2] Wu X. Intermetallics 2006;14(10–11):1114–22.
[3] Cheng T, Loretto MH. Acta Mater 1998;46(13):4801–19.

[4] Lapin J. Intermetallics 2006;14(2):115–22.
[5] Recina V, Lundström D, Karlsson B. Metall Mater Trans A 2002;33(9):2869–81.
[6] Agote I, Gutiérrez M, Sargsyan A, Coleto J, García de Cortazar M, Lagos MA, et al. Kovové Materiály 2008;46(2):87–95.
[7] Agote I, Coleto J, Gutiérrez M, Sargsyan A, García de Cortazar M, Lagos MA, et al. Intermetallics 2008;16(11-12):1310–6.
[8] Agote I, Lagos MA. Intermetallics 2013;36:51–6.
[9] Froes FH, Suryanarayana C, Eliezer D. ISIJ Int 1991;31(10):1235–48.

Aluminum Nitride

Vladimir V. Zakorzhevsky

Aluminum nitride was identified and obtained as an individual chemical compound in 1862, though a ceramic community was not really attracted to it until 1980s. Characteristics such as high thermal conductivity, excellent electrical resistance, and low coefficient of thermal expansion enabled the use of aluminum nitride as a thermally conductive dielectric ceramics. Rapid development of microelectronics since 1970s has made aluminum nitride the leading material in the production of hybrid microcircuits because it is a cheap and safe alternative to beryllium oxide.

Aluminum nitride synthesis in the combustion mode is based on aluminum powder combustion in nitrogen with additional regulating agents, for example, aluminum nitride as an inert diluent or salts (NH_4Cl, NH_4F, $MgCl_2$, AlF_3, $AlCl_3$, Na_3AlF_6, $(NH_4)_3AlF_6$).

$$2Al + N_2 \xrightarrow{\text{combustion, AlN}} 2AlN$$

$$2Al + NH_4Cl + N_2 \xrightarrow{\text{combustion, AlN}} 2AlN + HCl + 0.5N_2 + 1.5H_2$$

Production of aluminum nitride by the SHS began in the 1980s. At first, AlN was obtained as a powder (for subsequent sintering) in a pilot reactor [1]. The optimum AlN dilution of the green mixture and initial nitrogen pressure were determined, and the positive role of the gasifying additive NH_4Cl was identified [1].

Nitrides, including aluminum nitride, can be obtained through the use of solid nitriding components [2]. In this case, sodium azide was used as the main source of nitrogen. The synthesis is carried out in a nitrogen atmosphere according to the following scheme:

$$3Al + NaN_3 + NH_4Cl + N_2 \xrightarrow{\text{combustion}} 3AlN + NaCl + 1.5N_2 + 2H_2$$

As previously mentioned, it is also possible to use NH_4Cl, NH_4F, AlF_3, Na_3AlF_6, $(NH_4)_3AlF_6$, and so on as salt additives to the green mixture. They contribute to the development of the optimum temperature regime and participate in the structure formation mechanism.

Control of particle morphology. When aluminum nitride synthesis involves the participation of fluorides, in particular ammonium fluoride, spherical grains of AlN are formed, as shown in Fig. 1 [3]. The effect of NH_4F on the particle shape is connected to the participation

Concise Encyclopedia of Self-Propagating High-Temperature Synthesis
http://dx.doi.org/10.1016/B978-0-12-804173-4.00007-7

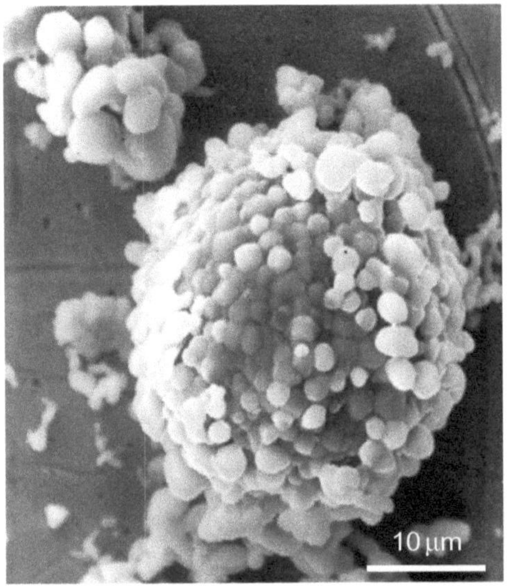

Fig. 1 Microstructure of AlN particles synthesized with participation of ammonium fluoride.

of AlF_3 (formed by the reactions $NH_4F \rightarrow HF + NH_3$; $Al + 3HF \rightarrow AlF_3 + 1.5H_2$) in the structure formation mechanism during the synthesis process.

AlN spherical particles appear because of the formation of an AlF_3 liquid film on liquid aluminum drops during the nitriding process.

When chlorides are used (NH_4Cl, $AlCl_3$), needle-like AlN particles are formed, as shown in Fig. 2 [4]. In this case, the gas-phase mechanism of structure formation with participation of HCl, NH_3, $AlCl_3$ is observed.

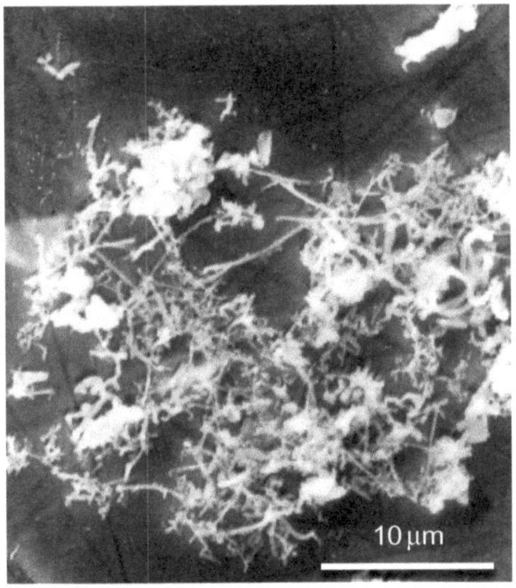

Fig. 2 Microstructure of AlN particles synthesized with participation ammonium chloride.

It has been shown that decreasing the temperature results in size reduction of aluminum nitride particles.

Synthesis of submicron aluminum nitride by SHS is based on the combustion components in thin films, which allows liquid aluminum to wet the surface of aluminum nitride. In such a case, the inert diluent (AlN) is not only a regulator for the combustion temperature but also a substrate for melt, that is, combustion of the liquid aluminum film occurs on the AlN surface; as a result, submicron aluminum nitride particles of 200–500 nm in size are formed [5]. When 0.3–0.5 mass% of AlF_3 or Na_3AlF_6 was introduced into the green mixture, aluminum nitride powders of $d=50$–200 nm with specific surface area up to 20 m^2/g were obtained.

REFERENCES

[1] Prokudina VK, Shestakova TV, Borovinskaya IP, Kuznetsova IG, Gracheva NA, Nedelko EI. Materials of the 3rd USSR conference on technological combustion, vol. 2. Chernogolovka; 1981. p. 5–8.
[2] Kosolapov VT, Shmelkov VV, Levashov AF, Merzhanov AG. Auth. cert. no. 658084, 1978; USSR.
[3] Merzhanov AG, Borovinskaya IP, Makhonin NS, Savenkova LP, Zakorzhevsky VV. Method of metal nitride production. Patent RF 2061653; 1996.
[4] Merzhanov AG, Borovinskaya IP, Zakorzhevsky VV, Savenkova LP, Ignatieva TI. Method of aluminum nitride production. Patent RF 2091300; 1997.
[5] Zakorzhevsky VV, Borovinskaya IP. Inorg Mat 2015;51(6):566–71.

Applications of Reactive Multilayer Foils

Timothy P. Weihs

Over the past two decades, many scientific studies have shown that one can control the ignition thresholds, the heat of reactions, and the velocities of self-propagating formation reactions in multilayer foils through the careful selection of chemistry, bilayer thickness, and total thickness [1–3]. Given this control, the nanolayered foils are ideal local heat sources for joining or sealing components, for heating devices or chemicals, for initiating other chemical reactions, and for producing optical signals. Here we consider their commercial use as heat sources for soldering, and then we briefly consider some of these other applications that are under development.

Reactive foils were developed as local heat sources for soldering and brazing components starting in 1995 and became commercially available as the NanoFoil product in 2003 [4,5]. Today they are actively used for soldering together small and large components. The process consists of placing a foil between two components that have been coated with solder or wetting layers; applying pressure to the sandwich structure; and then igniting a reaction within the foil, which produces a rapid burst of heat, melts the solder, and forms a bond, as shown schematically in Fig. 1A. By controlling the chemistry and thickness of the foils, the total heat released by the formation reaction can be tuned to ensure that the solder layers melt but the bulk of the components remain cool. Typically, only 30–40 μm of solder are melted on either side of reactive foil, as shown in Fig. 1B, and millimeter thick components rise only a few degrees in temperature [5]. With such small changes in temperature, the components do not undergo any significant expansion or contraction during bonding, even if their coefficients of thermal expansion (CTE) differ significantly [2,5]. Thus bonding with reactive multilayer foils enables the formation of relatively stress-free, metallic bonds between materials with dissimilar CTEs. This ability is particularly beneficial in bonding together large components such as sputtering targets and backing plates [6], and it is commercially available as the NanoBond process. More recent developments show an ability to bond Si wafers together using thinner but hotter Pd/Al patterning multilayer foils utilizing standard IC processing techniques [7].

The very localized heating offered by reactive multilayer foils helps in the joining of temperature-sensitive microelectronic components [2,5]. An exciting example, shown schematically in Fig. 2, is the bonding of chips and LED packages to heat sinks. To enhance heat dissipation from the chips or LEDs, one seeks very conductive interfaces

Concise Encyclopedia of Self-Propagating High-Temperature Synthesis
http://dx.doi.org/10.1016/B978-0-12-804173-4.00008-9

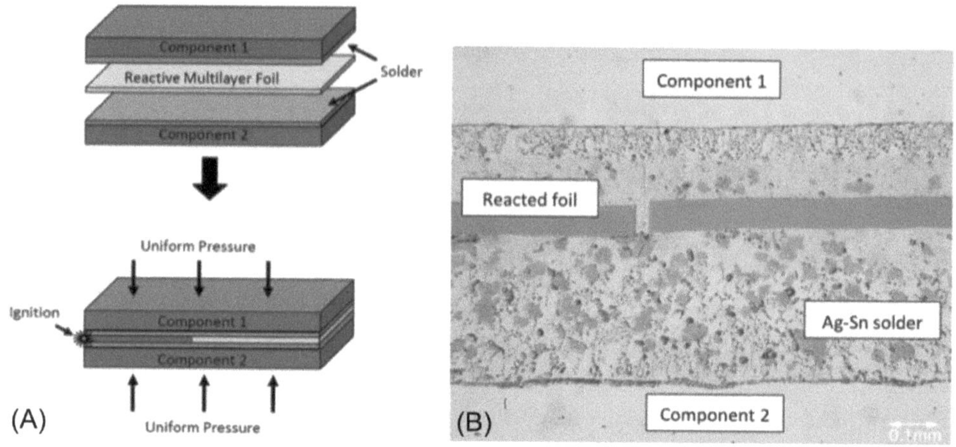

Fig. 1 (A) Schematic of bonding with reactive multilayer foils, (B) cross-sectional image of a bond of two components.

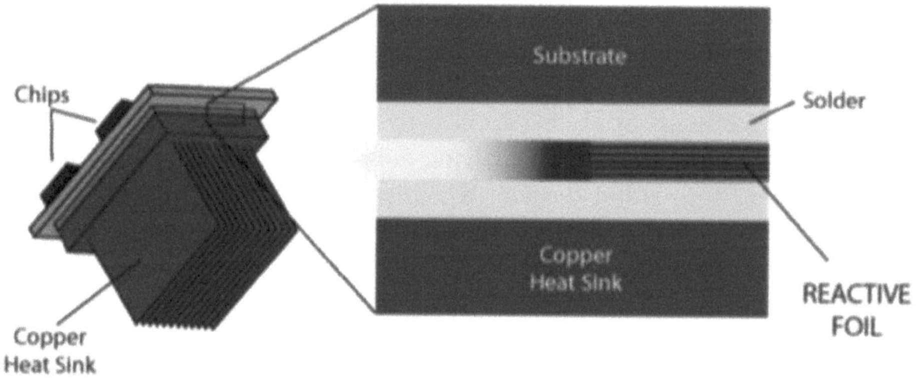

Fig. 2 Schematic of bonding a chip module to a heat sink using reactive multilayer foils and solder.

between the chip and the heat sink. Often this is a thermal paste, but well-bonded, metallic interfaces can dissipate heat more effectively. However, not all microelectronic components can be furnace soldered to achieve a conductive metallic interface, particularly at the later stages of assembly. Bonding with NanoFoil limits the thermal exposure of the front side of silicon dies to 100°C for less than a second [8]. The resulting interfaces are very conductive and could prove advantageous for CPUs, GPUs, solar cells, and LED devices [8]. Another example is the sealing of packages containing microelectronic components. The localized heating also allows one to form a metallic or polymeric seal along the perimeter of a package without heating the contents of the package [7,9]. This has been demonstrated at the micron and centimeter scale.

Beyond joining and sealing applications, the heat and light released by reactive multilayer foils can be used for a range of commercial and defense applications. In the area of pyrotechnics and explosives, one can ignite flares [10] or initiate reaction trains that lead to air bag deployment or explosive detonation [11]. As a stable and rapid source of energy, electrolytes in thermal batteries can be heated [12] or can act as a heater for other devices. The controllable burning rates can be utilized in chemical time delays [13], and the controllable ignition threshold can be beneficial in novel munitions [14]. All of these applications rely on the ability to store chemical energy for long periods of time and then release it in a tunable and precise manner. Additional investigations into the basic properties and reaction mechanisms within these novel reactive materials will help expand their utilization as controlled sources of heat and light.

REFERENCES

[1] Rogachev AS, Mukasyan AS. Comp Expl Shock Waves 2010;46:243–66.
[2] Weihs TP. In: Barmak K, Coffey KR, editors. Metallic films for electronic, magnetic, optical and thermal applications. Structure, processing and properties. Swaston: Woodhead Publishing; 2014. p. 160–243 [chapter 6].
[3] Adams DP. Thin Solid Films 2015;576:98–128.
[4] Makowiecki D, Bionta R. US Patent 5381944; 1995.
[5] Duckham A, Levin J, Weihs T. Adv Sci Technol 2006;45:1578–87.
[6] Duckham A. Vacuum technology & coating 2007;65–9.
[7] Braeuer J, Gessner T. J Micromech Microeng 2014;24:115002.
[8] Van Heerden D, Rude T, Newson J, Knio O, Weihs T, Gallus D. In: Proceedings of the twentieth annual IEEE semiconductor thermal measurement and management symposium; 2004. p. 46–9.
[9] Van Heerden D, Deger D,Weihs T, Knio O. US Patent 7,143,568; 2006.
[10] Nielson D, Tanner R, Dilg C. US Patent 7,690,308; 2010.
[11] Baginski T, Parker T, Fahey W. US Patent 6,925,938; 2005.
[12] Ding M, Krieger F, Swank J, Poret J, McMullan C, Chen G. In: 43rd power sources conference, Philadelphia, July 7–10; 2008.
[13] Kelly P, Tinston S, Arnell R. Surf Coat Technol 1993;60:597–602.
[14] Hugus G, Sheridan E. US Patent 7,955,451, June 7; 2011.

Auto-Oscillations

Konstantin G. Shkadinsky, Alexander S. Rogachev

Auto-oscillations are sustained oscillations in a nonlinear system, supported by an energy source [1]. They are determined by the properties of the medium, neglecting the initial stage, characterized by amplitude, frequency, as well as their oscillation spectrum. Nonlinear models describe self-oscillations, the nonlinearity does not allow fluctuations to grow indefinitely and regulates the power supply and removal of energy. The phase space is commonly used for analysis of oscillations. The geometric image of the oscillation is a steady attractor—the trajectory or set of trajectories in a limited region of phase space, attract all nearby trajectories. Examples of attractors in the phase plane (coordinates of the maximum rate of chemical interaction and velocity of the combustion wave movement) for self-oscillation in the tubular reactor are shown in Fig. 1.

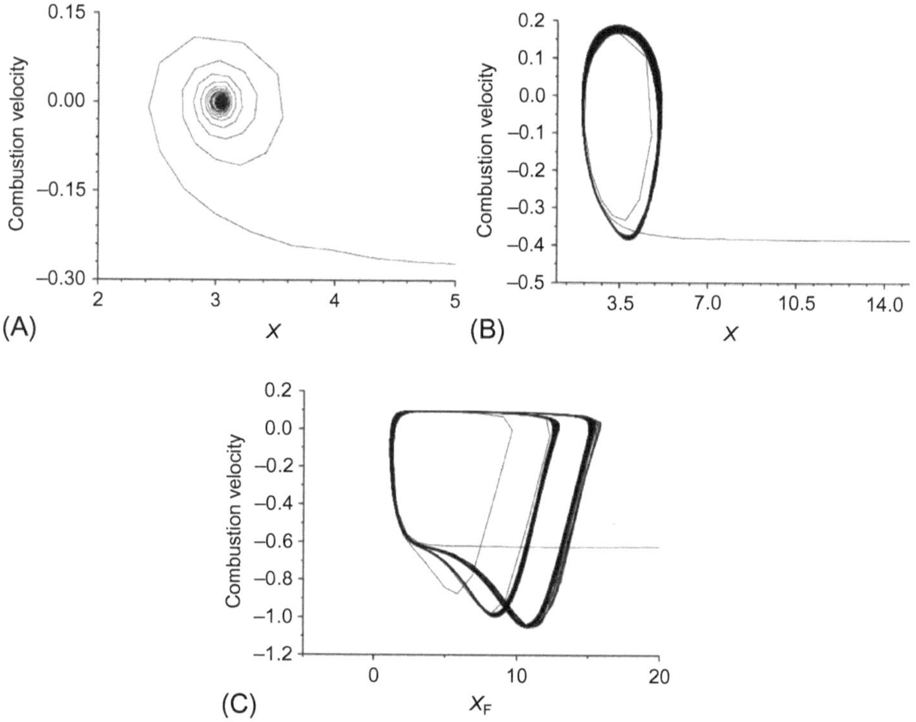

Fig. 1 Types of oscillations in phase space. (A) Stationary combustion; (B) Limit cycle; (C) Limit cycle in multidimensional space.

Concise Encyclopedia of Self-Propagating High-Temperature Synthesis
http://dx.doi.org/10.1016/B978-0-12-804173-4.00010-7

In two-dimensional phase space, the stationary state is characterized by a point (Fig. 1A). The simplest nontrivial attractor—a closed trajectory—is called a limit cycle (Fig. 1B). In distributed systems with many degrees of freedom, the structure and dynamics of oscillation become more complicated and can form, for example, a limit cycle in a multidimensional space (Fig. 1C shows its projection on the plane). The theory of self-oscillations is actively growing, but it is still far from complete.

Auto-oscillations of SHS waves exhibit themselves as periodic accelerations and stagnations of the combustion front. During the acceleration period, the local temperature increases, while during depressions it decreases. Therefore conditions of synthesis become different in different points in the sample. The influence of oscillating combustion can be found in solid products, as illustrated in Fig. 2. The sample can be easily fragmented into thin disks; each disk was formed during the active period of pulsation [2].

The nonhomogeneous product, which occurs as a result of pulsations, often becomes more uniform because of postcombustion reactions.

Fig. 2 Product of pulsating SHS [2].

REFERENCES

[1] Kuramoto Y. Chemical oscillations, waves, and turbulence. Berlin/Heidelberg/New York/Tokyo: Springer-Verlag; 1984.

[2] Merzhanov AG. Self-propagating high-temperature synthesis: twenty years of research and findings. In: Munir Z, Holt JB, editors. Combustion and plasma synthesis of high-temperature materials. New York, NY: VCH; 1990. p. 1–53.

Azide-Based Technologies

Alexander P. Amosov, Georgy V. Bichurov

Azide technology of SHS, which is abbreviated as SHS–Az, is based on the use of a solid powder of an azide as a nitriding reagent in the SHS process instead of gaseous or liquid nitrogen for synthesis of nitrogen-containing compounds (nitrides or carbonitrides). Azides are the salts of hydrazoic acid HN_3. Most azides have explosive properties, which limits their application in SHS–Az technology. When studying the SHS–Az processes, the azides of ammonium (NH_4N_3), lithium (LiN_3), and barium (BaN_6) were used. However, the practical application of SHS–Az technology has found sodium azide (NaN_3), which is produced on an industrial scale, to be nonexplosive, nonhygroscopic, and nonvolatile but highly toxic.

The scientific research in the field of the SHS–Az started in 1970 under the guidance of Professor V.T. Kosolapov in Samara State Technical University (SSTU) in Russia, formerly the Kuibyshev Polytechnic Institute in the USSR [1,2]. The first investigations were focused on the process of obtaining nitrides of transitive metals (TiN, ZrN, HfN) in a burning mode. Sodium azide (NaN_3) was recommended as the most effective product from the point of view of safety, chemical activity, and degree of purity of the final nitrides or carbonitrides. As a result of the initial mixture's decomposition after ignition, nitrogen with a high content of active atomic nitrogen and metallic sodium are formed: $NaN_3 \rightarrow Na + N_2 + N$. This result promotes the reaction of nitriding, which can be represented by the equation

$$3El + NaN_3 \rightarrow 3ElN + Na$$

where El = Ti, Zr, Hf, Nb, Ta, B is the initial element to be nitrided.

Sodium vapors also play a positive role by reducing the oxides on the surface of the powder of the initial element and thereby purifying it. However, sodium is a fire hazard that can lead to spontaneous combustion of the samples of nitrides after removing them from the SHS–Az reactor. Therefore a halide salt, for example, NH_4Cl, is added to the initial mixture of powders to fix the generated sodium into a neutral halide salt:

$$4El + NaN_3 + NH_4Cl \rightarrow 4ElN + NaCl + 2H_2 \uparrow$$

The hydrogen produced is burned at the outlet of the SHS–Az reactor, and the salt NaCl can be easily separated from the desired product by washing with water. The result is a high-purity nitride of the element with a high degree of nitriding. Because of the low

Concise Encyclopedia of Self-Propagating High-Temperature Synthesis
http://dx.doi.org/10.1016/B978-0-12-804173-4.00011-9

combustion temperature and the presence of gaseous products, these nitrides are highly dispersed and poorly caked.

In 1983 Dr. J.B. Holt at the Lawrence Livermore National Laboratory in the United States, proposed a similar idea on the use of a metal azide, preferably NaN_3, as a solid source of nitrogen in the combustion synthesis of transition metal nitrides following the SHS–Az technology developed at SSTU [3,4]. Instead of the initial system's "transition metal-sodium azide" he considered the system's "transition metal oxide-metal reducing agent—sodium azide." The mixture of azide, Ca, or Mg and metal oxide is ignited and undergoes self-sustaining combustion, producing the metal nitride with a high-percent conversion of metal to nitride at a low pressure of gaseous nitrogen.

In the further development of the SHS–Az technology at SSTU, the role of the halide salt was greatly expanded [5]. Originally it was introduced mainly to neutralize the fire hazard of sodium, but further development found that the addition of halides into the initial powder mixture allows obtaining the final refractory powders with higher purity and a specific surface area. The halide salt contains in its composition element to be nitrided or element to be nitrided and alkaline metals, for example, AlF_3 or Na_3AlF_6. For synthesis of nitrides such as Si_3N_4 and BN, complex halide salts are used containing, in addition to the halogen and element to be nitrided, an ammonia: $(NH_4)_2SiF_6$ and NH_4BF_4. In the latter cases, the use of halide salts in the original charge of SHS–Az is not less effective than the use of sodium azide. For example, in the case of $(NH_4)_2SiF_6$ in the synthesis of silicon nitride, the fluoride ions catalyze the nitriding process, the ammonia manifests as the nitriding and reducing actions, and silicon, having no oxide film, is nitrided in the gas phase, which favors the formation of Si_3N_4 fibers. Such systems using halide salts were found to be best for practical implementation. As a result, the technology of SHS–Az is based on the combustion of the initial mixture of powders' "element to be nitrided + sodium azide + halide salt." Technological processes have been developed to obtain nitrogen-containing powders such as the nitrides of titanium, zirconium, hafnium, aluminum, boron, and silicon; titanium carbonitride; composites of Si_3N_4–SiC, Si_3N_4–AlN; and others. Here are examples of some synthesis reactions:

$$4Ti + NaN_3 + NH_4Cl \rightarrow 4TiN + NaCl + 2H_2\uparrow$$

$$8Al + 3NaN_3 + AlF_3 \rightarrow 9AlN + 3NaF$$

$$8B + 3NaN_3 + KBF_4 \rightarrow 9BN + 3NaF + KF$$

$$14Si + 6NaN_3 + (NH_4)_2SiF_6 \rightarrow 5Si_3N_4 + 6NaF + 4H_2\uparrow$$

$$12Ti + 2NaN_3 + 6C + 2NH_4Cl \rightarrow 12TiC_{0.5}N_{0.5} + 2NaCl + N_2\uparrow + 4H_2\uparrow$$

$$19Si + 5C + 6NaN_3 + (NH_4)_2SiF_6 \rightarrow 5(Si_3N_4 - SiC) + 6NaF + 2H_2\uparrow$$

The refractory powders of the SHS–Az technology have the following characteristics. The powder of silicon nitride can contain from 40% up to 95% α-Si_3N_4 phase as anisotropic whiskers by a diameter $\ll 1$ μm. The particles of boron nitride are ultrafine, have a disc shape with the linear size about 20 nm. The crystal lattice of BN has turbostrate deformations. The powders of the nitrides of Al, Ti, Zr, and Hf differ from similar powders of traditional technologies of synthesis by a more branched structure, which is typical for the formation of compounds in a gas phase. With a specific surface area of 4–8 m^2/g, they are micropowders and nanopowders. The high purity, usually no less than 97–98%, is typical for all SHS–Az powders.

In recent years, much attention has been paid to the application of SHS–Az technology for obtaining nanosize powders of nitrides [5]. The main approach has been to use complex halides to be nitrided as precursors of these elements only with the exception of the powders of these pure elements from the initial powder mixture, that is, the use of the "halide-sodium azide" system:

$$(NH_4)_2TiF_6 + 6\,NaN_3 \rightarrow TiN + 6\,NaF + 4\,H_2\uparrow + 9.5\,N_2\uparrow$$

$$3\,(NH_4)_2SiF_6 + 18\,NaN_3 \rightarrow Si_3N_4 + 18\,NaF + 12\,H_2\uparrow + 28\,N_2\uparrow$$

$$NH_4BF_4 + 4\,NaN_3 \rightarrow BN + 4\,NaF + 2\,H_2\uparrow + 6\,N_2\uparrow$$

$$AlF_3 + 3NaN_3 \rightarrow AlN + 3NaF + 4N_2\uparrow$$

Thus the titanium nitride powder was synthesized in the form of nanostructured particles: fibers 50–100 nm in diameter and crystallites with an average size of 100–200 nm. Silicon and aluminum nitrides were crystallized in the form of nanofibers about 100 nm in diameter, while boron nitride was crystallized as a mixture of microsize and nanosize particles.

The same approach was used to produce a nanostructured composite powder of TiN-BN using precursors of both elements to be nitrided: Na_2TiF_6 and KBF_4 or composite powder of AlN-BN from precursors AlF_3 and KBF_4 or NH_4BF_4. The main problem with this approach is the presence of hard-to-remove impurity phases in the synthesis of the composite powders.

REFERENCES

[1] Kosolapov VT, Shmelkov VV, Levashev AF, Markov YuM. Synthesis of nitrides of aluminum, titanium, zirconium and hafnium in the combustion mode. In: Second all-union conference on technological combustion: abstracts. Chernogolovka: Branch of Institute of Chemical Physics, USSR Academy of Sciences; 1978. p. 129–30 (in Russian).
[2] Amosov AP, Bichurov GV, Bolshova NF, Erin VM, Makarenko AG, Markov YuM. Azides as reagents in SHS processes. Int J SHS 1992;1(2):239–45.
[3] Holt JB. Synthesis of refractory materials. Patent US 4,459,363, filed: Aug. 16; 1983.
[4] Holt JB. Exothermic process yields refractory nitride materials. Ind Res Dev 1983;25(4):88–91.
[5] Bichurov GV. Halides in SHS azide technology of nitrides obtaining. In: Gromov A, Chukhlomina L, editors. Nitride ceramics: combustion synthesis, properties, and applications. Weinheim: Wiley-VCH; 2015. p. 229–63.

Ball Milling-Induced SHS

Laszlo Takacs

When a highly reactive powder mixture is activated in a ball mill, the grain size decreases, the chemical components are mixed on an increasingly fine scale, and chemically active lattice defects are created. Consequently, the initiation and propagation of a self-sustaining reaction becomes easier, thus prior mechanical activation promotes SHS. It can also happen that the reaction starts in the running mill, thereby becoming a mechanically induced self-propagating reaction (MSR). Typical examples are transition metal-metalloid systems (e.g., Ti + C, Mo + Si), thermites (e.g., CuO + Al) and metal-chalcogen mixtures (e.g., Zn + S) [1].

Fig. 1 shows the temperature of the milling container recorded during a typical MSR milling of a mixture of Zn and Se powders in a SPEX 8000 Mixer Mill [2]. During the initial activation period, the temperature of the mill increases gradually because of the mechanical energy dissipated in the mill, while mixing and activation take place, but little chemical transformation occurs. At a well-defined moment (after about 2300 s of milling in this case), the powder mixture reaches a critical state; a self-sustaining reaction is initiated, and it propagates through the entire powder charge. Most of the reactants are consumed within seconds, followed by cooling toward a temperature determined by the balance between mechanical energy input and heat dissipation. The time to

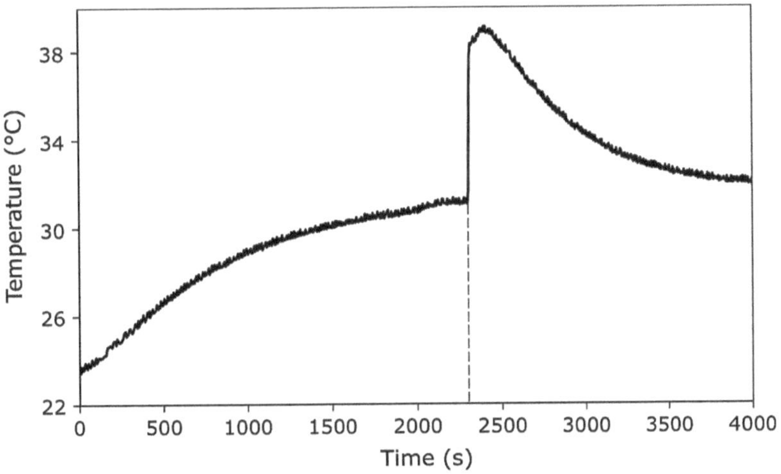

Fig. 1 Temperature of the container recorded while milling a Zr-Se powder mixture.

Concise Encyclopedia of Self-Propagating High-Temperature Synthesis
http://dx.doi.org/10.1016/B978-0-12-804173-4.00012-0

ignition, called the ignition time, is an important attribute of the process. It can vary from a few seconds to several hours depending on the system and the milling conditions. A model proposed by McCormick [3] characterizes the reactivity of the powder charge in terms of an ignition temperature, T_{ig}. Activation decreases T_{ig}, and when it becomes lower than the highest local temperature between the colliding balls, ignition occurs.

The variation of the ignition time provides a useful empirical method to explore the role of the physico-chemical properties of reacting systems. It also depends on the type of mill (vibratory, shaker, planetary) and the process parameters, such as the intensity of the mill, ball-to-powder ratio, and so on. In fact, comparing ignition times is a simple means to compare the efficiency of different mills. Whether MSR can happen in a particular reactive powder mixture depends mainly on its adiabatic temperature. Metal-metalloid and thermite reactions require at least 1800 K to become self-sustaining; lower adiabatic temperature is sufficient for metal-chalcogen reactions and reactions involving halides [1]. Compared to other forms of self-sustaining reactions, the importance of the mechanical properties has to be emphasized as they strongly affect the milling efficiency and consequently the ignition time. For example, adding an inert material to a reactive mixture is expected to increase the ignition time and eventually suppresses ignition entirely. That is indeed observed when soft stearic acid is added to a Zn-S mixture. Addition of a hard oxide slightly shortens the ignition time, as the abrasive oxide particles increase the efficiency of mechanical activation, and this change is more important than the simultaneous decrease of the adiabatic temperature [4]. The kinetics of MSR were studied in Ti-C mixtures as a function of composition, collision energy of the milling ball, and the volume of the powder charge [5]. It was shown that mechanical activation requires a minimal collision energy of about 0.027 J. Interestingly, only gradual mechanochemical transformation occurs, if the collision energy is greater than 0.155 J or the mass of the powder charge is below critical value.

The products of many MSR reactions are technologically important materials, thus MSR is considered a simple and efficient method to prepare small quantities of refractory materials such as $MoSi_2$ and TiB_2. Although the product immediately after MSR can be nonuniform and mixed with residual reactants and other unintended phases, fully reacted fine powder can be obtained simply by continuing the milling process. Scaling up to large quantities is difficult as the abrupt release of a large amount of heat in the mill can be problematic. A possible solution is the combination of mechanical activation followed by traditional SHS, called the mechanically activated SHS (MASHS) process.

REFERENCES

[1] Takacs L. Prog Mater Sci 2002;47:355–414.
[2] Takacs L. J Mater Synth Process 2000;8:181–8.
[3] McCormick PG. Mater Trans JIM 1995;36:161–9.
[4] Takacs L, Soika V. Mater Sci Forum 2001;360:427–32.
[5] Delogu F, Takacs L. Acta Mater 2014;80:435–44.

Bifurcation

Florence Baras

A bifurcation corresponds to the loss of stability of a reference state when a suited parameter, the bifurcation parameter, crosses a critical value, i.e., the bifurcation point. A bifurcation gives rise to occurrence of various complex behaviors: multiplicity of states, oscillations, period-doubling sequence, chaos, and spatio-temporal patterns [1]. Bifurcations can occur only in nonlinear systems subjected to nonequilibrium constraints. In nonisothermal systems, the nonlinear dynamics are directly associated with the thermal feedback of the highly nonlinear Arrhenius law. The most common nonlinear phenomena that occur in SHS processes are multiple stationary states, induction phenomena, auto-waves, and front velocity oscillations.

To illustrate the concept of bifurcation, let us consider the case of *thermal explosion mode of Combustion Synthesis* described by the Semenov model. The evolution equation reads as a balance between the heat generated by the reaction, q_+, and the heat exchanges with the environment, q_-. The nonlinear character of the function q_+ is at the origin of a positive feedback. Different regimes of combustion can coexist: a steady-state regime, close to the ambient temperature, or an explosive regime leading to an extremely high temperature value. We can represent this feature from the generic point of view of bifurcation theory as depicted in Fig. 1. For $\lambda < \lambda_c$, the state diagram shows a unique steady state, while for $\lambda > \lambda_c$, two stable stationary states coexist; λ_c corresponds to a limit-point bifurcation. Even in the monostable region, the proximity of the bifurcation (T) has a

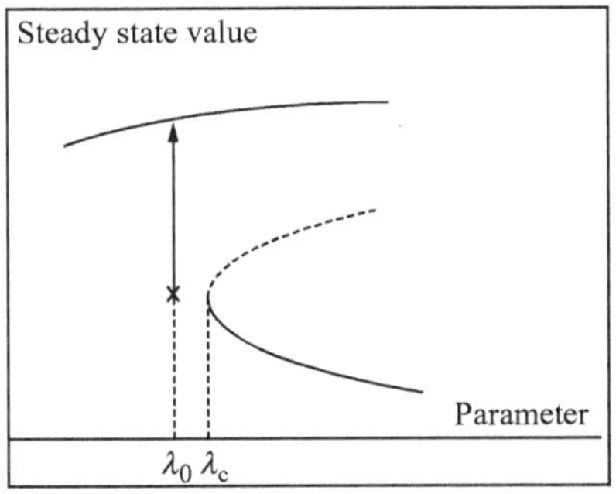

Fig. 1 Parameter range associated with explosive behavior in a state diagram in which the steady state is plotted against a control parameter.

Concise Encyclopedia of Self-Propagating High-Temperature Synthesis
http://dx.doi.org/10.1016/B978-0-12-804173-4.00014-4

dramatic influence on the system evolution. It is convenient to follow the kinetics in terms of the kinetic potential, which reads

$$U(T) = -\int_0^T dT'[q_+ - q_-] \tag{1}$$

for the Semenov model. In the subcritical region ($\lambda < \lambda_c$, explosive regime), the potential is characterized by a flat region around the reservoir temperature value, T_0, followed by a sharp decrease toward the unique minimum, as illustrated in Fig. 2. The induction period corresponds to the slow evolution of the system in the flat region, whereas the explosion coincides with the sudden switching toward the stationary state. For values near the turning point, $\lambda \to \lambda_c$, the kinetic potential becomes more and more flat with a significant increase of the induction period. This phenomenon is known as the critical slowing down of the dynamics near bifurcations. A straightforward calculation gives the induction time τ_i as a function of the distance from the critical point: $\tau_i = 1/\sqrt{1 - \lambda/\lambda_c}$.

Another important example of bifurcation occurring in nonisothermal systems is associated with the loss of stability of a self-propagating stationary front. The unstable combustion mode may give rise to spin combustion or auto-oscillating combustion. Matkowsky and Sivashinsky [2] have shown analytically that a solution exhibiting a periodically pulsating, propagating reaction front arises as a Hopf bifurcation from a solution describing a uniformly propagating front. The bifurcation parameter α was found as the product of

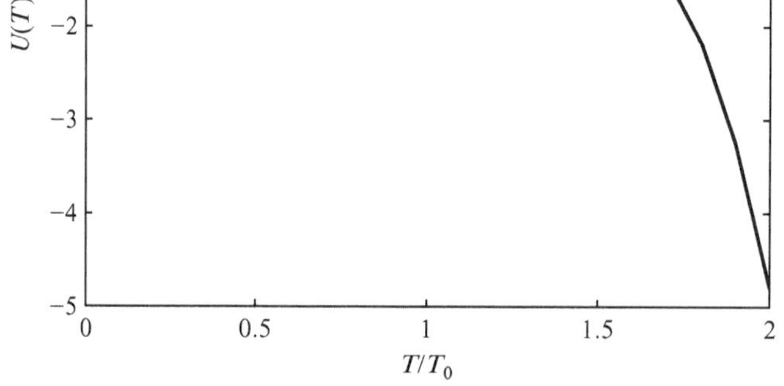

Fig. 2 Kinetic potential *U*(*T*) of the Semenov model exhibiting a plateau around the initial temperature.

dimensionless activation energy and a factor that is a measure of the difference between the dimensionless temperatures of the green mixture and the combustion products

$$\alpha = \frac{1}{2}\frac{E}{RT_b}\left(1 - \frac{T_0}{T_b}\right) \tag{2}$$

where T_0 and T_b are the temperature of the reactants and of the burned material and E the activation energy. If the bifurcation parameter crosses its critical value, $\alpha > \alpha_c = 2 + \sqrt{5}$, the combustion becomes pulsating. The Shkadinskii-Khaikin-Merzhanov criterion for steady-state combustion was estimated by computer simulation [3]:

$$9.1\frac{RT_b^2}{E(T_b - T_0)} - 2.5\frac{RT_b}{E} > 1 \tag{3}$$

REFERENCES

[1] Nicolis G, Progogine I. Exploring complexity. New York, NY: Freeman; 1989.
[2] Matkowsky BJ, Sivashinsky GI. SIAM J Appl Math 1978;35:465–78.
[3] Shkadinsky KG, Khaikin BI, Merzhanov AG. Combust Expl Shock Waves 1971;7:15–22.

Binary Systems

Inna P. Borovinskaya

For application of combustion processes, some chemical systems in which SHS processes can be carried out were determined. The compounds with high heat of formation (Q) can be obtained by the SHS method because in their cases the processes are characterized by high heat evolution and high combustion temperatures (800–4000°C). Also, high values of Q predetermine an excellent strength of chemical bonds in the compound and its stability at high temperatures, that is, the features that are inherent to refractory compounds. So the most important area of SHS process application is synthesis of simple binary compounds of various classes [1,2]. From a chemical viewpoint, the scheme of self-propagating, high-temperature synthesis of binary compounds is simply

$$aX + bY = X_aY_b + Q$$

where

X = Ti, Zr, Hf, V, Nb, Ta, W, Cr, Mo, etc.; Y = H_2, C, B, N_2, Si, O_2, S, Se, etc.

X_aY_b—hydrides, carbides, borides, nitrides, silicides, etc.

Q—thermal effect of the reaction.

In most combinations of X and Y, a great amount of heat is evolved, which allows synthesizing various compounds. The terms of formation of simple binary compounds—carbides, nitrides, borides, silicides, hydrides, chalcogenides, phosphides of II–VI Groups—are well studied and their application fields are determined. Synthesis of berillides, tellurides, and other compounds based on rare earth metals is very promising. Practically all known binary refractory compounds can be obtained by SHS. For low-exothermic compounds with low bond energy, some variations of conjugated SHS processes with the use of heat of high exothermic reactions were developed, and the methods of joint synthesis of high and low exothermic compounds were elaborated. Nonstoichiometric compounds (with metal and nonmetal vacancies in crystal lattice) and solid solutions of binary compounds are of particular interest, too (Table 1).

The production technology of these compounds is rather simple: the compound of the required phase composition can be obtained by finding the optimum ratio of initial components and organizing heat removal from the burning system properly. Multiphase systems are more sensitive to heat removal from the reaction mass than simple ones. In multiphase systems such as silicides, borides, carbides of Group V–VI, the departure from the optimum reagents ratio initiates undesirable intermediate products. The compound

Concise Encyclopedia of Self-Propagating High-Temperature Synthesis
http://dx.doi.org/10.1016/B978-0-12-804173-4.00015-6

Table 1 Binary and nonstoichiometric SHS compounds

Classes	Compounds
High exothermic compounds	
Simple binary compounds	
Carbides	TiC, ZrC, HfC, NbC, VC, TaC, Ta_2C, Si_3N_4, BN, AlN
Borides	TiB_2, ZrB_2, V_3B_2, VB_2, NbB_2, TaB_2
Nitrides	Mg_3N_2, BN, AlN, Si_3N_4, TiN, ZrN, HfN, VN, NbN, Ta_2N, TaN_{hex}, TaN_{cub}, LaN, ScN
Silicides	$ZrSi_2$, $MoSi_2$, Ti_5Si_3, $TiSi_2$
Chalcogenides	MgS, $NbSe_2$, $TaSe_2$, MoS_2, $MoSe_2$, WSe_2
Hydrides	TiH_2, ZrH_2, NdH_2, CsH_2, PrH_2, IH_2
Intermetallics	NiAl, CoAl, TiNi, NbAl, NbGe, TiFe, TiCo, CuAl, VAl
Phosphides	AlP, CuP, NbP, Co_2P, MnP, TiP
Low exothermic compounds	B_4C, Al_4C_3, SiC, Mo_2C, $TiSi_2$, MoB_4, MoB_2, MgB_2, Mo_3Si, WB, W_2B_5, WB_4
Nonstoichiometric compounds (x, low level of homogeneity)	TiN_x ($x=0.5$), ZrN_x ($x=0.56$), HfN_x ($x=0.74$), NbN_x ($x=0.5$), TaN_x ($x=0.5$), TaC_x ($x=0.6$), TiC_x ($x=0.5$)

purity that is connected with the reaction completeness degree and contamination level is not lower than that of the initial reagents since there is practically no contamination during the synthesis, and moreover, self-purification (evaporation of volatile impurities, oxide film recovery, etc.) occurs because of high temperatures. Under the optimum terms of the synthesis, SHS products are of equilibrium with the lattice parameters being similar to those mentioned in literature and have distinct peaks in diffraction patterns. Some of these products can be considered as a reference. Their crystal structures correspond to those of analogs obtained by other methods. But there are some exceptions. The most remarkable previously unknown examples of SHS of binary compounds at application of conventional methods were the production of cubic modification of δ-TaN, discovery and selection of ε-Ti_2N phase in the system of Ti-N, direct synthesis of superhardened modification of BN_{cub} without applying superhigh pressures in diamond presses, development of metastable phases of various compounds, for example, BN amorphous phase, metastable $BN_{0.75}$, MoC, ZrB, and so on [3]. At the directed organization of SHS under nonequilibrium terms, numerous nanostructured nitrides, carbides, sulfides, and other compounds were synthesized; they are characterized by a great variety of crystal lattice and morphology, including tubular structures of various shapes and nanotubes

REFERENCES

[1] Merzhanov AG, Borovinskaya IP. A new class of combustion processes. Combust Sci Technol 1975;10(5–6):195–201.

[2] Merzhanov AG, Borovinskaya IP. Self-propagating high-temperature synthesis in chemistry and technology of refractory compounds. J Mendeleev Russ Chem Soc 1979;24(3):223–7.

[3] Borovinskaya IP. Chemical classes of the SHS processes and materials. Pure Appl Chem 1992;64(7):919–40.

Black Ceramics and AlONs

Inna P. Borovinskaya

Among ceramic materials synthesized by SHS, ceramics with a matrix based on silicon nitride and silicon carbide and dispersion-strengthened by titanium nitride particles, as well as γ-AlON ceramic powders and ceramic items are of particular interest. Both materials are characterized by a high mechanical strength; γ-AlON can transmit light in visible, UV, and IR spectra and is considered to be a transparent ceramic that can substitute sapphire.

Black ceramics were obtained by one-stage direct combustion synthesis of initial reagents in a SHS gasostat at gaseous nitrogen pressure of up to 350 MPa. These ceramics were the first nonporous ones to be obtained (its porosity is close to 1%) and used for combustion synthesis of various items of complicated shapes for ceramic engines, ball bearings, and so on without any subsequent machining (net shape production) [1,2]. The conventional methods of γ-AlON powder production include furnace solid-phase nitriding of $Al + Al_2O_3$ mixtures and carbon-thermal reduction of Al_2O_3 with black carbon in nitrogen medium. There are some comparatively new methods—sol-gel and aluminum combustion in air [3]. The SHS method is also used for AlON synthesis [4]. Compact transparent ceramics of γ-AlON were first obtained by the American scientist McCauley by hot pressing AlN mixtures with Al_2O_3. This method is the main one used for ceramics synthesis [5]. High-temperature sintering and HIP processing are also used for this purpose.

Aluminum oxynitrides (AlONs) are within the scope of the double system xAlN-$y$$Al_2O_3$. Single-phase solid solutions can exist within a 28 to 41 mole percentage and temperatures from 1700°C to 2170°C. Powdery γ-AlON was obtained as a result of little-known variations of SHS, such as $Al + Al_2O_3$ combustion in gaseous nitrogen ($P_{N_2} = 5 - 100$ MPa) with the use of coupled combustion reactions (including "chemical furnaces") or with the introduction of energy additives to the low-exothermic reaction of γ-AlON production from the mixtures of aluminum with aluminum oxide [6].

When black ceramics were made in gasostats, the well-known peculiarities of the SHS process in formation of ceramic macrostructures and microstructures were taken into consideration, such as the volume effect of nitriding reactions, connection of combustion parameters with ceramics structure, final particle morphology, combustion modes (surface and layer-by-layer), and so on. These investigations allowed determining the optimum terms of production of excellent ceramics, including materials and items

Concise Encyclopedia of Self-Propagating High-Temperature Synthesis
http://dx.doi.org/10.1016/B978-0-12-804173-4.00016-8

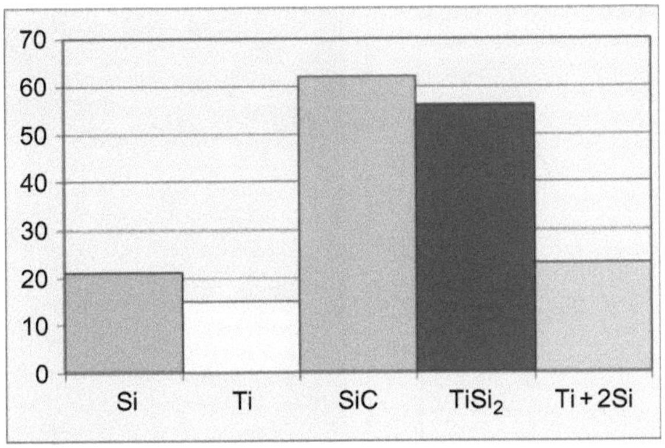

Fig. 1 Volume effect of nitriding.

with preset geometry, unique structure, and high-operation characteristics. The diagram in Fig. 1 demonstrates the volume effect (%) of the nitriding reaction that allowed choosing the optimum raw materials for dense black ceramics. According to this diagram, the main components are titanium disilicide and silicon carbide as an additive.

$$TiSi_2 + SiC \xrightarrow[P_{N_2}<350\ \text{MPa}]{N_{2(gas)}} Si_3N_4 - SiC - TiN(C)$$

In the case of a 70% mass of $TiSi_2$, the linear shrinkage of the sample is 6%. With a decrease in $TiSi_2$ content, the volume shrinkage decreases to 2%–0%. Mechanical strength of ceramic materials is very important for their application. It can be increased through dispersion-strengthening by creating strains at the micro level and hindrances on the way of cracking. Fig. 2A and B demonstrates the dependence of mechanical properties of black ceramics on the content of titanium nitride particles and the picture of cracking in the ceramics volume. Vickers Hardness (H_v) and crack resistance coefficient (K_{1C}) increased with an increase in TiN content. Under various terms of Si_3N_4-SiC-TiN synthesis, a material with the following characteristics was obtained:

Density: 3.05–3.40 g/cm^3
Porosity: 1–15%
Rockwell hardness: 85–93
Vickers hardness: 14.6–21.0
Bending strength: constant up to 1500°C (σ_{bend}=650 MPa)
Crack resistance: K_{1C}=6.0 MN/m$^{3/2}$
Thermal resistance (H_2O 1250±10°C): 60 cycles
Friction coefficient: 0.2–0.3 (dry), 0.02–0.03 (wet)

Fig. 3 shows some items of black ceramics made by one-stage synthesis of prepressed samples of preset geometry without any machining. It should be noted that further improvement of properties because of the formation of nanostructured ceramics with SiC_xN_y

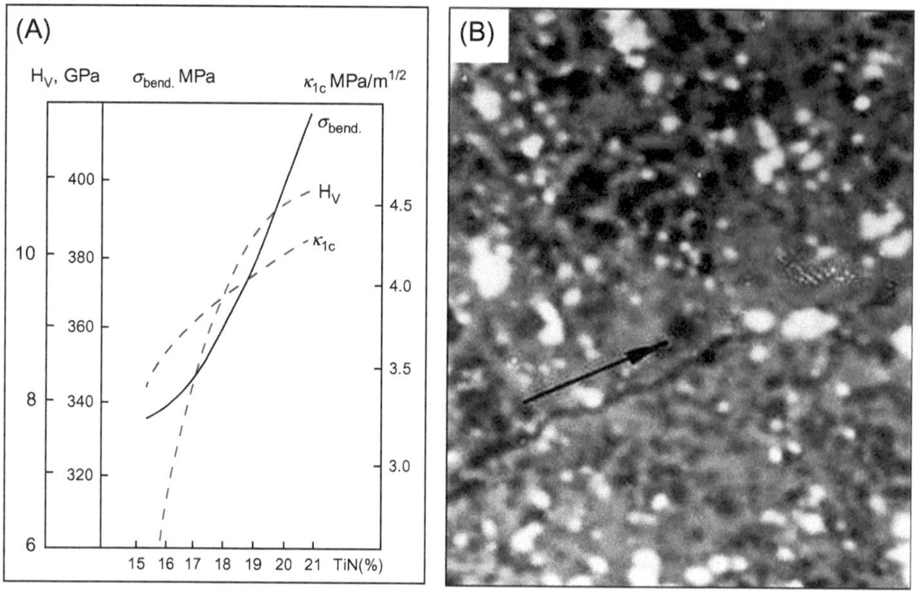

Fig. 2 (A) Strengthening by dispersion; (B) crack arresting.

Fig. 3 Some machine parts made of black ceramics: (1) pushers; (2) pusher bushings; (3) bushings of intermediate gear; (4, 8) friction pairs; (5) ball bearings; (6) tips on pusher bar; and (7) wear-resistant plates.

matrix and dispersion-strengthened particles TiC_xN_y was achieved by optimization of SHS-gas-statting parameters. The materials of $\sigma_{bend} = 830$ MPa and Rockwell hardness HRA = 94 were obtained.

Another important ceramic material (based on γ-AlON) as a compact with practical theoretical value of density (3.7 g/cm) was obtained from SHS γ-AlON powders by sintering at high temperatures and hot pressing at 1900°C. γ-AlON powders had been synthesized by thermally or chemically coupled combustion reactions of high-caloric and low-caloric mixtures. The main difference in the previously described thermal-coupled combustion reactions (chemical furnaces) using Al-Al_2O_3 + TiC mixtures (as compounds with high thermal effect) for γ-AlON synthesis is that in in these reactions, the Al + Al_2O_3 mixture was spatially separated from the chemical furnace that was a source of additional heat [7,8]. γ-AlON powders with energy additives (such as $Mg(ClO_4)$, $KClO_4$, and so on disengaging oxygen at their decomposition that reacts with aluminum and increases the exothermic effect within the first stage of combustion) possess some remarkable properties. They are easily milled, the particle structure of the SHS material is analogous to that of conventional materials (nanoplates of ~20 nm in thickness gathered in agglomerates; sometimes mosaic structures can be formed), with internal microstructure of the layers presenting packages of self-ordered nanostructures of geometrically complicated particles, as shown in Fig. 4A and B. Study of γ-AlON microstructure dependence on different combustion parameters allowed synthesizing γ-AlON powders with particles of various shapes, materials with glasslike transparent sections and γ-AlON melts. According to the analysis of optical properties of γ-AlON powders, they can be used as raw materials for manufacturing shortwave light-transmitting transparent ceramic materials and articles.

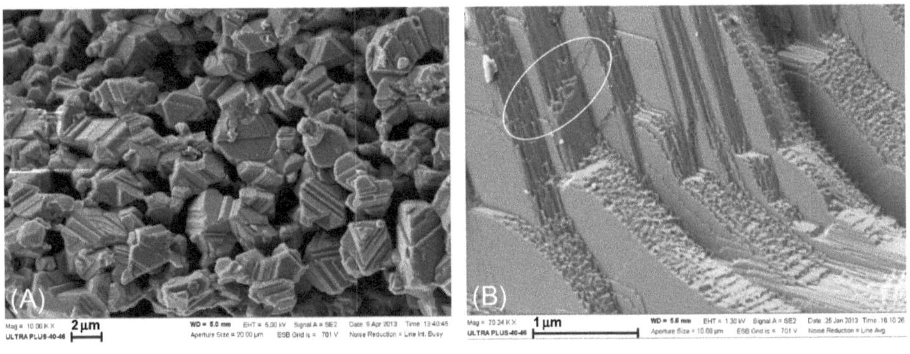

Fig. 4 (A) Microstructure of γ-AlON powder particles; (B) nanostructure of γ-AlON particle layers.

REFERENCES

[1] Merzhanov AG, Borovinskaya IP, Loryan VE, Blinov MYu, Mukasyan AS. A method of ceramic items obtaining. Internatoinal Patent WO90/11261, 1990;7082(English), 7092 (French); 1990.

[2] Borovinskaya IP, Loryan VE, Mukasyan AS. Gas-static technology of ceramic materials. In: Technology, equipment, materials. vol. 1. 1988. p. 16–20.
[3] Gromov A, Ilyn A, Ditts A, Vereshchagin V. Combustion of Al-Al_2O_3 mixtures in air. J Eur Ceram Soc 2005;25(9):1575–9.
[4] Lee J, Lee I, Chung H, Ahn J, Kim D, Kim B. Self-propagating high-temperature synthesis for aluminum oxynitride (AlON). Mater Sci Forum 2006;510–511:662–5.
[5] McCauley JW, Patel P, Chen M, et al. AlON: a brief history of its emergence and evolution. J Eur Ceram Soc 2009;29(2):223–36.
[6] Merzhanov AG. Thermally coupled processes of self-propagating high-temperature synthesis. Russ Dokl Akad Nauk 2010;434(4):489–92.
[7] Borovinskaya IP, Ignat'eva TI, Semenova VN, Chemagina EA. Aluminum oxynitride by SHS in chemical furnace. Int J SHS 2015;24(3):142–7.
[8] Borovinskaya IP, Akopdzhanyan TG, Chemagina EA. Aluminum oxynitride by SHS in chemical furnace. Int J SHS 2016;25(2):119–24.

Borides

Onuralp Yücel, Murat Alkan, Ahmet Turan

Metal borides are one of the subgroups of refractory hard materials and can be used in various applications, including the defense, nuclear industry, and high-temperature industries. Relatively high melting and stability temperatures; higher hardness and better electrical and thermal conductivities than oxide ceramics; and corrosion, wear, and thermal shock resistance are the main characteristic properties of the metal borides [1].

SHS is one of the common production method for metal borides. The first study on the investigation of solid-flame propagation to synthesize refractory metal borides (TiB_2 and ZrB_2) was made by Merzhanov et al. [2]. The main purpose of their study was the development of a new production process that would reduce energy consumption via a simple and fast route and that would enable the production of higher quality products. TiB_2 and ZrB_2 products were synthesized with 90% efficiency starting from elemental powders. Metal borides can be synthesized via SHS both by direct reaction of their elements and by using metal oxide mixtures with the presence of a reducing agent [3]. The SHS reactions can be shown as

$$\nu_1\ \mathrm{Me} + \nu_2\ \mathrm{B} \rightarrow \mathrm{Me}_x\mathrm{B}_y + \mathrm{Q} \tag{1}$$

$$\nu_1\ \mathrm{MeO} + \nu_2\ \mathrm{B_2O_3} + \mathrm{R} \rightarrow \mathrm{Me}_x\mathrm{B}_y + \nu_3\ \mathrm{RO} + \mathrm{Q} \tag{2}$$

where Me is a metal (Ti, Zr, W, Cr, etc.), R is a reducing agent (Mg, Al, Ca, etc.), Q is the generated heat during reaction, and ν_1, ν_2, ν_3, x, and y are molar amounts of the compounds.

The elemental system (Eq. 1) is the ideal one for the investigation and realization of solid flame phenomenon because of the extremely high heat generated during the reactions and the high melting temperature of the final products. The study by A. G. Merzhanov and coworkers provided the new modes of combustion wave propagation and a new approach to solving the problems related to combustion theory [2,4]. In addition, the particle size, compacted density of the raw materials, as well as the initial stoichiometric ratios, amounts of functional additives, and initiation temperature on the production of SHS borides are important parameters[5]. Initiation temperature and initial stoichiometric ratios of the raw materials directly influence the adiabatic reaction temperature (T_{ad}) and properties of the final products. Thermodynamic calculations obtained by a thermochemistry database

Concise Encyclopedia of Self-Propagating High-Temperature Synthesis
http://dx.doi.org/10.1016/B978-0-12-804173-4.00017-X

software (FactSage, HSC, Thermo, etc.) using the Gibbs Energy Minimization Method can give a prediction of T_{ad} and possible products. It was, for example, found that the stoichiometric reaction for TiB_2 production by magnesiothermic SHS was not suitable for obtaining only TiB_2 and MgO [6]. The reaction mechanism, structural formations, and productivity of SHS reactions are directly related to the T_{ad} to melting temperature ratio of the main products. At the end of the SHS reactions, three possibilities can be shown: both of the main products can be solid; one of the products can be melted, containing particles of the other product; or a mutually soluble heterogeneous melt can be formed. In the magnesiothermic process, MgO in the SHS product can be removed easily by acidic or alkali leaching, and high-quality B_4C, BN, TiB_2, and ZrB_2 powders can be produced on industrial scales. In the aluminothermic process, $B_4C+Al_2O_3$, $TiB_2+Al_2O_3$, etc., composite powders can be produced. The production cost of the selected SHS powders was calculated and compared with the commercially produced powders, and it was found that magnesiothermic SHS processes can produce TiB_2 powders for a 40% lower cost [6]. With the use of the SHS method, metal boride powders can be obtained at nanoscale without grinding processes. So contamination of the final product can occur at a lower degree than with use of the commercial carbothermic process. The advanced ceramic powders (B_4C, TiB_2, and ZrB_2) were produced by using a mixture of B_2O_3 and reducing Mg powders with C, TiO_2, or ZrO_2 powders. In the SHS studies, the effect of different

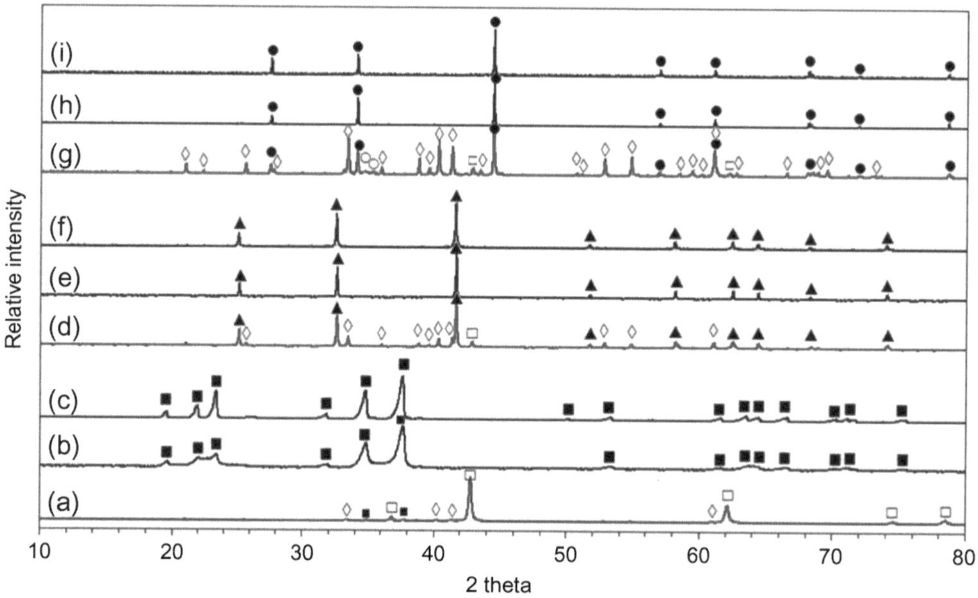

Fig. 1 XRD analysis of (a) unleached B_4C, (b) leached B_4C, (c) commercial B_4C, (d) unleached ZrB_2, (e) leached ZrB_2, (f) commercial ZrB_2, (g) unleached TiB_2, (h) leached TiB2, and (i) commercial TiB_2 powders. (▪: B_4C, ▲: ZrB_2, •: TiB_2, □: MgO, ○: $Mg_3B_2O_6$, ◊: $Mg_2B_2O_5$).

initial ratios on the final product's quality was investigated, and SHS products were cleaned with diluted HCl acid solutions. High-purity submicron size SHS borides consisting of porous agglomerated particles were obtained [7–9]. The comparative XRD analyses of the SHS products are given in Fig. 1.

REFERENCES

[1] Weimer AW. Carbide, nitride, and boride materials synthesis and processing. London: Chapman & Hall; 2007.
[2] Merzhanov, A.G., Shkiro, V.M., Borovinskaya, I.P., U.S.S.R. Inventor's Certificate 255221, 1967 (U.S. Patent 3726643, 1973).
[3] Borovinskaya IP. Pure Appl Chem 1992;64(7):919–40.
[4] Merzhanov AG. 40 Years of SHS: a lucky star of a scientific discovery. Sharjah: Betham e-Books; 2012.
[5] Munir ZA, Anselmi-Tamburini U. Part II, Ch. 2 In Handbook of ceramic hard materials. Wiley-VCH: Weinheim; 2000.
[6] Merzhanov AG. J Mat Pro Tec 1996;56:222–41.
[7] Akkas B, Alkan M, Derin B, Yücel O. Adv Sci Tech 2010;63:251–6.
[8] Alkan M, Sonmez MS, Derin B, Yücel O. Solid State Sci 2012;14(11–12):1688–91.
[9] Turan A, Bugdayci M, Yücel O. High Temp Mat Pr 2015;34(2):185–93.

Boron Carbide

Dmitry Y. Kovalev, Sergey V. Konovalihin

Boron carbide was first synthesized in 1889 by Henry Moissan [1]. Boron carbide crystals are shaped in the form of thin lamellas or prisms and have a black color (Fig. 1). A variety of chemical, physical, and mechanical properties of boron carbide ensure its wide range of applications in modern technology. High hardness (boron carbide is third in the list of the hardest compounds, the first two being diamonds and cubic boron nitride) makes the application of boron carbide possible as an abrasive and cutting material. In nuclear reactors, boron carbide is used as a material for rods that controls the kinetics of nuclear fusion. Boron carbide is highly stable chemically in various aggressive environments.

A high melting temperature (2000–2450°C) makes the use of boron carbide possible as a thermal insulation coating. Boron carbide is a high-temperature semiconductor and is used as a material for thermocouple electrodes utilized at elevated temperatures in inimical environments where metallic electrodes decay rapidly.

Boron carbide crystals have the hexagonal crystalline system, space group $R3m$. Boron carbide is a phase with variable composition. Crystal lattice parameters depend on carbon content within the boron carbide. It was found that boron carbide composition varies from $B_{13.5}C_{1.5}$ to $B_{12}C_3$. Crystal lattice parameters lie within the intervals: $5.594 \leq a \leq 5.672$ Å, $11.977 \leq c \leq 12.314$ Å [2]. The boron carbide structure is based on icosahedron with composition $B_{12-x}C_x$ $(0 \leq x \leq 1)$ (Fig. 2). Icosahedrons are connected one with another by six covalent bondings B—B. Six other atoms of icosahedron are connected by covalent bondings with groups Y—Q—Z (Y,Q,Z = B,C).

The constitution of these groups is still in discussion [3]. Atoms of Y and Z groups form three covalent bondings, each with a different icosahedron. Therefore all atoms of an icosahedron are connected with neighboring structural elements by covalent bondings. That's why boron carbide possesses such high hardness and thermal stability. The place of incorporation of carbon atoms into the structure is still a disputed topic.

The most supported version claims that in the composition of Y—Q—Z group at least one boron atom must be presented. The carbon atom could be incorporated randomly in any apical position of an icosahedron (Fig. 2) [2,4].

Various methods are used to produce boron carbide, including direct synthesis from elements, carbon reduction of boron anhydride B_2O_3 or boron acid H_3BO_3, reduction of boron anhydride by magnesium in the presence of carbon, reduction of boron halogenides in the presence of carbon, sedimentation from the vapor phase, and plasma chemical synthesis.

Concise Encyclopedia of Self-Propagating High-Temperature Synthesis
http://dx.doi.org/10.1016/B978-0-12-804173-4.00018-1

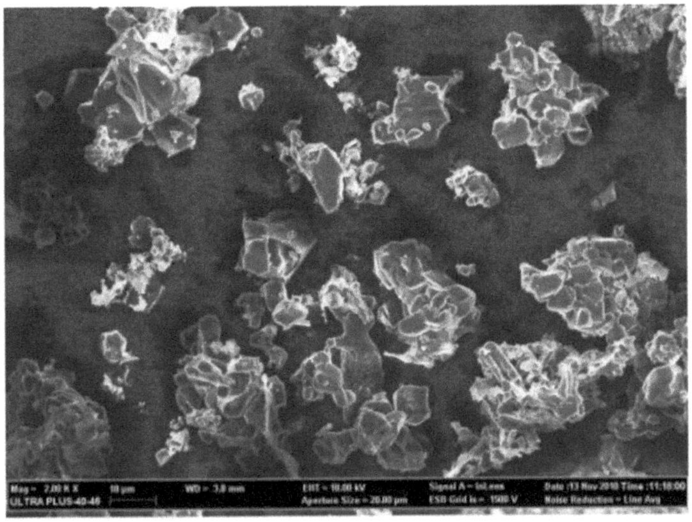

Fig. 1 Polycrystalline boron carbide.

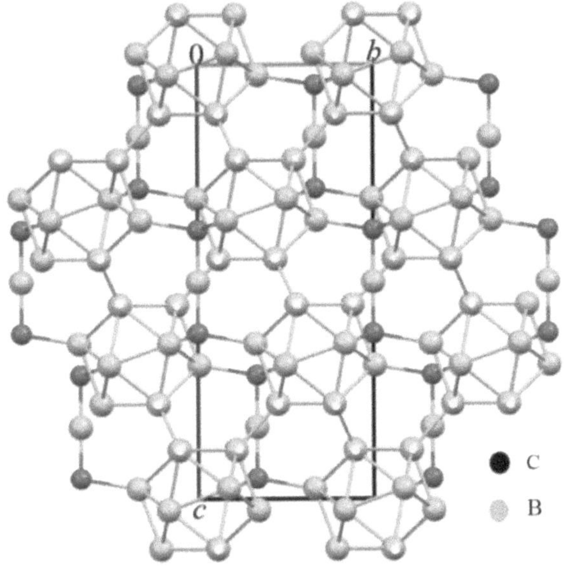

Fig. 2 Crystal structure of boron carbide with composition $B_{13}C_2$. Atoms of Y and Z within YQZ group are marked as black spheres. Apical position of icosahedron are marked as white points.

One of the most used methods is direct SHS synthesis [5,6], which is achieved by magnesium-thermic reduction of boron oxide in the presence of carbon via the reaction:

$$2B_2O_3 + 6Mg + xC \rightarrow 4B + xC + 6MgO \rightarrow B_4C_x + 6MgO$$

Synthesis is performed in a SHS reactor with 3 MPa pressure of argon atmosphere. The reaction is initiated by heating a tungsten coil. Without special burning additives, the temperature of combustion is approximately 2000–2300°C, along with a combustion velocity of 2 cm/s. If the carbon content within the reaction mixture exceeds 20%, then an independent carbon phase occurs in the final product. The SHS technique's eminence lies in its high productivity, the absence of energy spent for heating to high temperatures, and the simplicity of its production equipment.

Recently, a new derivative of boron carbide was synthesized. The obtained compound has the composition $B_{25}C_4Mg_{1.42}$ [7]. Creating new derivatives of boron carbide with peculiar physical and chemical properties using the SHS technique is feasible.

REFERENCES

[1] Moissan H. Preparation et properties de borure de carbone, CR 1899;118:556–60.
[2] Ponomarev VI, Kovalev ID, Konovalikhin SV, Vershinnikov VI. Crystallography 2013;58(3):420–5.
[3] Domnich V, Reynaud S, Haber RA, Chhowalla M. J Am Ceram Soc 2011;94(11):3605–28.
[4] Balakrishnarajan MM, Pancharatna PD, Hoffmann R. New J Chem 2007;31:473–85.
[5] Merzhanov AG, Borovinskaya IP. Dokl Acad Sci USSR 1972;214(2):366–9.
[6] Merzhanov AG. Russ Chem Rev 1976;45:827–47.
[7] Ponomarev VI, Konovalikhin SV, Kovalev ID, Vershinnikov VI, Borovinskaya IP. Mendeleev Commun 2014;24:15–6.

Boron Nitride

Alexander S. Mukasyan

Boron nitride (BN) is a chemical compound that involves equal numbers of boron and nitrogen atoms. Boron nitrate exists in various crystalline forms that are isoelectronic to carbon. The hexagonal (h-BN) form, which corresponds to graphite, is the most stable and soft among polymorphs. The cubic polymorph (c-BN) is analogous to diamond. The rare wurtzite BN modification is similar to lonsdaleite and may be harder than c-BN. BN also exists in the exotic fullerene-type molecules, as well as nanotubes and atomic layers.

h-BN has a set of unique properties, including high dielectric breakdown strength; it is not wetted by most molten metals, glasses, or salts, and hence has a high resistance to chemical attack and high-volume resistivity. Based on the preceding properties, a variety of important applications can be outlined: high-temperature electrical insulators; crucibles and reaction vessels; molds and evaporating boats [1]. h-BN powder is made primarily by the reaction between boron trioxide or boric acid with ammonia (or sometimes urea) in a nitrogen atmosphere. There are also several methods to produce BN-based ceramics, such as hot pressing and reaction-sintering. In all cases, long-term high-temperature conditions are required, which makes the materials expensive.

There are two different chemical routes to produce h-BN powder by using the combustion-based approach. One is the CS from the elements:

$$2B + N_2 + yBN = 2(1 + y)BN + Q_1 \quad (1)$$

The other one is CS by reduction reaction:

$$B_2O_3 + 3Mg + N_2 = 2BN + 3MgO + Q_2 \quad (2)$$

Both reactions are adequately exothermic to proceed in a self-sustained manner. The reduction route (2) assumes the leaching stage for complete elimination of MgO phase and is used for the synthesis of technical and ultrapure BN powders, for which the characteristics are shown in Table 1.

The benefits of CS are fully exploited in the direct production of ceramic materials (Fig. 1) in the combustion wave [3]. Some properties of CS BN-based ceramics are summarized in Table 2. For special applications (of MHD generator; erosion-residing sleeve in cutting torches for air-plasma cutting of metals and alloys), which allow the

Concise Encyclopedia of Self-Propagating High-Temperature Synthesis
http://dx.doi.org/10.1016/B978-0-12-804173-4.00019-3

Table 1 Some characteristics and properties of BN powders [2]

	CS product		Furnace product	
Content, wt.%	**Ultrapure**	**Technically pure**	**ORPAC Grade 99**	**Denka (Japan)**
Boron nitride, BN	>99.5	97.3	98–99	>98
Nitrogen, N	>55.7	54.9	54–55	54.5
Oxygen, O	<0.5	1.5	1.5	1.5
Carbon, C	<0.01	0.3	N/A	N/A
Metal impurities (Fe, Mg)	<0.2	0.3	N/A	N/A
Specific surface area, m^2/g	11.0	8–14	10	N/A

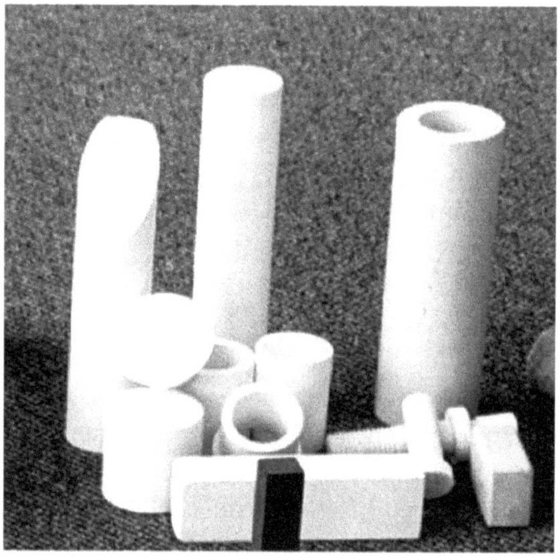

Fig. 1 Different net-shape articles (tubes, cylinders, screws, blocks, etc.) produced directly in the chemical reactor after minor finishing treatment.

Table 2 Some properties of BN-based Ceramics [4]

Properties	**BN**	**BN + SiO_2**
Chemical composition, wt %	N = 55; O < 0.5; B_{free} < 0.5; C < 0.5	BN = 74; SiO_2 = 26
Density, g/cm^3	1.5	1.85
Dielectric strength, kV/mm	25	19
Dielectric permeability	3.1	6.0
Tangent of dielectric loss at 1 MHz	0.0034	0.036
Resistivity, Ω	2×10^8	$1.3° \times 10^8$
Thermal conductivity, W/mK		
T = 400 K	20	N/A
T = 900 K	8	8
Bend strength, σ_b, MPa	25	50

presence of a SiO_2 phase, one may significantly increase the mechanical strength of the ceramics without changing the material's thermal conductivity.

REFERENCES

[1] Ertuğ B. Powder preparation, properties and industrial applications of hexagonal boron nitride. In: Sintering applications. Rijeka: InTech; 2013.
[2] Merzhanov AG. Particulate materials and processes: advances in powder metallurgy and particulate materials. Princeton, NJ: Metal Powder and Industries Federation; 1992. p. 341.
[3] Mukasyan AS. Nitride ceramics: combustion synthesis and applications. In: Gromov A, Chukhlomina L, editors. Weinheim: Wiley, VCH; 2014.
[4] Varma A, Rogachev AS, Mukasyan AS, Hwang S. Adv Chem Eng 1998;24:79–226.

Casting of Alloys

Vladimir N. Sanin, Vladimir I. Yukhvid

Based on the thermite-type reactions, the processes of SHS metallurgy gave birth to the technology named SHS production of cast alloys [1] yielding intermetallics, carbides, borides, silicides, and so on. The utmost advance was reached in the fabrication of cast polymetallic alloy and intermetallics [2,3].

Because vigorous metallothermic reactions are accompanied by splashing and loss of material, the process can be made much more effective when carried out under conditions of high artificial gravity [2], as in centrifugal machines. This approach helps to improve the yield of the targeted product, facilitate the removal of gaseous products, diminish the grain size in the product, and make the product more uniform. Fig. 1 shows sketches of a process chart and a centrifugal SHS machine.

The scheme of metallothermic SHS for fabrication of cast polymetallic alloys can be represented as

$$(\mathrm{Ox}_1 + \mathrm{Ox}_2 + \mathrm{Ox}_3 + \cdots + \mathrm{Ox}_n) + \mathrm{R} \rightarrow [\text{polymetallic alloy}] + \mathrm{R}_k\mathrm{O}_l + \mathrm{Q}$$

where Ox_n stands for the oxides of Ni, Co, Ti, Cr, Nb, Mo, W, V, Mn, and so on, R is the metal reducer (Al, Ti, K, etc.), and Q thermal is the effect of reaction.

Because of highly different densities, the melted phases undergo gravity-assisted phase segregation and subsequent crystallization. Product macrostructure markedly depends on the completeness of phase separation. At low overloads a ($<30 \times g$), a synthesized cermet represents an oxide matrix with uniformly distributed metallic drops. For $a = 50$–$100 \times g$, two-layer products comprising the desired metallic alloy as a lower layer and corundum as an upper layer are obtained. When phase separation is complete, the lower metallic layer is easily detached from the upper oxide layer.

The preceding technology is also known as centrifugal SHS metallurgy. To date, this technique has been successfully applied to fabricate a number of promising multicomponent alloys based on Ni, Co, Cr, Mo, Nb, Ti, Fe, and so on, including a new class of high-entropy alloys. SHS-produced cast alloys have the advantage of low production cost, fine-grained and uniform structure, and metallurgical heredity. Such alloys can be applied in the aerospace industry, nuclear power engineering, implant surgery, and so on. Implementation of metallothermic SHS can be expected to improve the quality of modern high alloys.

Concise Encyclopedia of Self-Propagating High-Temperature Synthesis
http://dx.doi.org/10.1016/B978-0-12-804173-4.00020-X

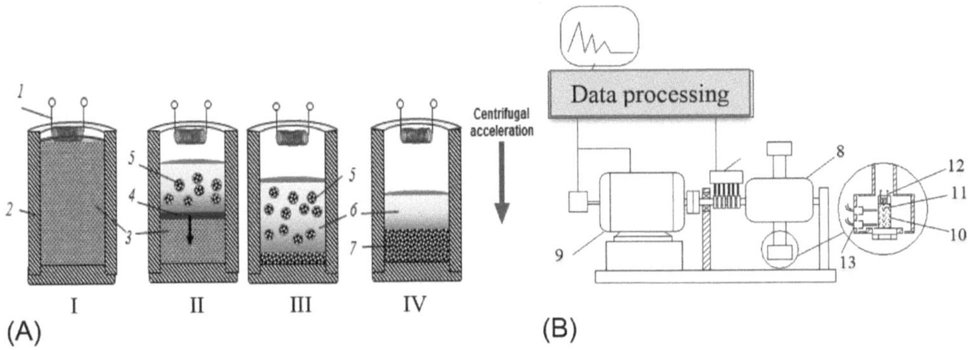

Fig. 1 Schematic of the SHS process (A) for fabrication of cast alloys and (B) sketch of centrifugal SHS machine: (I) charge preparation, (II) combustion, (III) phase separation, (IV) pattern formation. Here (1) stands for igniting coil; (2) refractory mold; (3) green composition; (4) combustion front; (5) metal drops; (6) oxide phase (Al_2O_3); and (7) cast alloy (ingot), (8) rotor, (9) electromotor, (10) mold (graphite, silica), (11) reactive mixture, (12) igniting coil, and (13) photodiodes.

For further successful development, research on the new cost-effective process for fabricating cast alloys is needed to address the problem of deviation of nominal/calculated alloy composition as well as for searching optimal cavity conditions for synthesis to assist more complete chemical conversion and, as a result, increase the yield of target products (capacity).

REFERENCES

[1] Yukhvid VI. Pure Appl Chem 1992;64(7):977–88.
[2] Sanin V, Andreev D, Ikornikov D, Yukhvid V. Open J Met 2013;3:12–24.
[3] Sanin VN, Ikornikov DM, Andreev DE, Yukhvid VI. Russ J Non-Ferr Met 2014;55(6):613–9.

Catalysts

Galina G. Xanthopoulou

The self-propagating high-temperature synthesis (SHS) method is being developed for the production of a new class of active catalyst materials based on metals, alloys, metal oxides, borides, carbides, and spinels for various applications [1]. The high velocity of the combustion wave, followed by rapid cooling, aids the synthesis of catalytically active structures because of the formation of defect-filled structures of intermediate and non-stoichiometric compounds. A wide range of materials have been produced and characterized by a variety of physicochemical and mechanical tests. A number of catalytically active SHS materials all over the world have been identified that offer promise for applications in a broad range of processes, such as highly active catalysts for exhaust emission control [2], methane oxidation [3], dehydrodimerization of methane [4], hydrogenation, dehydrogenation, pyrolysis [5], dry reforming of methane [6], soot oxidation [7], Fischer-Tropsch synthesis [8], hydrodesulfurization [8], and many other processes.

It was found that the SHS method feasibly allows preparation of new, active ceramic catalysts and carriers with compositions, structures, and properties that satisfy the stringent requirements of many applications. Interest in this well developed technology is growing every year, and many countries are working intensively with SHS and combustion synthesis catalysts, including Russia, United States, Kazakhstan, Japan, Armenia, Greece, Brazil, Spain, Portugal, Korea, India, China, Puerto Rico, and Malaysia. The SHS method is more effective than traditional preparation methods because it provides very high heating and cooling rates (10^{3}°C/s to 10^{6}°C/s) and very short completion times, on the order of minutes, which lead to formation of a highly defected structure that promotes high catalytic activity. This method is attractive for industrial production not only because it consumes much less energy than traditional production methods, introduces the possibility of "just-in-time" manufacturing, high productivity, and cheap catalysts, but also because of its simplicity, scalability, control over physicochemical properties of the products, applicability to a large range of new catalyst materials, and production of a broad range of structural forms (granules of different sizes, blocks of honeycomb-like structures, and varied geometric forms). Despite the low surface area (in the range 0.5–5 m^2/g) of SHS catalysts, they remain highly active because of their unique composition, as well as their very high density of atomic-scale defects. SHS catalysts are used in the industrial pyrolysis of naphtha (ethene and propene production) [9].

Concise Encyclopedia of Self-Propagating High-Temperature Synthesis
http://dx.doi.org/10.1016/B978-0-12-804173-4.00021-1

Table 1 Comparison of activity of several SHS catalysts with traditional catalysts

PROCESS/conditions	SHS catalysts/ characteristics	Traditional catalyst/ characteristics [10]
CO oxidation	Cu-Cr spinel [2]	0.05%Pd/Al_2O_3
Temperature at 50% conversion	410 K	500 K
CH_4 deep oxidation of methane	K-8 [3]	Pd on carrier
At 820–870 K	60%	50%
Octane dehydrogenation	Cobalt-based catalyst [9]	Ni/ZrO_2
H_2 yield	100%	95%
Synthesis gas combustion	Co-Al-O catalyst [9]	$CuCr_2O_4$, CuO, Cr_2O_3
Temperature at 100% CO conversion:	(spec. area 1.4 m^2/g) 790 K	(spec. area 123 m^2/g) 780 K
Diesel fuel pyrolysis	Co-based catalyst [5]	KNO_3 on mullite-corundum carrier
Ethylene yield without vapor:	38% (coke 0.86)	28% (coke 3.2)
Methane oxidative dehydrodimerization	Mn-based catalyst [4]	LiCl-MnO_2
Ethylene yield:	20%	15%

Table 1 compares the activity of several SHS catalysts to traditional catalysts.

One of the new directions spawned from the SHS method is solution combustion synthesis (SCS) [11]. SCS, which is used produce supported catalysts (oxides, spinels, metals, alloys), involves impregnation of the reaction solution within the porous structure of the inert solid support, followed by reaction initiation. This method permits the synthesis of supported catalysts with extremely high surface areas (up to 200 m^2/g). SCS has been used to fabricate active catalysts for dry reforming of methane, hydrogenation [12] reforming, CO, and propane [13] oxidation, and other processes.

REFERENCES

[1] Xanthopoulou G. Adv Sci Technol 2010;63:287–96.
[2] Xanthopoulou G, Vekinis G. Appl Catal B Environ 1998;19:37–44.
[3] Xanthopoulou G, Vekinis G. Appl Catal A Gen 2000;199(2):227–38.
[4] Xanthopoulou G. Appl Catal A Gen 1999;185:L185–92.
[5] Xanthopoulou G. Appl Catal A Gen 1999;182(2):285–95.
[6] Karanasios K, Xanthopoulou G, Vekinis G, Zoumpoulakis L. Int J SHS 2014;23(4):222–31.
[7] Hirano T, Tosho T, Watanabe T, Akiyama T. J Alloys Compd 2009;470(1-2):245–9.
[8] Borshch V, Pugacheva E, Zhuk S, Sanin V, Andreev D, Yukhvid V, et al. Kinet Katal 2015;56(5):690–7.
[9] Xanthopoulou G, Vekinis G. Eurasian Chem-Technol J 2010;12(1):17–21.
[10] Gladoun-Xanthopoulou G. Int J SHS 1994;3:51–8.
[11] Patil KC, Aruna ST, Mimani T. Curr Opinion Solid State Mater Sci 2002;6(6):507–12.
[12] Dinka P, Mukasyan A. J Phys Chem B 2005;109:21627–33.
[13] Hirano T, Purwanto H, Watanabe T, Akiyama T. J Alloys Compd 2009;470(1–2):263–6.

Cathodes-Targets

Yury S. Pogozhev, Artem Yu. Potanin, Evgeny A. Levashov

The technology of forced SHS pressing makes it possible to obtain a wide range of composite cathodes-targets, based on ceramics, metalloceramics, and intermetallide compounds, for ion-plasma deposition of functional multicomponent nanostructural coatings. The cathodes-targets obtained via forced SHS pressing have high relative density, homogeneous or gradient structures, and high chemical purities along with necessary mechanical properties and thermal and electrical conductivities. Composite targets are used for magnetron sputtering allowing the substance to be transferred with a homogenous flow of metallic (Ti, Ta, Cr, Mo, Al, Zr, etc.) and nonmetallic (Si, C, O, N, P) atoms and ions, since the cathode contains all the elements required for the formation of the coating. Both disk and planar-extended segmented cathodes-targets (Fig. 1) can be used in magnetron installations.

In recent years, many compositions of cathodes-targets, intended for the deposition of functional coatings onto cutting tools, parts of aircraft and rocket engines, internal-combustion engines, medical devices, and so on have been developed. Multicomponent composite cathodes-targets in the TiB_2-Ti_5Si_3, TiC_xN_y-Ti_5Si_3-$TiAl_3$, CrB_x-Cr_5Si_3-Cr_4Al_{11}, $MoSi_2$-MoB, Mo_5SiB_2, ZrB_2-$ZrSi_2$, SiC-B_4C, and so on systems have been synthesized for the deposition of heat-resistant coatings with an enhanced thermal stability [1–3]. SHS cathodes-targets made of $TiC_{0.5}$-CaO, TiC_xN_y-Si_3N_4-CaO, $TiC_{0.5}$-Ti_3PO_x-CaO, and (Ti,Ta)C_x-Ti_3PO_x-CaO compositions [4,5] have been used for

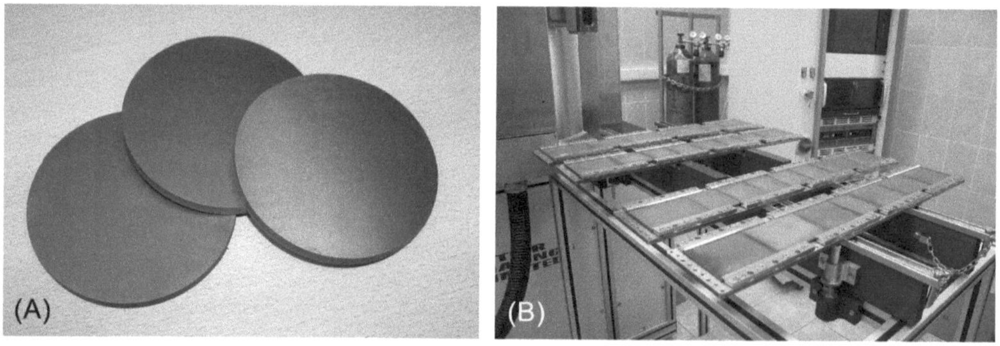

Fig. 1 Composite disk (A) and planar-extended segmented (B) ceramic-based targets obtained via forced SHS pressing.

Concise Encyclopedia of Self-Propagating High-Temperature Synthesis
http://dx.doi.org/10.1016/B978-0-12-804173-4.00022-3

the deposition of biocompatible bioactive coatings onto orthopedic and dental implants, as well as implants for cranio-maxillofacial surgery, fixation of the cervical and lumbar spine, and so on.

The SHS process combined with preliminary mechanical activation (MA) of the reactionary mixture is successfully applied for obtaining $M_{1-n}AX_n$-phase targets based on the following compounds: Ti_3AlC_2, Ti_2AlC, Ti_3SiC_2, Cr_2AlC, and $Ti_{1-x}Cr_xAlC$ [6]. They are applied in the deposition of corrosion- and heat-resistant coatings. In addition, MA SHS made it possible to obtain cathodes-targets on the basis of CrB_x and MoB borides, as well as of Ti_2CrB_2 and Cr_4Ti_9B [6] ternary compounds for depositing hard wear-resistant coatings.

Resistance of the cathodes-targets to thermal cycling is increased because of the application of functionally gradient materials (FGMs) [7] when a ceramic sputtering layer (for example, $TiC_{0.5}$-Ti_3PO_x-CaO) is fastened to a load-bearing layer, which has high thermal and electrical conductivity (for example, one made from titanium). Additionally, the difference between thermal expansion coefficients of the ceramic and metallic layers is smoothed because of the intermediate gradient layers. The working life of such FGMs cathodes-targets is limited by the thickness of the sputtered layer. The second way to improve thermal stability is to reinforce the material with a refractory component (Ta, W, Mo, etc.) in the form of continuous fibers or meshes, which are inserted into the reactionary mixture at the stage of the charge briquette obtaining prior to SHS.

The SHS cathodes-targets have been successfully tested and applied in different installations for ion-plasma sputtering: in direct-current magnetron systems, high-frequency and pulse magnetron systems, magnetron systems with additional inductively coupled plasma, arc evaporators, and so on.

REFERENCES

[1] Pogozhev YuS, Levashov EA, Zamulaeva EI, Potanin AYu, Vlasova AYu, Novikov AV, Kochetov NA. Galvanotechnik 2015;2:371–6.
[2] Levashov EA, Pogozhev YuS, Potanin AYu. Ceram Int 2014;40:6541–52.
[3] Fedotov AF, Amosov AP, Ermoshkin AA, et al. Russ J Non-Ferr Metals 2014;55:477–84.
[4] Potanin AYu, Levashov EA, Pogozhev YuS, Shvindina NV, Kovalev DYu. Ceram Int 2015;41:8177–85.
[5] Levashov EA, Rogachev AS, Epishko YuK, Kochetov NA. Russ J Non-Ferr Metals 2007;48(6):496–505.
[6] Levashov EA, Pogozhev YuS, Kurbatkina VV, et al. Advances in ceramic—synthesis and characterization, processing and specific applications. In: Sikalidis C, editor. Rijeka, Croatia: InTech Open Access; 2011. p. 3–48.
[7] Levashov EA, Larikhin DV, Shtansky DV, Rogachev AS, Grigoryan HE, Moore JJ. J Mater Synth Proc 2003;10(6):319–30.

Cemented Carbide Tools

Igor' Yu. Konyashin

WC-Co cemented carbides are composite materials consisting of a hard phase, tungsten carbide, and a ductile phase, cobalt-based binder. Because of the unique combination of different properties, including high hardness, fracture toughness, abrasion resistance and strength, WC-Co cemented carbides are widely employed for fabricating different types of tools for metal-cutting, mining, construction, and so on. Such tools are used for exploitation drilling of oil and gas, geotechnical works, coal and ore mining, and different construction works, such as tunneling, road planing, trenching, soil stabilization, and so on [1,2].

Tools for metal-cutting comprise carbide inserts, which are either mechanically attached or brazed to steel tool holders.

Mining tools for the two following major fields of application of cemented carbides can be distinguished: (1) drilling holes and wells by use of drilling bits, and (2) cutting rock, asphalt, and concrete by use of picks.

Cemented carbide tools for drilling holes and wells can be divided into the four following application ranges: (1) rotary drilling and overburden drilling by use of cutters or bits; (2) percussive drilling; (3) rotary-percussive drilling; and (4) rotary drilling by use of roller-cone bits.

Tools for rotary drilling comprise carbide plates or inserts brazed to steel bodies, which rotate without percussion and cut the rock or coal. In this case, rock or coal is destroyed mainly by compression, crushing, and cutting. A certain value of force is applied to the bit needed to overcome the rock toughness and get a certain rate of penetration into the rock. After a certain wear of carbide plates is obtained, they have to be reground to maintain the penetration rate. Bits for overburden drilling, which are mainly employed for geotechnical drilling, work similarly.

Percussive drilling is employed for degradation of rocks with medium to very high hardness and is performed by creating high-impact loads on the rock surface as a result of percussion. A bit for percussive drilling strikes the rock and rotates at a certain speed to move the position of carbide insert or plate to another part of rock. Typical percussive drilling bits comprise either carbide plates or inserts. The major mechanism of rock degradation in percussion drilling is the formation of micro- and macro-cracks in the rock under high-energy stroke and high compressive stress leading to rock crushing and chip

Concise Encyclopedia of Self-Propagating High-Temperature Synthesis
http://dx.doi.org/10.1016/B978-0-12-804173-4.00023-5

formation. Tools for rotary-percussive drilling combine characteristics of the tools for both rotary drilling and percussive drilling.

Rotary drilling performed by roller-cone bits is employed for drilling holes of large diameters, which are needed for oil and gas drilling or surface drilling of large holes, for example, in iron ore quarries. Roller cones rotate together with the drill rod and cut the rock without percussion under pressure provided by the drill rod from the surface. Carbide inserts situated on the surface of the roller cones penetrate into the rock under pressure and crush it.

Rock-, coal-, asphalt- and concrete-cutting is performed by picks with cemented carbide inserts or tips. Picks for cutting coal and rock as well as for trenching and soil stabilization are designed in such a way that they rotate either during or after each cut, leading to continuous wear propagation in the whole volume of the carbide insert. Cutting of rocks and asphalt leads to the formation of high-impact loads, intensive abrasion, and high cutting temperatures, resulting in strong thermal shock and extremely severe thermal fatigue.

Percussive drilling bits for mining can be successfully produced with inserts made of WC-Co cemented carbides fabricated from WC powders manufactured by the SHS method [3,4]. A medium-coarse carbide grade with 6% Co made from the SHS WC powder has a homogeneous microstructure with no abnormally large WC grains and is characterized by high values of TRS, hardness, and fracture toughness. Its microstructure and properties are comparable with those of the standard grade for percussive drilling. The performance of the drilling bits with the inserts made of the cemented carbide manufactured from the SHS WC powder is similar to that of the bits with the cemented carbide manufactured from the conventional WC powder. Therefore, it was established that if the SHS WC powders do not contain Mg and additives employed in the self-propagation high-temperature synthesis process and are stoichiometric, cemented carbides on their basis and tools produced with inserts of such cemented carbides on their basis are characterized by high performance properties.

REFERENCES

[1] Panov VS, Chuvilin AM, Falkovsky VA. Technology and properties of sintered cemented carbides and articles thereof. Moscow: MISIS; 2004.
[2] Konyashin I. Cemented carbides for mining, construction and wear parts. In: Sarin V, editor. Comprehensive hard materials. USA: Elsevier Science and Technology; 2014. p. 425–51..
[3] Zaitsev AA, Vershinnikov VI, Konyashin I, Levashov EA, Borovinskaya IP, Ries B, et al. High-quality cemented carbides on the basis of near-nano and coarse-grain WC powders obtained by self-propagating high-temperature synthesis (SHS). Int J Self-Propag High-Temp Synth 2015;24:152–60.
[4] Zaitsev AA, Veshinnikov VI, Konyashin I, Levashov EA, Borovinskaya IP, Ries B, et al. Cemented carbides from WC powders obtained by the SHS method. Mater Lett 2015;158:329–32.

Cemented Carbides

Igor' Yu. Konyashin

The invention of WC-Co cemented carbides by K. Schröter in Germany in 1922 was one of the most important aspects of the technological revolution during the last century. Cemented carbides are metal-ceramic composites consisting of a ceramic phase, tungsten carbide, and a metal binder, usually cobalt. WC-Co cemented carbides are used widely in various industries, including mining, construction, oil-and-gas-drilling, and so on because of their high level of hardness, their resistance to wear, and their toughness and strength.

Since their discovery, the WC-Co cemented carbides have not changed dramatically. The main improvements relate to variance in the WC grain size, Co content, and the employment of inhibitors of WC grain growth during liquid phase sintering [1]. It is well known that one can vary the hardness, wear-resistance, strength, and toughness of cemented carbides in a wide range by changing the Co content and WC mean grain size; however, the hardness and wear resistance can be increased only at the expense of fracture toughness.

Fine, ultrafine, or near-nano carbide grades with the WC mean grain size of 0.2–0.8 μm are usually employed for the fabrication of wear parts and cutting tools. The majority of industrial WC-Co carbide grades having a WC mean grain size of 2–5 μm are employed for mining and construction; they are fabricated from medium-coarse, coarse, or ultra-coarse WC powders. Cemented carbides for cutting steel also comprise grains of (Ti, Ta, W)C or (Ti, Ta, Nb, W)C mixed carbides.

Cemented carbides can be successfully fabricated by the use of WC powders produced by the self-propagating high-temperature synthesis (SHS) method [2–5].

WC powders with different grain sizes varying from near-nano to coarse-grain were produced by the SHS method [3–5]. Near-nano WC powders with mean grain sizes of about 200 nm were prepared by the SHS method including the reduction of WO_3 by Mg in the presence of carbon and regulating additives. The chemical leaching and refinement of the SHS reaction products allowed one to obtain stoichiometric WC containing only traces of oxygen and magnesium. The thermal reduction of WO_3 and V_2O_5 by magnesium in the presence of carbon resulted in obtaining two carbide phases of WC and complex carbide (W,V)C with the fcc crystal lattice having a grain size of less than 300 nm. It was established that the tungsten oxide reduction by magnesium in the presence of carbon cannot be used to synthesize coarse-grain WC powders. Coarse-grain

Concise Encyclopedia of Self-Propagating High-Temperature Synthesis
http://dx.doi.org/10.1016/B978-0-12-804173-4.00024-7

WC powders were obtained using a W + C mixture heated to high temperatures by a simultaneous exothermic reaction of interaction between magnesium perchlorate $Mg(ClO_4)$ and magnesium. The coarse-grain WC powder synthesized in such a way is nearly stoichiometric and consists of sintered round-shaped agglomerates with an average grain size of up to 16 μm and containing only traces of magnesium and oxygen. The agglomerates comprise WC single crystals of roughly 1–8 μm in size.

In Refs. [4,5] it was established that microstructures of a submicron grade with 5 wt.% Co made from the near-nano WC powder obtained by the SHS method and a medium-coarse WC-6 wt.% Co grade obtained from the SHS medium-coarse WC powder are similar with those of cemented carbides made from conventionally produced WC powders. Physical, mechanical, and performance properties of the cemented carbides made from the SHS WC powders are also comparable with those of the corresponding cemented carbides made from the conventionally produced WC powders. The microstructure and properties of cemented carbides were found to be dependent on the parameters of the SHS process, as they lead to noticeably various WC mean grain size of the original WC powders. However, if the SHS WC powders do not contain impurities of Mg and regulating additives and are stoichiometric, usually high-quality WC-Co cemented carbides can be fabricated on their basis.

REFERENCES

[1] Bose A. A perspective on the earliest commercial metal-ceramic composite. Cemented tungsten carbide. Int J Powder Metall 2011;47:31–50.

[2] Zhang J, Lee JH, Maeng DY, Won CW. Synthesis of tungsten monocarbide by self-propagating high-temperature synthesis in the presence of an activative additive. J Mater Sci 2001;36:3233–8.

[3] Zaytsev AA, Borovinskaya IP, Vershinnikov VI, Konyashin IYu, Patsera EI, Levashov EA, Ries B. Near-nano and coarse-grain WC powders obtained by the self-propagating high-temperature synthesis and cemented carbides on their basis. Part I: structure, composition and properties of WC powders. Int J Refract Met Hard Mater 2015;50:146–51.

[4] Zaytsev AA, Vershinnikov VI, Konyashin IYu, Levashov EA, Borovinskaya IP, Ries B, et al. High-quality cemented carbides on the basis of near-nano and coarse-grain WC powders obtained by self-propagating high-temperature synthesis (SHS). Int J Self-Propag High-Temp Synth 2015;24:152–60.

[5] Zaitsev AA, Vershinnikov VI, Konyashin I, Levashov EA, Borovinskaya IP, Ries B. Cemented carbides from WC powders obtained by the SHS method. Mater Lett 2015;158:329–32.

Centrifugal Casting

Osamu Odawara

It is a conventional practice that various kinds of pipes and vessels of metals used for transportation of fluids or for handling chemical substances in carrying out chemical reactions are provided with a lining of a ceramic material in order to improve their resistance against corrosion, abrasion, and heat. There are several methods for providing such a ceramic lining to a hollow body, including coating and binding methods. In a coating method such as a plasma spray coating, the inward surface is coated with ceramic materials (if necessary, by baking) with limitations in thickness. In the binding method, another hollow body of a ceramic material is shaped in advance with somewhat smaller dimensions than the housing one where the ceramic body is inserted, followed by the filling of narrow interstices with a lower melting substance.

Self-propagating high-temperature synthesis (SHS) has been used under centrifugal force for in situ applications of ceramic linings inside metal pipes and vessels, the synthesized product densifications, and/or directionally controlled uniform composite formations. As a technological application, the SHS process could successfully produce more than 6 m long ceramic-lined pipes with significant features for the production by using a thermite reaction under the influence of centrifugal force. According to the centrifugal casting process with the thermite reaction under centrifugal forces, the molten metal and ceramic synthesized by the reaction of thermite mixtures are separated in layers by virtue of their density differences under centrifugal force; the ceramic forms the innermost layer, and the metal forms the intermediate layer between the ceramic and the wall of the outer metal, resulting in strong bonding with the ceramic layer through the solidification of the ceramic layer upon cooling.

The main characteristics of the present centrifugal casting are due largely to the reaction heat generated by the SHS process, the centrifugal force applied to the reaction products, and the behavior of the reaction propagation under the influence of centrifugal force, which were confirmed through long-sized ceramic lined pipe development [1]; the SHS process proceeds rapidly along the inner hollow surface of the reactant formed under the centrifugal effect first and then into the reactant layer in the radial direction, resulting in a homogeneous quality in the direction of the pipe length under the conditions of the proper amount of thermite powders and centrifugal force. The latter reveals unique mechanical properties of the present ceramic-lined pipes; the innermost ceramic layer is compressed by the intermediate metal and outer steel pipe layers, and the stress in

Concise Encyclopedia of Self-Propagating High-Temperature Synthesis
http://dx.doi.org/10.1016/B978-0-12-804173-4.00025-9

the outer steel pipe layer is graded from tension to compression toward larger radial direction in both pipe circumference and length. With the intermediate metal layer, in which the tensile stress mainly remains, the ceramic-lined pipes produced by the process can reveal high potential as compared to other conventional ceramic lining techniques.

The quality of the produced ceramics is affected mainly by the heat and the product contents during the SHS process under centrifugal forces [2]. The effect of centrifugal force is considered to be evident in producing uniform layers and enhancing properties such as porosity and strength obtained by a squeezing test. With respect to the ceramic layer composition obtained by the reaction of the thermite powders of Al and Fe_2O_3, the phase of corundum dendritically grew in the pipe radial direction, and the hercynite (FeO-Al_2O_3 spinel) was distributed at the former boundary region, which content was about 12 wt%. Although the hercynite would play a role in improving mechanical properties of a ceramic layer such as toughness and thermal shock resistance, it should be reduced for improving its corrosion resistance, because the FeO component would be attacked in high acidic conditions.

The formation of layers was improved with increasing centrifugal force, and undesirable entanglement of the ceramic behind the product metal was decreased by more than 100 G. When the centrifugal force reached 200 G, the layers of ceramic and metal were uniformly formed. The ceramic layer produced was considerably dense, although some small cracks were observed. With increasing centrifugal force, compressive strength and density are improved, but compressive-shear stress and thermal shock resistance are not necessarily improved. Therefore the effect of the centrifugal force is considered to be evident in producing uniform layers and enhancing properties of materials such as the porosity and the strength in a squeezing test. However, the excessive centrifugal force would decrease the binding force between layers and deteriorate the internal uniformity of the ceramic, and it would give no advantage to properties such as compressive-shear stress and thermal shock resistance.

The centrifugal force also plays an important role in reaction propagation and product formation, which is realized through the control of grain sizes and compositions of the reactants. The centrifugal force affects movements of intermediate materials during the SHS process, i.e., spreading of melts, coalescence of liquid drops, and material separation by the difference of the component's specific gravity. The reaction propagation rate during the SHS process under inverse centrifugal effect toward the reaction front propagation is faster than that under the other conditions [3]; the product grains become smaller in size. Application of loads up to 1000 G to the SHS process of cast multicomponent alloys and metal-matrix composites [4] improved the yield of the designed product, removed gaseous byproduct, diminished product grains in size, and made the product composition uniform.

REFERENCES

[1] Odawara O. J Am Ceram Soc 1990;73:629–33.
[2] Odawara O. Key Eng Mater 1996;122–124:463–76.
[3] Miyazaki E, Odawara O. Centrifugal effects on combustion synthesis of (Ti-B-C) compound system. Mater Res Bull 2003;38:1375–86.
[4] Sanin V, Andreev D, Ikornikov D, Yukhid V. Open J Metals 2013;3:12–24.

Chalcogenides

Valentina K. Prokudina

Chalcogenides (compounds of metals and sulfurum, selenium, and tellurium) possess a number of valuable characteristics, one of them is a high tribotechnical (greasing) ability because of the lamellar structure of these compounds.

One of the most conventional solid greases is natural disulfide of molybdenum (MoS_2), but its performance under air atmosphere is flawed by the double increase in wearing and in its frictional coefficient.

Thus another chalcogenide, tungsten diselenide, attracts attention because of the stability of its tribotechnical characteristics, even in cases of its performance in the air. That is why the study on chalcogenides synthesis in SHS regime started from synthesis of this compound.

SHS of chalcogenides requires a special approach, because S and Se, being easy melting elements, have a boiling point lower than the combustion temperature of powder mixtures, which leads to intense evaporation of these elements from the combustion zone and, therefore, to its insufficiency in synthesized compounds. To prevent the evaporation of S and Se from the combustion zone, a variety of methods are used, including dilution of a reactive mixture with the products of reactions, addition of extra S or Se over the stoichiometry into reactive mixture, synthesis under high argon pressure, pressing the reactive mixture into briquettes, and their special arrangement within the SHS reactor.

Synthesis of WSe_2 is performed in the medium of argon in the hermetic multipurpose reactor SHS-2, which was designed to work with gas pressures up to 40 MPa.

A mixture of W and Se is used as a raw material for synthesis. To prevent the evaporation of Se, all the previously mentioned methods are used, except for dilution with the final product.

Dilution of powder mixtures that contain tungsten is not needed because of its high thermal properties (temperature of melting that is equal to 3650 K and high thermal resistivity).

To prevent heat loss through the walls of the reactor, powder briquettes are covered with thin tungsten foil and arranged on ring mandrels in the center of the reactor.

A model of the arrangement of four powder mixture briquettes within the reactor is shown in Fig. 1.

Optimal conditions for the SHS synthesis of WSe_2 are as follows: Se particles with a size less than 250 μm; excess Se over the stoichiometry, 6%; and argon pressure, 8 MPa.

Concise Encyclopedia of Self-Propagating High-Temperature Synthesis
http://dx.doi.org/10.1016/B978-0-12-804173-4.00026-0

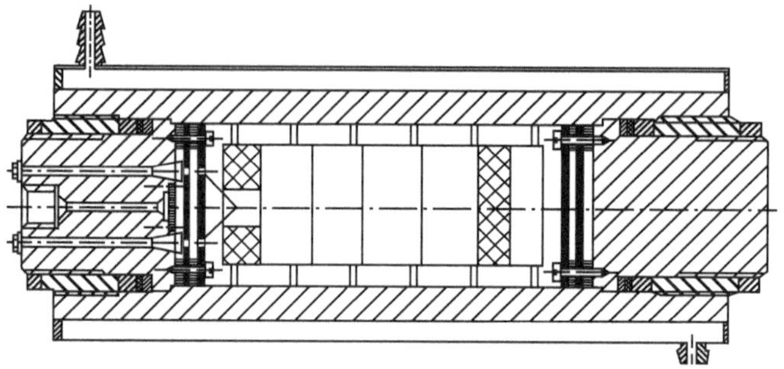

Fig. 1 Schematics of loading of powder mixture briquettes to obtain chalcogenides in universal (multipurpose) SHS reactor.

A comparison between the properties of SHS WSe_2 and its analogue, obtained in metallic (quartz) ampules through long-time heating in a furnace, showed the identity of both powders by chemical composition, crystal lattice parameters, and specific electro resistivity, physical, mechanical, and frictional properties [1].

As shown in Fig. 2, particles of both WSe_2 powders obtained in furnace and by SHS have hexahedral shape and size of 5–10 μm.

The production of SHS-sulfides was engendered from MoS_2 and WS_2 [2].

Commercial MoS_2 is usually produced from natural mineral molybdenite, but products obtained by this route contain abrasive particles (SiO_2 mostly) and oils.

Attempts to produce synthetic MoS_2 have been made for decades, but none of the technologies used allowed the procurement of MoS_2 that could compete property-wise and pricewise with the natural form.

Tungsten disulfide is also found in the Earth's cortex (in the form of tungstenite), but much more scarcely than MoS_2. The methods for synthesis of both compounds are the same.

The main challenge for SHS of MoS_2, besides preventing the sulfur evaporation, is obtaining the monophasic product. In the Mo-2S system, an undesired Mo_2S_3 compound with abrasive properties also exists.

In the W-2S system only one compound (WS_2) exists. The calculated adiabatic temperature of combustion of a Mo + 2S mixture is close to the melting temperature of the

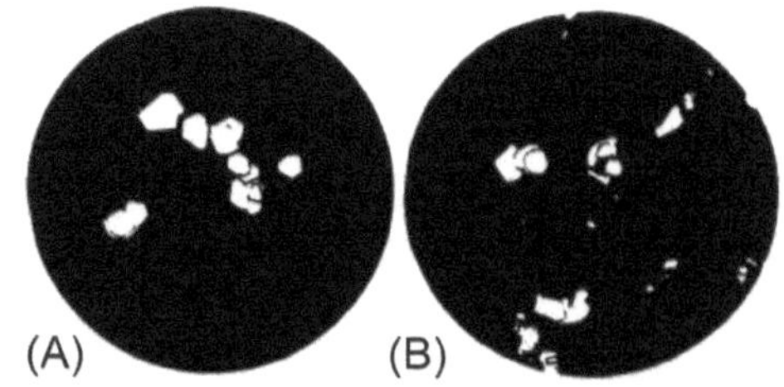

Fig. 2 Shape of WSe_2 particles obtained by ampoule synthesis (A) and SHS (B), magnification 900 ×.

molybdenum. Sulfur at this temperature exists in a liquid-vapor phase, which leads to considerable sintering of reaction mixture and the impossibility of obtaining a monophasic MoS_2. However, it is possible to obtain monophasic MoS_2 by reduction of the combustion temperature (through the method of dilution of reactive mixture by final product) and by increasing the argon pressure.

Optimal conditions for WS_2 synthesis are the same as for MoS_2 except for the need for dilution of the reactive mixture by the final product.

As synthesized, the SHS powder body easily dissipates to powders with lamellar form and peculiar greasing abilities. Comparison studies of MoS_2 powders, obtained by different methods, showed that "natural" powder has inferior properties in terms of dissolved impurities. Frictional properties of "natural" and synthesized powders are equal.

In order to produce chalcogenides with controlled electro resistivity by means of SHS, partial replacement of molybdenum by niobium in MoS_2 was used [3].

It was shown that a 5% replacement of Mo to Nb within the MoS_2 leads to a drastic decrease in specific electro resistivity of the compound down to $(1 \div 2) \times 10^{-3}$ Ω cm.

Besides $Mo_{1-x}Nb_xS_2$, other ternary compounds containing sulfur and selenium were synthesized: $Mo_{1-x}Nb_xSe_2$, $W_{1-x}Nb_xS_2$, $W_{1-x}Nb_xSe_2$.

All these compounds were produced by specific technologies that were approbated on a pilot industrial scale with an annual amount of chalcogenide production equaling 5 tonnes.

Commercially produced SHS chalcogenides were supplied to electro and radio engineering factories to be used as solid lubricants in dry friction junctions (bearings, electrobrushes, electrocontacts), in commuting electrocontacts, and in small nonwire variable resistor.

On the basis of SHS MoS_2, prospective antiwear additives "MOLUCAR" were developed and produced that were designed to be used in motor oils, polishes, paints, and emulsions for friction surfaces.

These additives differ from comparable "natural" MoS_2 because they are nontoxic and noninflammable, they overcome the analogue in frictional properties, and they are considerably cheaper.

SHS technology was developed also for zinc chalcogenides. Produced experimental batches of ZnS and ZnSe were used in thin optical films, deposited by vacuum evaporation-condensation and for obtaining optical ceramics materials with enhanced performance.

As the trials showed, SHS-produced ZnS and ZnSe exceeds the world's top commercial analogs in gas emission and light transmission.

Extra-pure SHS ZnS, ZnSe, and CdTe (with oxides content 0.02–0.04 wt.%) were used as optical materials with increased transparency in visible and infrared fractions of spectrum.

As was shown by investigations of a wide range of oxisulfides of rare-earth elements [4,5], combustion processes could be used successfully to produce chalcogenides of rare-earth elements and the continuous increase of the quality of related materials that are later used in luminophores, semiconductors, superconductors, and laser materials.

REFERENCES

[1] Merzhanov AG, Borovinskaya IP, Ratnikov VI, Prokudiva VK, Memelov VL, Gluskin YaA. Neorg Mater 1977;(5):811–4 [Russ.].

[2] Prokudiva VK, Kalikhman VL, Golubnichnaya AA, Borovinskaya IP, Merzhanov AG. Porosh Metall 1978;(6):48–52 [Russ.].

[3] Kalikhman VL, Golubnichnaya AA, Gladchenko EP, Prokudiva VK, Schepinova LP. Porosh Metall 1982;(10):56–9 [Russ.].

[4] Kazarbina TV, Maksimov YuM, Serebrennikov VV, Merzhanov AG. Neorg Mater 1980;16(11): 2014–7 [Russ.].

[5] Mishenina LN, Kazarbina TV, Kozik VV. Neorg Mater 1994;30(6):788–91.

Chaotic Regimes

Konstantin G. Shkadinsky, Alexander S. Rogachev

Chaotic refers to complex, aperiodic changes in the physical system with respect to either space or time. Chaos occurs when random processes occur within a system. Examples of random, chaotic processes are Brownian motion and turbulent motion of liquid or gas. The complex irregular behavior, possessing a basic random process, is found in many nonlinear dynamic systems. In recent years, significant progress has been made in the theory of chaotic processes; new concepts, such as the turbulent structures, turbulent crystal, and so on were developed.

Chaotic regimes of SHS were reported for the combustion of 3Ni + Al in argon [1], Ti-N_2 at low nitrogen pressure [2], and the thermite-type mixture, Cr_2O_3-Al-Al_2O_3, in air [3]. In the first two cases, chaotic combustion looks like irregular motion of hot spots (reaction sites, ~1 mm in diameter each) on the surface of porous samples. Direction of spot motion changed sporadically. In the latter case, nonperiodic oscillations were observed, and Fourier analysis confirmed the absence of any period of the oscillations [4]. The chaotic regimes of SHS were considered in the book [5].

REFERENCES

[1] Wenning LA, Lebrat JP, Varma A. Some observations of unstable self-propagating high-temperature synthesis of nickel aluminides. J Mater Synth Process 1994;2(3):125–32.

[2] Mukasyan AS, Vadchenko SG, Khomenko IO. Combustion modes in the titanium-nitrogen system at law nitrogen pressure. Combust Flame 1997;111:65–72.

[3] Strunin DV. On chaotization of the combustion front oscillations of condensed systems with melting. J Comput Math Math Phys (Zh Vychisl Mat Mat Fiz) 1991;31:543–50.

[4] Strunin DV, Strunina AG, Rumanov EN, Merzhanov AG. Chaotic reaction waves with fast diffusion of activator. Phys Lett A 1994;1925(6):361–3.

[5] Merzhanov AG, Mukasyan AS. Solid-flame combustion. Moscow: Torus Press; 2007. p. 180–9.

Concise Encyclopedia of Self-Propagating High-Temperature Synthesis
http://dx.doi.org/10.1016/B978-0-12-804173-4.00027-2

Chemically Assisted Combustion Synthesis

Suren L. Kharatyan

The primary condition for the implementation of the SHS is the ability to generate a sufficient amount of energy that makes the progression of reactions in combustion mode possible. But often it is difficult to initiate the reactions in combustion mode because of kinetic/diffusion barriers [1].

To overcome these limitations, that is, to activate low-caloric systems and/or overcome kinetic or diffusion barriers in solid-state combustion reactions, a unique approach known as chemically assisted combustion synthesis (CACS) was developed. The approach is based on the use of some minor additives playing multifunctional roles, such as

- Thermal activation of the combustion process
- Catalytic activation through the forming active intermediates
- Purification of the reagents from protective oxide films
- Improvement of the infiltration of gaseous reagent to the combustion zone, and so on

Another significant advantage of the application of minor additives in combustion synthesis is the creation of methodologies that facilitate the design of materials with required micro(nano)structures and relevant properties.

The CACS approach includes two main aspects: thermal and kinetic.

In the first case, the main factor contributing to the reaction proceeding under controlled combustion mode is ensuring a certain thermal regime ($T_{ad} > T_{ad}(min)$).

In the latter case, a high reaction rate is achieved by changing the mechanism of overall reaction without a significant change to its thermal regime.

In most cases, these two types of activation modes occur together, and in this case, we deal with the thermo-kinetic activation of reactions.

Accordingly, for chemical activation of SHS processes, the following three activation types are distinguished:

- Purely thermal activation
- Kinetic activation
- Thermo-kinetic activation

Thermal activation often is used to synthesize compounds with low enthalpy of formation. Examples of thermal limitations occur in the Si-C, B-C, W-C, Mo-C systems, and

Concise Encyclopedia of Self-Propagating High-Temperature Synthesis
http://dx.doi.org/10.1016/B978-0-12-804173-4.00028-4

so on. For these systems adiabatic combustion temperature for direct elemental synthesis is too low to generate a self-sustaining combustion wave. Therefore high temperature, needed for combustion reaction, is provided by adding a high-caloric mixture to the base reaction system. (Mg + Teflon) mixture is widely used for this purpose [2,3]. For example, adding 8–10 wt.% of this high-caloric mixture enables an increase in the adiabatic combustion temperature for the W + C system from 700°C to ~2500°C [2] and the 2Mo + C system T_{ad} increases from 700°C to 2000°C [3].

SEM examinations have shown that the morphology of produced materials is determined by the amount of the additive. For example, the Mo_2C produced at temperatures below 2200°C (less than 10% additive) is characterized by a fine particulate microstructure (≤1 μm). A higher additive content (e.g., 14 wt.%) produces coarse (15–20 μm) sintered particles. The reason for such microstructure differences was attributed to the presence of a molten phase, as the reaction temperature exceeds the eutectic temperature (~2200°C) of the Mo-Mo_2C system [3].

Kinetic activation is an effective approach to activate self-sustaining reactions in a number of other carbide systems (Ti-xC, Zr-xC, Nb-C, SiO_2-Al-C, etc.) [4].

Many halogen-containing organic (e.g., polyvinylchloride—PVC—or Teflon—PTFE) and inorganic (e.g., NH_4Cl and NH_4F) compounds are used as minor additives to provide highly reactive intermediates and thus activate rapid reactions [4]. In the Ta + C system, iodine or carbon dioxide gas are used as activators [5]. This approach was also shown to be very effective in terms of tuning the microstructure and composition of products.

It was shown that Teflon is the most suitable additive among others for the activation of combustion synthesis of TiC and ZrC below the melting point of the metal, as well as for synthesis of niobium carbides, SiC/Al_2O_3 composite, and so on. The reason is that Teflon contains the most active halogen and that the minimum amount required to activate the reactions is about five times less than for other halogen-containing polymers. For example, to activate a combustion reaction in the Nb-C and SiO_2-Al-C systems, it is sufficient to introduce only 1–2 wt.% Teflon. The activity of PTFE is so high that it may even facilitate a self-propagating reaction in millimeter-size titanium particles [4].

Typical examples of thermo-kinetic activation are the activation of (i) Si-C system by various additives (Teflon, KNO_3, $KClO_3$, etc.), (ii) SiO_2-Al-C system by the Al + KNO_3 mixture, (iii) Mo-Si-C system by Teflon for obtaining $MoSi_2$-SiC composite with high content of SiC at which the products with different microstructures are formed.

Thus based on the results obtained for a variety of low-caloric systems, it can be concluded that using the appropriate activated additive in the appropriate quantity, it is possible not only to achieve the SHS process but also to regulate the phase and chemical composition as well as the microstructure of products.

REFERENCES

[1] Merzhanov AG. Combustion processes that synthesize materials. J Mater Process Technol 1996;56:222–41.
[2] Abovyan LS, Nersisyan HH, Kharatyan SL. Combustion of tungsten with carbon under thermal activation conditions and synthesis of tungsten carbide. Chem Phys Rep 1995;13(10):1740–7.
[3] Nersisyan HH, Kharatyan SL. Combustion-wave propagation and phase formation mechanism in the molybdenium-carbon-active additive system. Combust Explos Shock Waves 2002;38(1):60–4.
[4] Kharatyan SL, Nersisyan HH. Chemically activated SHS in synthesis of refractory carbide powders, progress in self-propagating high-temperature synthesis. Key Eng Mater 2002;217:83–92.
[5] Kim T, Wooldridge MS. Catalytically assisted self-propagating high-temperature synthesis of tantalum carbide powders. J Am Ceram Soc 2001;84:976–82.

Chromium Borides

Evgeny I. Patsera, Victoria V. Kurbatkina, Evgeny A. Levashov

Chromium borides draw the attention of developers of heat-resistant and hard structural materials because of their unique combination of resistance to high-temperature oxidation and hardness, resistance to wear, and electrical and thermal conductivities. Chromium boride-based ceramics are in demand particularly for ion-plasma (magnetron) sputtering of ultrahard wear- and heat-resistant nanostructural coatings.

SHS-powder, SHS-metallurgy, and SHS-compacting technologies are the main ways of producing chromium boride. The former two technologies use, as a rule, reactions that reduce chromium oxide and boric anhydride mixtures with magnesium or aluminum [1]. For SHS compacting, mixtures of the elemental powders chromium and boron [2,3] are taken as initial reagents.

Obtaining CrB_2 through forced SHS-compacting technology is somewhat difficult because of the low exothermicity of the forming reaction $\left(\Delta H^0_{298} = -125\,\text{kJ/mol for } CrB_2\right)$, the relatively low adiabatic combustion temperature (1941 K), the high ignition temperature, and the low combustion rates. That is why, in order to use SHS, it is necessary to bring additional energy into the reaction zone, for example, by preliminary heating of the reaction mixture or by activating it mechanically (MA).

Research on the influence of MA for the 70.6%Cr + 29.4%B-composed reaction mixture's structure, as well as on the heat release and rate during combustion reaction have shown that MA leads to pulverizing of the chromium particles and increasing the structure defect density. The dependence of the absolute value of heat released during the combustion process on the MA time goes to the maximum. The increased MA time, at the first stage, increased the heat release rate and the mixture reactivity. Then the increased MA time reduced the heat released because of the mechanochemical reactions forming chromium borides within the mill drums.

It is known that MA can significantly influence the combustion wave propagation characteristics. For example, a transition from the stationary combustion mode (if the mixture is not exposed to the MA) to an unusual combustion wave propagation mode—a combination of spin and auto-oscillating modes [2] was discovered in Cr-B mixtures. It has been established that, in the MA mixture with maximum rate of exothermicity, the combustion front propagates layer by layer from the lateral surface to the center (Fig. 1), perpendicular to the sample axis and in the general combustion direction, in the form of periodically repeated sources.

Concise Encyclopedia of Self-Propagating High-Temperature Synthesis
http://dx.doi.org/10.1016/B978-0-12-804173-4.00030-2

Fig. 1 A frame-by-frame video filming (12 frames/min) of the combustion process of 70.6% Cr+29.4% B mixture mechanically activated for 21 min. Top view, the initial temperature $T_0=293$ K.

The spin combustion mode (from the lateral surface to the center), which is characteristic of solid-gas systems, is often connected with the leading role of the gas reagent when there are filtration impediments for bringing the gas to the internal sectors of the briquette that have not reacted. The auto-oscillating mode appears in weakly exothermal mixtures when the loss of heat thwarts the stationary combustion process. When the Cr-B mixture was mechanically activated (MA) that a SHS wave variety simultaneously combined the features of the auto-oscillating and spin combustion modes, which is not typical for solid-solid systems.

The SHS process can be realized within the mixture activated for 21 min even at room temperature, whereas for the weakly activated charge mixture ($t_{Ma}=1$ min), it is possible only at the initial temperature of $T_0=525$ K. In both cases, there is a linear dependence of the combustion rate on T_0. A weak dependence of the combustion temperature on T_0 for both the powerfully and the weakly activated mixtures is peculiar to this system. If MA is performed in air, the reaction mixture accumulates not only energy but also oxygen, which is both adsorbed and bound within oxide films on the surface of the particles. This is why the combustion temperature is reduced when the mixtures are mechanically activated.

Based on the experimental results, along with the low values of the effective combustion activation energy (171 kJ/mol for 1 min of the MA and 142 kJ/mol for 21 min of the MA), a sequence of chemical reactions that will determine the combustion mechanism in general has been proposed. Within the preheating zone, B_2O_3 melts and volatile suboxide B_2O_2 is formed as a result of the reaction $B_2O_{3(liquid)}+B \rightarrow 3/2B_2O_{2(gas)}-Q$. Within the combustion zone, the gaseous suboxide is chemisorbed onto the surface of the chromium particles with the formation of chromium boride according to the reaction: $B_2O_{2(gas)}+Cr \rightarrow B_2O_{3(liquid)}+Cr_xB_y+Q_1$. In addition, within the combustion zone, there is a simultaneous interaction between the solid chromium and boron particles: $Cr+B_{(solid)} \rightarrow Cr_xB_y+Q_2$. Analysis involving the compositions of the synthesized products made it possible to conclude that the depth of the chemical reactions with the formation of the products according to the scheme $CrB \rightarrow Cr_3B_4 \rightarrow CrB_2$ depends directly on the MA time. When the MA time increases, the conversion depth also increases. The established combustion laws for the MA mixtures were taken into account when developing a technology for making compact items from chromium diboride and titanium-chromium borides.

REFERENCES

[1] Raimkhanova DS, Abdulkarimova RG, Mansurov ZA. J Mater Sci Chem Eng 2014;2:66–9.
[2] Kurbatkina VV, Levashov EA, Patsera EI, Kochetov NA, Rogachev AS. Int J SHS 2008;17(3):1–6.
[3] Patsera EI, Kurbatkina VV, Levashov EA, Kochetov NA, Rogachev AS, Umarov LM. Int J SHS 2012;21(2):110–6.

Combustion Mechanisms

Yury M. Maksimov

Investigations into the mechanisms of combustion and phase formation are of great importance in understanding the SHS processes and development of structural macrokinetics [1,2]. Fig. 1 shows the two main mechanisms of the final product formation during the SHS process: equilibrium and nonequilibrium mechanisms determined by the ratio between the characteristic times of the structure formation (τ_c) and reaction (τ_r).

For the equilibrium mechanism corresponding to the condition $\tau_c \ll \tau_r$, chemical destruction of a green mixture in the combustion wave and structure formation of a final product are considered to occur simultaneously. In this case, combustion and structure formation are strongly dependent on each other.

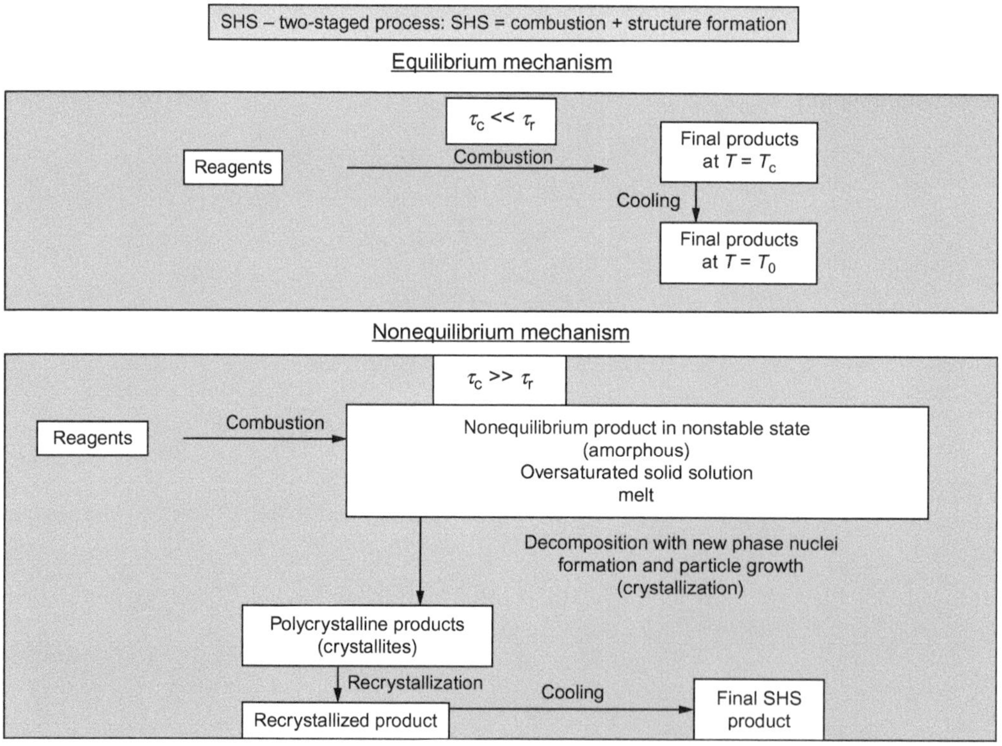

Fig. 1 Two threshold mechanisms of the final product formation during SHS processes.

Concise Encyclopedia of Self-Propagating High-Temperature Synthesis
http://dx.doi.org/10.1016/B978-0-12-804173-4.00031-4

The nonequilibrium mechanism corresponding to the condition $\tau_c \gg \tau_r$ describes a more complex structure. Here the combustion process does not depend on structure formation in the reaction products. Structure formation occurs far behind the front of the exothermic reaction.

The development of a concept for the mixed mechanism was of great importance for the development of structural macrokinetics when the mechanism of combustion was equilibrium for one of the intermediate products and nonequilibrium for another product.

It should be noted that the mechanism of combustion is a common and significant concept that includes the processes of heat and mass transfer, reaction kinetics, phase formation, as well as the combustion modes of SHS systems, the study of which allows us to create new promising materials with unique properties.

REFERENCES

[1] Merzhanov AG. Combustion and synthesis of materials. Chernogolovka: ISMAN; 1998. 512 p.
[2] Borovinskaya IP. Combustion and explosion/proceedings of the 4th all-union symposium on combustion and explosion. Moscow: Nauka; 1977. p. 138.

Combustion of Chromium in Nitrogen

Boris Sh. Braverman, Yury M. Maksimov, Ludmila N. Chukhlomina

Chromium and nitrogen form the two nitrides Cr_2N and CrN. For the reaction with formation of Cr_2N, the calculated adiabatic temperature is 1290°C, and for the reaction with formation of CrN, the calculated adiabatic temperature is 2063°C. CrN has a relatively low thermal stability. The CrN dissociation temperature is described by the equation T_d [°C] = 10,620 $(13.03 - \lg p)^{-1}$ [1] and equals 1050°C at atmospheric pressure and 1490°C at a pressure of 10 MPa. The dissociation temperature of CrN is below the calculated adiabatic temperature at pressures up to 10 MPa, which allows us to consider the CrN system to be a system with a dissociating product [2]. The maximum temperature reached during combustion of chromium in nitrogen depends on the pressure and should not exceed the dissociation temperature. The maximum combustion temperature reaches values close to the minimum temperature during the formation of liquid in the Cr-N system (eutectic at 1640°C) at low pressures. In this case, synthesis of Cr nitrides will take place without the formation of a significant quantity of a liquid phase, and the sample will preserve gas permeability close to the initial one.

Properties of chromium nitrides determine the characteristics for the combustion of chromium in nitrogen. These characteristics were experimentally investigated using a powder containing 99 wt.% of chromium.

The reaction between chromium and nitrogen proceeded in the combustion wave in two stages. Cr_2N was formed at the first stage; CrN was formed at the second stage. The temperature reached the values close to the temperature of CrN dissociation at the second stage; then the reaction proceeded at a practically constant temperature, and the degree of conversion increased at a practically constant temperature because of the heat dissipation [2,3]. There was a possibility to obtain a product consisting only of Cr_2N at low pressures (less than 0.5 MPa) since the combustion temperature reached values less than the dissociation temperature of CrN.

The relatively low temperature for the combustion of chromium in nitrogen and preserving the gas permeability of the material allowed the surface combustion regime to be visually recorded first by use of quartz plates [4]. Fig. 1 presents the video images demonstrating the evolution of the surface fronts during the combustion of the chromium powder in a flat assembly with the gas-permeable lateral surfaces. The counter-movement of the surface fronts led to the increase in the temperature in the central part

Concise Encyclopedia of Self-Propagating High-Temperature Synthesis
http://dx.doi.org/10.1016/B978-0-12-804173-4.00032-6

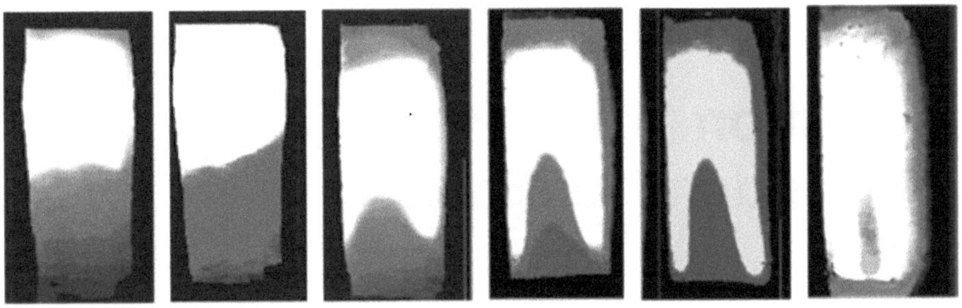

Fig. 1 Video images of the surface mode during propagation of the reaction zone.

of the sample, which might lead to melting of the combustion product. The transition of combustion into the surface mode resulted in the fact that the quality of the product depended on the size of the sample [5]. This phenomenon should be taken into account during self-propagating high-temperature synthesis (SHS) of chromium nitride.

REFERENCES

[1] Levinsky YuV. PTX phase diagrams of binary metallic systems. Reference book, vol. 1. Metallurgy: Moscow; 1990. p. 400.

[2] Aldushin AP. Nonadiabatic waves during combustion of condensed systems with dissociating reaction products. Combust Explos Shock Waves 1984;3:10–7.

[3] Braverman BSh, Ziyatdinov MKh, Maksimov YuM. Combustion of chromium in nitrogen. Combust Explos Shock Waves 1999;35(5):50–2.

[4] Braverman BSh, Ziyatdinov MKh, Maksimov YuM. Combustion of porous samples under conditions of multidimensional filtration. Combust Explos Shock Waves 2002;38(4):49–52.

[5] Braverman BSh. The scale effect during the self-propagating high-temperature synthesis of chromium nitrides. Chem Chem Technol 2009;52(8):124–5.

Combustion Wave Propagation Over Condensed Matter

Peter M. Krishenik

Propagation of reaction waves in condensed matter may proceed in two modes, either at supersonic or subsonic velocities. The former mode, referred to as detonation, involves rapid compression of matter within the shock wave and vigorous heating to temperatures that strongly accelerate the reaction; in turn, the heat released in the reaction is spent on sustaining a constant velocity of the shock wave so that it can propagate over a long distance. The theory of detonation is a large and important part of combustion/explosion science and is closely related to gas dynamics.

A subsonic mode of reaction wave propagation concerns the molecular processes of thermal conductivity and diffusion, in other words, the mode of flame propagation. The heat released in chemical reaction is transferred via conduction to the neighboring layer of unreacted material. The resulting increase in temperature thermally initiates a chemical reaction in this neighboring layer. Depending on their aggregate state, combustible condensed systems can be subdivided into two large groups: nongasifying and gasifying.

Complete or partial gasification during combustion of heterogeneous systems is caused not only by intense heat release from exothermic reactions, which are capable of sustaining the combustion wave, but also as a result of evaporation, sublimation, and thermal decomposition of condensed matter. As a rule, combustion of gasifying condensed systems is a complicated multistage process accompanied by heat release in both the condensed and gaseous phases.

SHS reactions in reactive heterogeneous media typically proceed in a mode of frontal combustion where initial reagents, some of the intermediate products, and combustion products behind the combustion wave are present in their condensed state. Some solid-state SHS reactions yield highly valued inorganic compounds or materials, and since SHS reactions are accompanied by a significant temperature rise, their extent of conversion is typically quite high. Combustion reactions in SHS systems are often multistage and kinetically controlled.

Frontal exothermic reactions yielding condensed products are normally described in terms of theoretical models that neglect the mobility of reactants and assume diffusion coefficients to be smaller than thermal diffusivity.

Concise Encyclopedia of Self-Propagating High-Temperature Synthesis
http://dx.doi.org/10.1016/B978-0-12-804173-4.00033-8

Combustion Wave Propagation Velocity

Peter M. Krishenik

Classical combustion theory [1] postulates that pulsed local thermal initiation of reactive matter gives rise to the propagation of a combustion wave with some burning velocity u. Thermal combustion waves are self-sustained by heat release from the combustion reaction and self-heating of matter. The propagation of planar combustion waves is described in terms of the model of gasless combustion, by using the equations of thermal conductivity and chemical kinetics

$$c\rho\frac{\partial T}{\partial t}=\frac{\partial}{\partial x}\left(\lambda\frac{\partial T}{\partial x}\right)+Q\rho\frac{\partial\eta}{\partial t}$$

$$\frac{\partial\eta}{\partial t}=k(1-\eta)\exp(-E/RT)$$

where T, is temperature; T_0, initial temperature; Q, heat of reaction; c, heat capacity; ρ, density; E, activation energy; and η, the degree of conversion.

A typical structure of combustion wave is given in Fig. 1.

Because of strong (exponential) dependence of burning velocity on temperature, the chemical transformation happens largely within the narrow zone in the vicinity of maximal combustion temperature T_m, which for adiabatic conditions is $T_m = T_0 + Q/c$. In the theory of frontal combustion, a key problem is finding out proper conditions for steady wave propagation at constant velocity u. Under such conditions, all model parameters depend only on the linear combination of coordinate and time: $\xi = x - ut$.

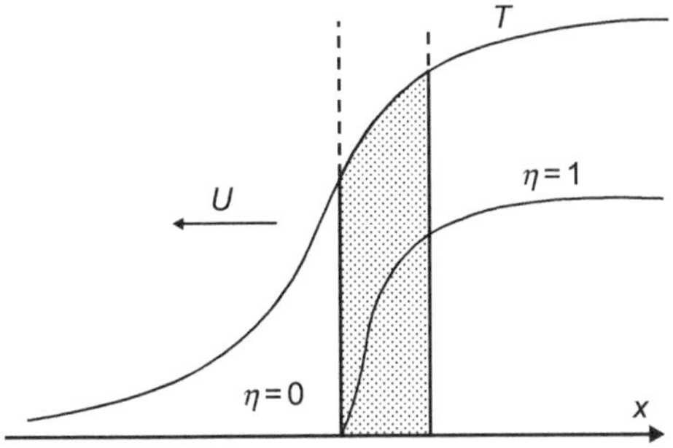

Fig. 1 Combustion wave structure: preheating zone (left), reaction zone (shadowed), and product zone (right).

Concise Encyclopedia of Self-Propagating High-Temperature Synthesis
http://dx.doi.org/10.1016/B978-0-12-804173-4.00034-X

Velocity u is sought as an eigenvalue of the problem formulated at the following boundary conditions: $x = -\infty$, $T = T_0$, $x = +\infty$, $T = T_m$. Just as for combustion of gases [1], it is convenient to subdivide the combustion wave into the three zones shown in Fig. 1. But instead of relying on the similarity of temperature and concentration profiles as used by Zeldovich and Frank-Kamenetskii [1], Novozhilov suggested [2] to make use of the similarity between the temperature gradient and the extent of conversion within the reaction zone: $\eta_{max} - \eta = (\lambda/\rho Q u)\frac{dT}{dx}$, where η_{max} is the extent of conversion attained at T_m. By matching solutions for the three zones, one obtains the expression for u:

$$u^2 = \frac{\lambda}{\rho Q}\frac{RT_m^2}{E}\exp(-E/RT_m)$$

REFERENCES

[1] Zeldovich YaB, Frank-Ksamenetskii DA. Zh Fiz Khim 1948;22:27–48.
[2] Novozhilov BV. Dokl Akad Nauk SSSR 1961;141:151–3.

Combustion Wave Velocity and Temperature Profile

Alexander S. Rogachev

Propagating velocity and temperature profile are basic characteristics of SHS wave. Strictly speaking, these characteristics are applicable for steady state combustion regime. Only in this case, stationary temperature profile $T(x)$ propagates with constant linear velocity u. The equation for propagating velocity was derived by Zel'dovich [1] as

$$u \approx \frac{1}{c\rho(T_C - T_0)} \sqrt{2Q\lambda k_0 \int_{T_0}^{T_C} e^{-\frac{E}{RT}} dT} \tag{1}$$

where c, ρ, and λ are heat capacity, density, and thermal conductivity of the medium; k_0 is preexponential factor in the Arrhenius law; Q is heat of reaction; T_0 is initial temperature; and T_C is maximum temperature of the combustion products. This equation was simplified by Frank-Kamenetskii [2] as

$$u \approx C \exp\left(-\frac{E}{2RT_C}\right) \tag{2}$$

where $C \approx \frac{1}{c\rho}\sqrt{\frac{2Q\lambda k_0 R}{E}}$ is a function that weakly depends on temperature compared with the exponent and thus can approximately (in a limited temperature range) be considered a constant.

Several types of the temperature profiles were suggested [3] and are presented schematically in Fig. 1. The "classical" profile (Fig. 1A) is called the "profile with narrow reaction zone." This profile appears if exothermic reaction and therefore heat release exist in the narrow zone near the maximum temperature T_C. The remaining temperature distribution can be described by the Mikhelson profile and is commonly referred to as the preheating zone

$$T(x) = T_0 + (T_C - T_0)\exp\left(\frac{ux}{a}\right)$$

where a is thermal diffusivity of the medium.

Concise Encyclopedia of Self-Propagating High-Temperature Synthesis
http://dx.doi.org/10.1016/B978-0-12-804173-4.00035-1

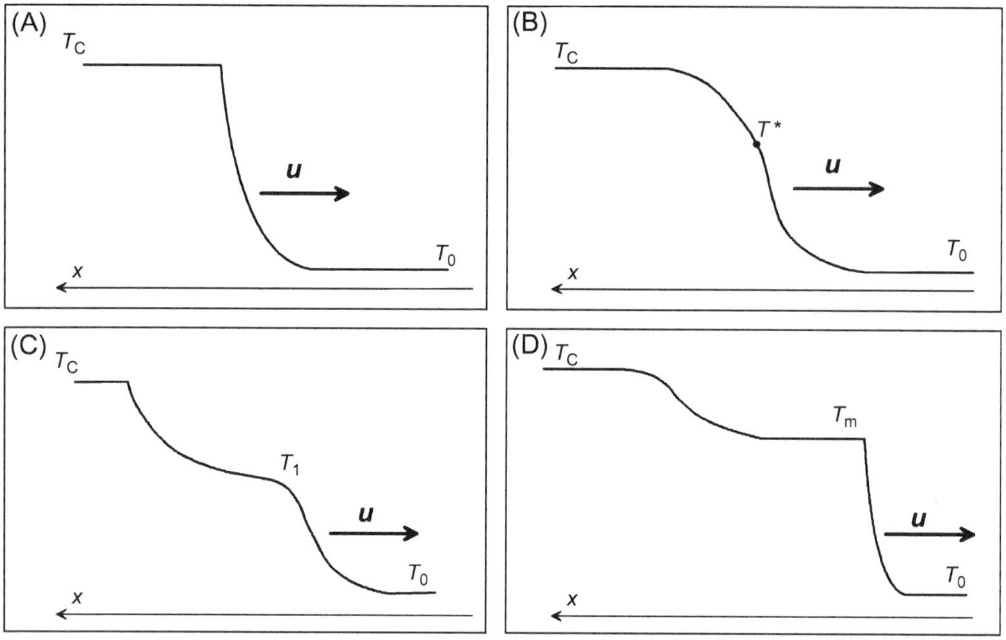

Fig. 1 Types of the temperature profiles: (A) combustion with narrow reaction zone; (B) combustion with broad reaction zone and postcombustion zone; (C) two-stage combustion; and (D) combustion with phase transition (melting).

Combustion with broad zones was described by Merzhanov and his school [3] (Fig. 1B). In this case, propagating velocity depends on an intermediate temperature T^* that must be substituted in Eqs. (1), (2) instead of T_C. The T^* corresponds to maximum temperature gradient dT/dx, or maximum rate of reactive heat evolution. Reaction behind T^* does not affect propagating velocity; this zone ($T > T^*$) is named postcombustion or postburning, zone. The broad zones appear when solid products retard the reaction. Therefore T-profile with a broad zone describes most of the SHS waves, which produce solid products, while the profile with a narrow zone better fits flames in premixed gases.

Two other profiles (Fig. 1C and D) were also suggested for combustion of condensed matters. In the case of a complex two-stage reaction, the propagation velocity may depend on the temperature of the first stage T_1. Phase transition, e.g., melting of some reactant(s) or intermediate product, in some cases leads to appearance of isothermal "shelf" on the temperature profile, and the propagating velocity depends on the temperature of melting T_m.

Real temperature profiles measured by thermocouples or pyrometers can be complex combination of the considered elementary profiles or can even present a new $T(x)$ curve. In cases when combustion is not steady state (e.g., oscillating mode), the propagating velocity and temperature profile undergo continuous changes during the combustion wave propagation.

REFERENCES

[1] Zel'dovich YaB, Barenblatt GI, Librovich VB, Makhviladze GM. Mathematic theory of combustion and explosions. New York, NY: Consultant Bureau; 1985.
[2] Frank-Kamenetskii DA. Diffusion and heat transfer in chemical kinetics. Princeton, NJ: Plenum Press; 1967.
[3] Merzhanov AG. Self-propagating high-temperature synthesis. In: Kolotyrkin YaM, editor. Physical chemistry: current problems. Moscow: Khimiya; 1983. p. 8–44.

Composite Materials

Alexander P. Amosov

The SHS process can be used to produce composites in two ways: (1) for the preliminary synthesis of components of composites with subsequent use for the manufacture of the composites by traditional technologies (ex situ); (2) for one-step synthesis of the composites from initial reagents by realization of the SHS process (in situ).

In the case of ex situ, SHS products such as porous skeletons, ingots, powders, and nanopowders can be used as components of composites. Application of such products may be justified either they facilitates the unique properties of composites or products are much cheaper compare with commercial one. For example, the performance of the process of SHS in vacuum in ISMAN, Chernogolovka, Russia, allowed researchers to obtain a porous ceramic skeleton of TiC, which was then infiltrated with the melt of high-temperature nickel alloy and formed a composite, the high-temperature strength of which was 2.5 times higher than the nickel alloy, whereas the skeleton, obtained by conventional sintering from a powder of TiC, was destroyed when impregnated with the alloy. Nanopowders, for which industrial production is not yet developed, are offered on the market at a very high price, and SHS nanopowders can be quite competitive as fillers in the manufacture of various types of composites.

The possibility of obtaining ceramic and ceramic-metal composites in situ in one step is the main advantage of the SHS process. SHS process significantly increases the productivity and efficiency of manufacturing composites compare with an industrial techniques: solid-phase sintering of mixtures or liquid-phase reinforcing metallic matrix or matrix metal infiltration directly to the porous skeleton. Rapid achievement of very high temperatures (on the order of 3000°C) and rapid cooling of combustion product allows sintering of hard-to-sinter ceramic powders, prevents the dissociation of the synthesized phases, and allows you to retain and use metastable phases in the composite, which is also an advantage of the process of SHS because it enables you to obtain composites that are impossible to obtain by traditional methods. High gradients of temperatures and concentrations facilitate the impregnation of finely dispersed ceramic phase even with nonwetted liquid phase through the implementation of the thermocapillary effect. One-step obtaining of composites by the SHS method provides intimate contact and good adhesion between the composite phases, as these phases are not introduced from the outside with inevitably contaminated surfaces, but are formed directly in the body

Concise Encyclopedia of Self-Propagating High-Temperature Synthesis
http://dx.doi.org/10.1016/B978-0-12-804173-4.00036-3

of the composite and have a fresh, clean surface. The SHS process can be carried out in the mode of layer-by-layer combustion or in the mode of volumetric thermal explosion at different heating methods, which also allows you to vary the composition and structure of the composite.

With one-step synthesis of the composite, the SHS process can be applied for the synthesis of dispersed phase or matrix phase, or both phases simultaneously. Relevant examples are composites of TiC-Ni, SD-NiAl, TiC-Al_2O_3, where the SD is the synthetic diamond (powder). During synthesis and immediately after synthesis, the phases (components) of the composite can be (a) fully solid; (b) solid and liquid (molten); (c) completely liquid. In the case (a), the porosity of the SHS product is typically more than the original powder mixture. A high level of porosity of the final product (about 50%) is one of the main disadvantages of SHS as a method of composite production. To obtain a dense porous-free composite, the application of external pressure (force consolidation) is required. This also applies to case (b) if the volume of the liquid phase is significantly less than the volume of the solid phase. Various methods of force SHS consolidation were developed for these cases (the cases of compacted SHS composites). If the liquid phase occupies most of the body or the entire body of the composite at synthesis temperatures, case (c), a dense porous-free composite can be obtained without force compaction (the case of cast SHS composite).

Fundamental studies on the application of the SHS process for the one-step production of composites were carried out in ISMAN headed by Merzhanov and Borovinskaya, where the main in situ technologies of SHS composites were developed [1]. SHS gasostatic technology allowed them to obtain nitride-containing ceramic composites at pressures of reactive gas of nitrogen up to 1000 MPa, for example, SiC-Si_3N_4 with a porosity of not more than 5%. The method of SHS pressing made it possible to produce a great quantity of ceramic-metal composites based on TiC, TiB_2, and intermetallic compounds for use primarily as hard alloys (cemented carbides). Particularly it should be noted that the effectiveness of the SHS pressing technology increases with an increase of the synthesized product dimensions, and in the case of large-size products with diameters more than 300 mm, SHS pressing is unique and is beyond comparison. SHS extrusion by extruding hot SHS products through a die allowed them to obtain long-length products made of ceramic and metal-ceramic composites with brittle and hard components. The combination of SHS with the action of shock waves and explosive loading led to dynamic consolidation of composites with significant comminution of the structural constituents. Cast ceramic and ceramic-metal composites of different compositions were obtained by SHS metallurgy on the basis of metal-thermite systems under gas pressure or in the field of centrifugal force. Adjusting the phase separation process of SHS liquid-phase products, they could obtain cast composites with matrix mixed with dispersed phase, and cast laminar composites with spatially separated phases, in particular, steel pipes with internal ceramic coating Al_2O_3. In recent years, new methods have been developed in ISMAN to produce multilayer and gradient ceramic-metal composites when SHS of intermetallides, carbides, silicides, and MAX-phases are implemented in layered systems; furthermore, intensively studies on the application of the previously listed SHS technologies to obtain nanostructured composites are conducted.

In Moscow SHS Center, Russia, SHS pressing has been used to obtain a significant number of ceramic and ceramic-metal composites from double, triple, and quadruple SHS systems, including the use of mechanically activated charge, for use as multicomponent targets for PVD-technologies of nanostructured coatings, as electrode materials for electric-spark alloying technology, as well as the diamond-bearing tool materials [2].

Other types of the force SHS consolidation for production of composites were developed in the US (SHS rolling and SHS impact die forging) and Japan (SHS with hot isostatic pressing).

In Samara SHS Center, Russia, in China and India, considerable attention is paid to developing ceramic-aluminum composites, primarily TiC-Al and TiB_2-Al. Skeleton composites containing up to 80% of the ceramic phase are obtained by combustion of appropriate powder mixtures, including aluminum powder, followed by the shock wave compaction or by the SHS pressing [3]. Aluminum matrix composites contain up to 20% of ceramic phase and are obtained either ex situ by mixing SHS ceramic powders in the melt of aluminum or in situ by carrying out the SHS process of ceramic particles directly in the melt of aluminum [4].

REFERENCES

[1] Merzhanov AG, Borovinskaya IP. Historical retrospective of SHS: an autoreview. Int J SHS 2008;17 (4):242–65.

[2] www.shs.misis.ru.

[3] Amosov AP, Fedotov AF, Latukhin EI, Novikov VA. TiC–Al interpenetrating composites by SHS pressing. Int J SHS 2015;24(4):187–91.

[4] Amosov AP, Luts AR, Ermoshkin AA. Nanostructured aluminum matrix composites of Al-10%TiC obtained in situ by the SHS method in the melt. Key Eng Mater 2016;684:281–6.

Diamond Containing Tools

Evgeny A. Levashov

The possibility of obtaining new diamond-containing materials with ceramic and metallic binders using SHS is based on the fact that, under certain conditions, a diamond grain withstands, without substantially changing its properties, a short-time exposure to a high-temperature chemical synthesis wave. The degree of diamond preservation within the combustion wave depends on a wide range of parameters, such as combustion temperature and rate, composition of the atmosphere, the mass content of diamond in the reaction mixture, dispersity and nature of the diamond, mutual solubility and interphase interaction of the diamond and the binder, intermediate and final SHS products, synthesis product cooling rate, and so on. Experimental and theoretical research was carried out for various reaction mixtures, such as Ti-B, Ni-AI, Ti-C-Ni, Ti-Mo-C, Ni-AI-Cu-Ti, Cu-AI-Ni-Sn, Ti-AI-C, Ti-Si_3N_4, Ti-BN, and Ti-Si_3N_4-Al [1–5]. An insignificant graphitization of the diamond grain surface was noted in highly exothermic systems. With an increase in the combustion temperature, for example, from 1960 to 2180 K, the degree of graphitization can increase from 3% to 6%. With the increase in combustion rate, the strength properties of the diamond are better preserved. The diamond's weakening within the volume under SHS conditions depends on the time of its exposure at high temperature rather than on the absolute value of the temperature [1,2].

As the combustion wave passes through, the adsorbed and diluted additives are emitted intensively as gases. The created reducing atmosphere restricts oxygen's access to the hot products of synthesis, which contributes to the diamond's preservation. Analysis of processes that affect the diamond within the combustion waves of various SHS systems made it possible to formulate a list of problems associated with the new method for production of diamond-containing materials and tools [3]:

1. Study of a diamond crystal's behavior within an SHS wave. Analysis of interactions of the diamond crystals with initial reagents, intermediate, and combusted products into the preheating and combustion zones. Comparison research on the strength and morphology of the diamond grains in different zones of the combustion wave. Analysis of the diamond's concentration affect on the combustion process parameters.
2. Development of technical methods to ensure competitive advantages of new diamond-containing materials.
3. Search for promising compositions of the SHS binder material to be used in various diamond tools.

Concise Encyclopedia of Self-Propagating High-Temperature Synthesis
http://dx.doi.org/10.1016/B978-0-12-804173-4.00037-5

4. Comparative laboratory and industrial testing of the obtained items.
Several ways to reduce a diamond's loss of strength within the combustion wave and enhancing the quality of the final products of synthesis have been developed [4–6].

The first way is a two-layer SHS process option (graded approach) that introduces the diamond into the reaction mixture of only one layer. This approach makes it possible to solve three tasks:

- To reduce the residence time of the diamond within the combustion wave because of the increasing combustion rate compared to the homogeneous option.
- To increase the maximum diamond concentration, because after the combustion reaction initiation, the diamond-free layer serves as a chemical oven, but the emitted heat is consumed for warming up and initiating the chemical reaction and sintering the diamond-containing layer.
- To reduce the consumption of the expensive diamond powder in the case of large-size tool production.

The second way is to add a gas-emitting substance with a dissociation temperature that is much lower than the combustion temperature into the reaction mixture in order to create a protective reducing atmosphere during the synthesis process. Titanium hydride TiH_2 is an example of such a substance. Within a hydrogen atmosphere, the diamond's resistance to high temperatures increases. In this case, the diamond's strength properties and structure change insignificantly within the combustion wave.

The third way is to switch to low exothermic compositions with low combustion temperatures and a relatively high linear combustion velocity. Metal-metal mixtures are an example of such compositions, for example, Ni-Al, Ti-Al, Cu-Al, Ti-Ni, and so on, with intermetallides and solid solutions being their products. During the Ni-Al combustion reaction, the diamond hardly changes its properties because of the high rate of combustion at a lower temperature within the reaction zone. The reaction sintering of the diamond-containing compositions is a technological method that consists of selecting low exothermal compositions with a relatively low combustion temperature. Optimization of the SHS sintering modes for different grades of diamond powder made it possible to create a technology for obtaining diamond tools: honing stones, grinding heads, and faceplates.

REFERENCES

[1] Padukov KL, Kost AG, Levashov EA, Borovinskaya IP, Bogatov YV. Int J SHS 1993;1(3):443–6.

[2] Levashov EA, Borovinskaya IP, Rogachev AS, Koizumi M, Ohyanagi M, Hosomi S. Int J SHS 1993; 2(2):189–201.

[3] Levashov EA, Borovinskaya IP, Rogachev AS, Koizumi M, Ohyanagi M, Hosomi S. Int J SHS 1994;2 (2):189–201.

[4] Ohyanagi M, Kanno M, Koizumi M. J Mater Synth Process 1993;1(5):311–22.

[5] Levashov EA, Borovinskaya IP, Ohyanagi M, Hosomi S, Koizumi M. In: Shiota I, Myamoto Y, editors. Proceedings of the 4th int. symposium on FGM, Tsukuba, Japan, Oct. 21–24; 1996. p. 283–8.

[6] Levashov EA, Akulinin PV, Sorokin MN, Sviridova TA, Hosomi S, Ohyanagi M, Koizumi M. Phys Met Metallogr 2004;97(1):41–9.

Discovery of SHS

Alexander G. Merzhanov, prepared by Inna P. Borovinskaya

Development of the method of self-propagating high-temperature synthesis SHS is closely connected with the "The Phenomenon of the Wave Localization of Solid-State Autoretarding Reactions" in 1967. [1]. In the 1960s of the XX century, the combustion mechanism of gasifying condensed systems such as explosives, gun powders, solid propellants, and so on was studied at the Research Center of the USSR Academy of Sciences (Chernogolovka, Moscow region). The aim of this investigation was to find out the role gas evolution played during combustion. A gasless thermite mixture of iron-aluminum diluted by alumina was used to check one of the hypotheses. The samples made from this mixture kept their initial weight after burning since all the reagents and products were condensed during the combustion process. Unlike the gasifying systems, gasless combustion was in accordance with the predictions of the primary combustion theory. Later, in 1970s, the search of the systems became purposeful and was extended with the use of exothermic mixtures in which refractory metals with low reactivity (niobium, tantalum, molybdenum, tungsten, etc.) and nonmetals (boron, carbon, etc.) acted as reagents while the products were refractory compounds (borides, carbides). The solid flame was found to be an efficient way for obtaining refractory materials, and the method worked out on its base was called SHS (self-propagating high-temperature synthesis) [2] and patented in the world's leading countries: USSR, the United States, the United Kingdom, France, Japan, and so on.

Solid-phase processes were carried out (solid reagents, solid products) in the first experiments at 1960s. They usually occurred rather slowly, but during our experiments we observed high velocities of the combustion front propagation. We quickly realized that the wave began to propagate even at a low conversion degree of the reaction because of the high values of its thermal effect and a significant kinetic retardation. It was not necessary to achieve the complete conversion of a substance (as it followed from the classical theory of the reaction wave propagation). SHS was related to combustion and was viewed as its variety. In fact, SHS has much in common with combustion (significant heat release at very high temperatures, front (or wave) nature of the reaction propagation, high rates of the substance heating, etc.). But there is an essential difference between SHS and combustion theory. For combustion theory, it is important to know what processes occur near the combustion front. The processes near the front determine the combustion pattern, namely the front velocity (or combustion velocity). These characteristics are also important for SHS but is insufficient. The processes behind the front are of great importance for

Concise Encyclopedia of Self-Propagating High-Temperature Synthesis
http://dx.doi.org/10.1016/B978-0-12-804173-4.00038-7

SHS, though they do not influence the front velocity. They are phase and structure transformations in primary combustion products. Therefore, on the one hand, the idea behind SHS is much broader than that of combustion because it includes not only macrokinetic concepts but also material science concepts. On the other hand, SHS theory is narrower than combustion theory because it does not consider extensive classes of combustion reactions that do not result in material formation, which leads to the conclusion

$$\text{Combustion} = \text{heat release} + \text{heat transfer}$$

$$\text{SHS} = \text{combustion} + \text{structure formation}$$

It should be noted that at first the authors of SHS wanted to call the process "combustion synthesis," taking into consideration its combustion origin. However, they finally rejected that idea when they understood the great role of structural factors.

Nearly all scientific discoveries have their own predecessor, and the SHS method is not an exception. The discovery of the SHS method was preceded by theoretical studies on the combustion mechanism of gasifying condensed systems on the one hand, and on such mechanisms of thermit systems on the other hand (Fig. 1).

The author has reviewed publications in which even classical processes known since the time of Beketov and Goldschmidt were considered as being specific variations of SHS. Today nearly all combustion processes resulting in valuable compounds and materials can be related to SHS. In addition, the combustion processes in open air were known. However, there was not a link to join the notions of combustion, synthesis,

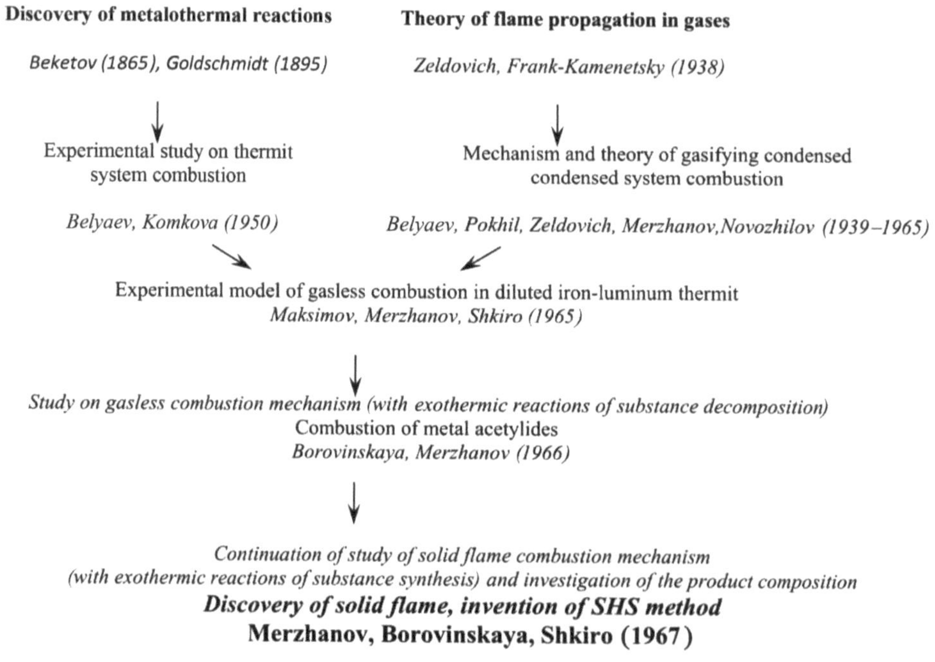

Fig. 1 Schematic concept of historical discovery of SHS.

and structure formation, which are the backbones of SHS and structural macrokinetics. SHS theory finally produced that link [3,4].

No wonder the SHS method did not appear until the 1960s of the XX century, despite numerous papers on the subject. The perception of SHS is different now. On the one hand, it was a scientific discovery; on the other hand, "everything had already been known." In his review, Professor Hlavacek [5] showed a rich collection of articles published in Western countries during the 19–20th centuries. The articles were dedicated to the exothermic effects of synthesis processes. However, these articles are not related to the history of SHS. Initially, they did not have a significant impact on the further development of the investigation; they were not connected with each other and did not result in the formulation of methodological approaches. The review by Prof. Hlavacek is not about the method's history; it is "a collection of works," though this fact does not diminish its value. Unfortunately, the review does not mention the analogous papers by the Russian scientists, particularly the article by Beketov [6], the inventor of aluminothermy, though it was published earlier than that of Goldschmidt. But there is an excellent review in the paper, "Self-propagating high-temperature synthesis: a Soviet method for producing ceramic materials" by Crider [7], who described the origin and development of SHS methods in the former USSR. Crider wrote the article at the request of McCauley, who issued 15 papers about SHS during the initial stage of its development in the United States [8] and is considered to be the pioneer of SHS in the United States. Later, the remarkable activity and excellent papers of Professors Munir (the United States) [9], Holt (the United States) [10], Odawara and Miyamoto (Japan) [11], Pampuch (Poland) [12], and other scientists from different countries made a great contribution to the development of SHS.

REFERENCES

[1] Merzhanov AG, Borovinskaya IP, Shkiro VM. The phenomenon of the wave localization of solid-state autoretarding reactions. USSR Discovery 287, registered 5.07.1967, No. 32; 1984. p. 51.
[2] Merzhanov AG, Borovinskaya IP. Self-propagating high-temperature synthesis of refractory inorganic compounds. Dokl Akad Nauk USSR 1972;204(2):366–9.
[3] Merzhanov AG. History and new developments in SHS. Ceram Trans: Adv Synth Proc Compos Adv Ceram 1995;56:3–25.
[4] Merzhanov AG, Rogachev AS. Structural macrokinetics of SHS processes. Pure Appl Chem 1992;64 (7):941–53.
[5] Hlavacek V. Combustion synthesis: a historical perspective. Am Ceram Soc Bull 1991;70(2):240–3.
[6] Beketov NN. Study on substitution of some metals by others. Thesis, Kharkov University; 1865.
[7] Crider JF. Self-propagating high temperature synthesis – a Soviet method for producing ceramic materials. Ceram Eng Sci Proc 1982;3(9–10):519–28.
[8] McCauley JW. A historical and technical perspective of SHS. Ceram Eng Sci Proc 1990;11 (9–10):1137–81.
[9] Munir ZA. Synthesis of high-temperature materials by self-propagating combustion method. Am Ceram Bull 1988;67(2):342–9.
[10] Holt JB, Danmead SD. Self-heating synthesis of materials. Annu Rev Mater Sci 1991;21:305–34.
[11] Odawara AO, Kaieda AY. A new synthesis method for ceramics. Self-propagating high temperature synthesis method. Sosei to Kako 1987;28(312):3–8.
[12] Pampuch R, Lis J, Stobierski L. Self-propagating high-temperature synthesis of ceramic powder. Sci Ceram 1988;14:15–26.

Dissolution-Precipitation

Florence Baras

Dissolution-precipitation is one of the reactive mechanisms occurring in the SHS process when the ignition temperature coincides with the melting temperature of one of the reactants. Dissolution is exothermic because of the heat released during mixing. More than 60% of heat production by SHS is due to dissolution [1]. The usual expression of the dissolution rate is as a linear function of the saturation degree of the solution

$$\frac{dC}{dt} = k(C_s - C) \tag{1}$$

where C and C_s are the concentration of the solid substance and its solubility, respectively. The explicit form of k depends on the surface of contact and other parameters such as the diffusion coefficient in the case of the "diffusion layer" model introduced by Nernst and Brunner [2]. The dissolution is supposed to be controlled by diffusion across an interfacial layer without any influence of diffusion in the liquid phase. The general form (1) is commonly used to describe the dissolution of micrometric particles into liquid [3]. However, the standard description (1) becomes questionable for reactive nanolayers because dissolution may interfere with other elemental processes such as melting, interdiffusion, or phase growth [4].

Precipitation corresponds to the occurrence and growth of a new solid phase from a saturated melt (Fig. 1). The new phase is supposed to be more stable than the liquid solution. According to the classical theory of homogeneous nucleation, the nucleation stage appears with the formation of new phase seeds, and the growth stage proceeds when the seeds exceed a critical size. The kinetics governing the transformation are either limited by diffusion in liquid or by interfacial barrier. In isothermal systems, the volume fraction $Y(t)$ occupied by the new phase follows the well-known KJMA law (Kolmogorov-Johnson-Mehl-Avrami) $Y(t) \approx 1 - \exp(-Kt^n)$, where the exponent n varies from 1 to 4 and K is related to temperature by the Arrhenius law [5]. When the volume fraction reaches its maximum value, the subsequent evolution will be associated with coarsening and coalescence of the grains (Fig. 2). The main question is the relevance of such modeling for describing reactive processes in the SHS wave that proceeds in nonequilibrium conditions (high heating rates and high combustion temperature) and that takes place in a heterogeneous material (powder mixture, stacked layers). In the case of metal aluminides, a modeling based on the coupling between macrokinetics (heat transport) and microkinetics (dissolution at the level of one particle) was carried out [6].

Concise Encyclopedia of Self-Propagating High-Temperature Synthesis
http://dx.doi.org/10.1016/B978-0-12-804173-4.00039-9

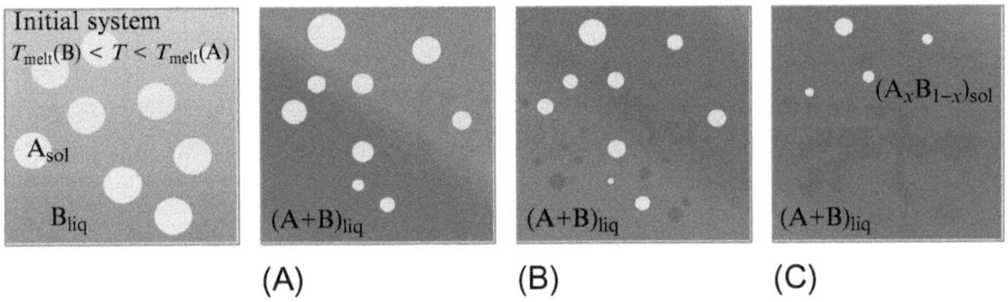

Fig. 1 Dissolution-precipitation process: (A) dissolution of the solid A particles in the melt B; (B) nucleation and growth of the new phase A_xB_{1-x}; (C) coarsening and coalescence.

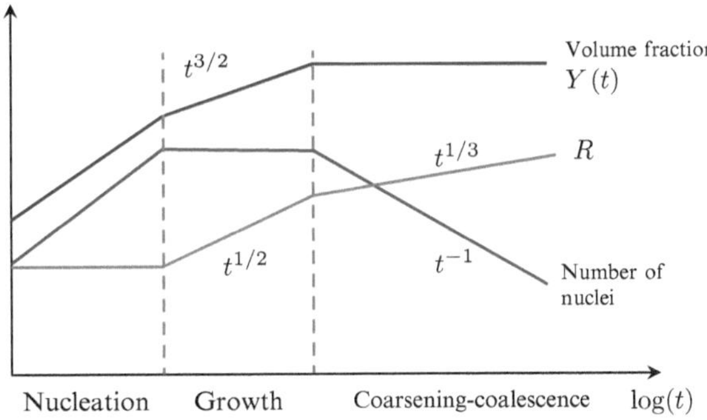

Fig. 2 Typical evolution of the volume fraction of the new phase, $Y(t)$, of the number and size of nuclei radius, R, in an isothermal system for a diffusion-controlled growth.

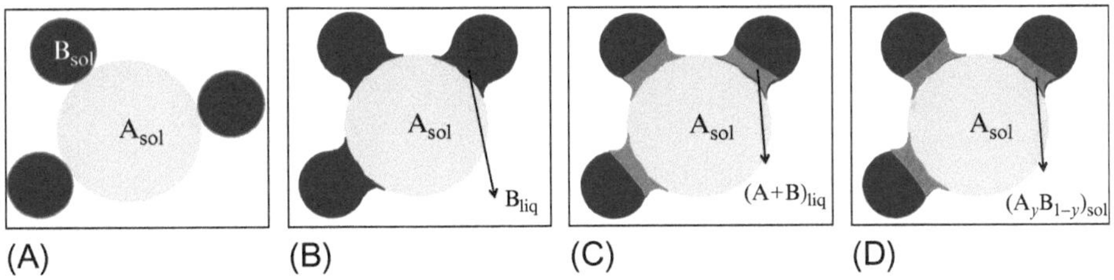

Fig. 3 First stages of the heterogeneous reaction between A and B particles: (A) initial system; (B) wetting of A solid particles by B liquid; (C) partial dissolution of A in B; (D) precipitates of B rich intermediate.

Experimental studies of nucleation are difficult to achieve mainly because of the extremely rapid phase transformations associated with combustion synthesis. The most common way to study the nucleation is by quenching the sample during the SHS propagation and analyzing the products formed along the arrested front. The available microscopic images demonstrate that the heterogeneity of the reactive powder is of primary importance. Fig. 3 shows one example of the complex mechanisms involved at the level of the particles: wetting of the solid particle A by the less refractory reactant B; partial dissolution of A into liquid B; precipitation of intermediate phases, rich in B, A_yB_{1-y}; continuation of A dissolution; formation of a shell of liquid B around. It is worth noting that some precipitate may be characteristic of an A_yB_{1-y}-B eutectic mixture. This microstructure has been observed experimentally during nickel aluminide formations in SHS wave, as an Al_3N-Al eutectic [7].

REFERENCES

[1] Grigorovich KV, Krylov AS. Thermochim Acta 1998;314:255–63.
[2] Dokoumetzidis A, Macheras P. Int J Pharm 2006;321:1–11.
[3] Baras F, Kondepudi D. J Phys Chem B 2007;111:6457.
[4] Turlo V, Politano O, Baras F. Acta Mater 2015;99:363–72.
[5] Avrami M. J Chem Phys 1939;7:1103–12. 8 (1940) 212-224; 9 (1941) 177-184.
[6] Gennari S, Anselmi Tamburini U, Maglia F, Spinolo G, Munir ZA. Acta Mater 2006;54:2243–351.
[7] Li Y, Nan Y, Guo W, Che H, Fan Q. Intermetallics 2010;18:179–80.

DTA/TGA-Based Methods

Alexander S. Mukasyan

Thermal analysis (TA) refers to a group of techniques that studies the properties of materials as they change with temperature [1]. Thermal analysis allows one to obtain a variety of properties in many compounds and reactive systems, including enthalpy, thermal capacity, mass changes, and the coefficient of heat expansion. Solid-state chemistry uses TA for studying reactions in the solid state, thermal degradation reactions, phase transitions, and phase diagrams. TA involves several methods that are distinguished by the measured property. The methods are thermogravimetric analysis (TGA), differential thermal analysis (DTA), differential scanning calorimetry (DSC), pressurized TGA, thermalmechanical analysis, dilatometry, and evolved gas analysis (EGA). Often different properties can be measured at the same time: TGA/DTA, TGA/EGA.

The TGA measures changes in weight in relation to changes in temperature. The measured weight loss curve may give information on changes in sample composition, its thermal stability, and kinetic parameters for chemical reactions. In DTA, the material under study and an inert reference are made to undergo identical thermal cycles, while recording any temperature difference between the sample and reference. The phenomena that may cause changes in heat include physical adsorption and desorption, a change in crystal structure, crystallization, melting, vaporization, sublimation, chemisorption, gas-solid reactions, and solid-state reactions. DSC shares much in common with DTA, except that DSC measures the difference in heat, which is needed to flow to the sample as compared to the reference to maintain both at the same temperature.

TGA/DTA/DSC methods are widely used to investigate the combustion synthesis reactions. Several issues can be addressed: (i) to study the characteristic temperatures (e.g., melting point, dissociation, decomposition) of the initial precursors; (ii) to find the onset temperatures for the reaction; (iii) to measure the heat released during the SC process; and (iv) to investigate the kinetics of the chemical reaction. TGA/DTA data on initial precursors can be found in almost all publications with material related to the *solution combustion synthesis*. For example, based on TGA/DTA data analysis, it was demonstrated that, in many such systems, onset reaction temperature is related to the decomposition temperatures of the fuel or oxidizer [2,3]. Another example of using TGA/DSC methods to analyze reactions in gas-solid system is shown in Fig. 1 [4]. Analysis of such data allows the authors to explain why the degree of nitridation, as well as the reaction

Concise Encyclopedia of Self-Propagating High-Temperature Synthesis
http://dx.doi.org/10.1016/B978-0-12-804173-4.00040-5

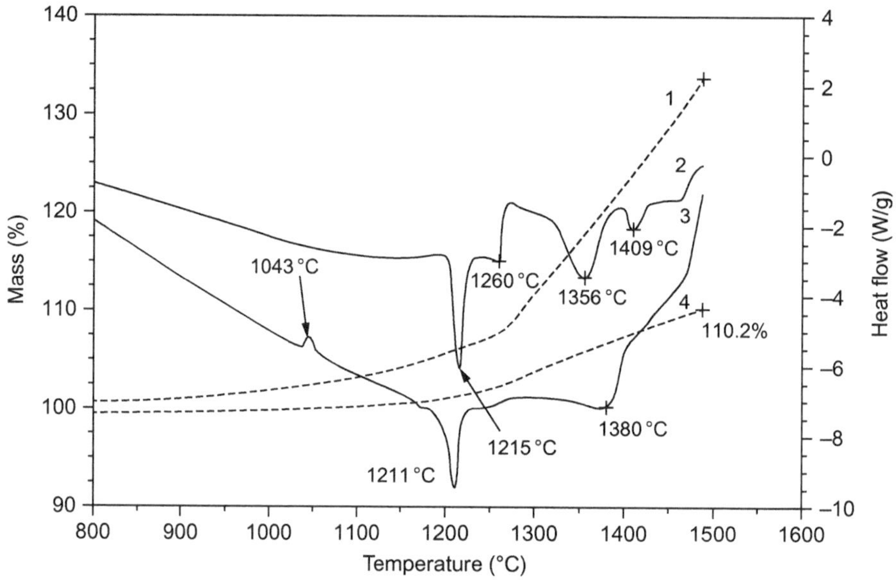

Fig. 1 TG (*1, 4*) and DSC (*2, 3*) curves of complex thermal analysis of ferro-silicon powder (*1, 2*) and powder of ferro-silicon with 30% ilmenite added (*4, 3*) in nitrogen.

rate, is lower for a ferro-silicon/ilmenite powder mixture (curve 2,4) than for ferro-silicon (curve 1,3).

A special task is obtaining kinetic data based on DTA/DSC data analysis, where an isoconversional approach is typically used [5]. The principal idea of the isoconversional method is based on two main assumptions: (i) rate of the processes is defined as a product of two independent functions, one of temperature the other of degree of conversion; (ii) The effective activation energy is obtained based on a set of kinetic trials from the dependences temperature vs heating rate, β, for incremental methods with linear heating rate. For example, for the Kissinger-Akahira-Sunose (KAS) method [6] the following relation can be obtained:

$$\ln\left(\frac{\beta}{T^2(\eta)}\right) = \ln\left(\frac{k_0 R}{E_{a,\eta} g(\eta)}\right) - \frac{E_{a,\eta}}{RT(\eta)}$$

Thus, for a constant degree of conversion, η, a plot of $\ln(\beta/T^2(\eta))$ vs $1/T(\eta)$ obtained from thermal curves recorded at several heating rates is a straight line, and its slope allows for evaluation of the apparent activation energy of the reaction, whereas the intercept allows the value of the preexponential factor to be obtained. This method has been widely used to obtain the effective activation energies for a variety of CS reactions [3,7–9].

REFERENCES

[1] Gabbott, editor. Thermal analysis: principle and applications. Oxford: Blackwell Publishing; 2008.
[2] Deshpande K, Mukasyan AS, Varma A. Chem Mater 2004;16(24):4896–904.
[3] Kumar A, Wolf EE, Mukasyan AS. AIChE J 2011;57(8):2207–14.
[4] Chukhlomina LN, Vitushkina OG, Vereshchagin VI. Glas Ceram 2011;67:9–10.
[5] Šimon P. J Therm Anal Calorim 2004;76:123–32.
[6] Kissinger HE. Anal Chem 1957;29:1702.
[7] González-Cortés SL, Imbert FE. Appl Catal A Gen 2013;452:117–31.
[8] White J, Reeves RV, Son SF, Mukasyan AS. J Phys Chem A 2009;113:13541–7.
[9] Horvitz D, Klinger L, Gotman I. Scr Mater 2004;50:631–4.

Dynamic Methods of Experimental Diagnostics

Alexander S. Rogachev, Alexander S. Mukasyan

Investigation of the variety of physical, chemical, and structural transformations, which take place during SHS, requires a comprehensive set of advanced diagnostics [1]. The use of a particular method depends primarily on the level (macro; micro; nano; atomic) of material structure and characteristic time of the investigated stages of structural transformation. The macrostructure can be examined by direct observation with a relatively small magnification, with the help of conventional video recording. The transformation of the microstructure can be investigated by the methods of optical and scanning electron microscopy (SEM). The evolution of the crystal structure and phase composition can be studied by different diffraction-based techniques, including X-ray diffraction (XRD), neutron diffraction, and/or electron diffraction. The nano and atomic level of structures are typically investigated by transmission electron microscopy (TEM), coupled with energy-dispersive X-ray spectroscopy (EDS). In addition to these structural levels, which are commonly recognized in material science, for SHS it is necessary to also study the thermal structure of the combustion wave. Such thermal structures can be examined by the methods of macro and micro video recording, as well as by micro thermocouples and infra-red (IR) thermal vision systems.

Depending on the characteristic time of the investigated stages of structural transformation, dynamic or static methods are used. The slower processes and structures that may be quenched to obtain products can be studied by static methods (XRD, SEM, TEM, EDS, etc.). Fast processes and short-lived structures are studied by dynamic or in situ (operando) methods.

The dynamics of changes in the macrostructure and thermal structure of SHS wave are studied by using laboratory setups, as schematically shown in Fig. 1.

The following diagnostics can be used to investigate both the combustion propagation and the thermal explosion modes:

- High-speed micro video recording [2], to study the microstructure of the combustion wave with time resolution up to 10^6 fps and scale resolution of 1–10 μm
- High speed IR thermal vision [3], to investigate 2D-temperature combustion maps with time resolution up to 1.5×10^3 fps, scale resolution of 10 μm, and temperature range from 300 to 2300 K

Concise Encyclopedia of Self-Propagating High-Temperature Synthesis
http://dx.doi.org/10.1016/B978-0-12-804173-4.00041-7

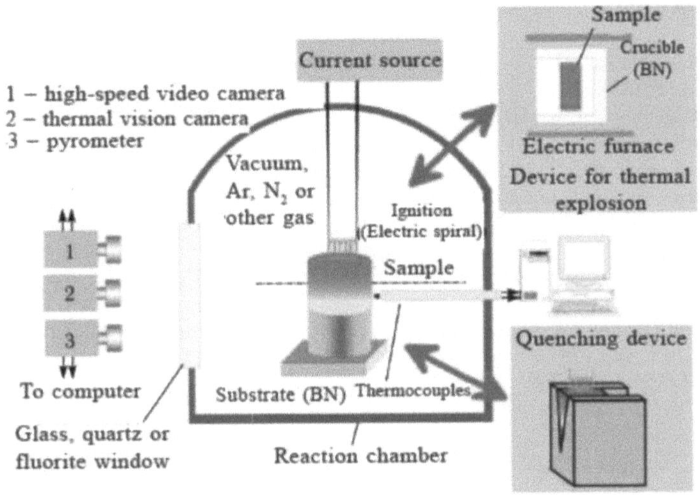

Fig. 1 Experimental setup for the study of the SHS process.

- Micro-thermocouple measurements, to explore the thermal structure of the combustion wave time resolution ~1 ms, scale resolution 7 mm, and temperature range 300–3000 K [4]
- 2D rapid optical pyrometer: spatial resolution 1 μm and time resolution 10 ms, cf. [5] or special resolution 100 μm and time resolution 1 μs, cf. [6]
- In situ mass spectroscopy, for investigation of gas phase evolution in the combustion wave [7]

To investigate structural transformations in the combustion wave on the micro; nano, and atomic scale several time-resolved diagnostics were designed and used:

- Time-resolved XRD, to obtain dynamic information on phase formation in the combustion wave, time resolution up to 50 ms using a synchrotron radiation source [8]. Use of synchrotron radiation is quite expensive, and therefore, for many tasks that do not require extremely high values of temporal resolution, it is convenient to apply the conventional X-ray tubes with time resolution ~0.1 s, cf. [9].
- Time-resolved TEM, to investigate the structural transformations with atomic scale resolution and time resolution ~3–15 ns [10]. Modern SEM and TEM electron microscopes may have special heating holders that allow controllable preheating of the samples. Using such a TEM holder, the structural transformations during the interaction between the nano-layers of titanium and aluminum in the multilayer nano-films was studied [11].

REFERENCES

[1] Rogachev AS, Mukasyan AS. Combustion for materials synthesis. New York, NY: CRC Press, Taylor & Francis; 2015. p.44.
[2] Varma A, Rogachev AS, Mukasyan AS, Hwang S. Proc Natl Acad Sci U S A 1998;95:11053.
[3] Rogachev AS, Mukasyan AS, Varma A. J Mater Synth Process 2002;10:31.
[4] Zenin AA, Merzhanov AG, Nersisyan GA. Combust Explos Shock Waves 1981;16:63.
[5] Anselmi-Tamburini U, Campari G, Spinolo G. Rev Sci Instrum 1995;66:5006.
[6] Garkol DA, Gulyayev PY, Evstigneyev VV, Mukhachev AB. Combust Explos Shock Waves 1994;30:72.
[7] Kamynina DC, Kidin NI, Kudryashov VA, Rogachev AS, Umarov LM. Combust Explos Shock Waves 2002;38:446.
[8] Rogachev AS, Gachon J-C, Grigoryan HE, Nosyrev AN, Tsygankov PA, et al. Nucl Inst Methods Phys Res A 2005;543:175.
[9] Merzhanov AG, Borovinskaya IP, Khomenko IO, Mukasyan AS, Ponomarev VI, Rogachev AS, et al. Ann Chim 1995;20:123.
[10] Kim J, LaGrande T, Reed B, Taheri M, Armstrong M, et al. Science 2008;321:1472.
[11] Gachon J-C, Rogachev AS, Grigoryan HE, Illarionova EV, et al. Acta Mater 2005;53:1225.

Electric Probes

Alexander S. Rogachev

Like other processes of combustion, SHS is accompanied by thermoelectrical and chemoelectrical phenomena. Two types of electrical probes have been used for investigation of electrical signals generated by SHS: Langmuir probes for diagnostics of gas phase and simple metal contacts or thermocouples inserted into the sample. Studies of the gas phase by an electrostatic probe located near the sample surface without contact with solid phase [1,2] have shown that low-temperature plasma forms close to the reaction zone, in which the concentration of charged particles is two to three orders of magnitude higher than the equilibrium value. The concentration of charge carriers is also one to two orders of magnitude higher than the possible concentration of charged particles because of thermal emission of the electrons from the surface of the hot condensed phase. Another approach was to measure the electric voltage that appeared between two wires, inserted into the sample, during propagating of SHS wave [3,4]. Relatively large signals (up to ~0.1–1.0 V) were detected, which were probably caused by the collective action of the Seebeck effect (thermal electromotive force), ionized gas plasma, chemical ionization, and so on. Further investigation [5] revealed very unusual effects, such as ultraviolet and even X-ray radiation from burning SHS compositions, as well as the emission of hot (100 eV) electrons. These observations coupled with the spectrometry data indicate that the reaction zone is capable of generating particles and radiation. The nature of these processes remains unclear, and further experimental research will help in understanding these phenomena.

REFERENCES

[1] Kudryashov VA, Mukasyan AS, Filimonov IA. Chemoionization waves in heterogeneous combustion processes. J Mater Synth Process 1996;4(5):353–8.
[2] Kamynina DC, Kidin NI, Kudryashov VA, Rogachev AS, Umarov LM. Ionization in a combustion wave. Combust Explos Shock Waves 2002;38(4):446–8.
[3] Morozov YuG, Belousova OV, Kuznetsov MV. Some approaches to collecting electric voltage generated by SHS reactions. Int J Self-Propag High-Temp Synth 2011;20(3):208–10.
[4] Shcherbakov VA, Barinov VYu. SHS under pressure: I. Thermal electromotive force arising during combustion of Ti + C mixtures. Int J Self-Propag High-Temp Synth 2011;20(1):33–5.
[5] Kirdyashkin AI, Salamatov VG, Maksimov YuM, Sosnin EA, Tarasenko VF, Gabbasov PM. Optical spectrum of emission from the combustion zone. Combust Explos Shock Waves 2010;46(1):117–20.

Concise Encyclopedia of Self-Propagating High-Temperature Synthesis
http://dx.doi.org/10.1016/B978-0-12-804173-4.00043-0

Electro-Composites Using SCS Powders

Singanahally T. Aruna

Electrodeposition of metal matrix composites is an important surface-coating technique. Such metal matrix coatings are being developed to meet the particular demands of low-friction coefficients and high-wear and corrosion resistance for current advanced technological applications such as aerospace, defense, automobile, and nuclear power industries. Electroplating composite coating is an effective method to prepare electro-composite coatings through the codeposition of metallic, nonmetallic, or polymer particles with metal to improve properties such as corrosion resistance, hardness, and wear performance [1]. This method has advantages such as low cost, low temperature, and a single-step process without additional thermal treatment.

Among different electrodeposited metals, nickel deposition is very popular for electrodeposition because it shows good mechanical properties, excellent corrosion resistance, high electrical conductivity, and good thermal conductivity and magnetic properties [2]. However, the mechanical properties deteriorate during severe environmental and stress conditions. To improve this situation, the metal matrix is reinforced with inert particles [2]. Such coatings are produced by codeposition of fine ceramic or polymer particles in a metal matrix from electrolytic baths. Among the composite coatings, Ni-SiC is the most widely used wear-resistant coating [2]. However, the interfacial reaction between nickel and SiC has limited the application of this composite for high-temperature applications (~600°C), and oxide particles are being investigated as reinforcement [2].

With the advent of nanomaterials, there has been an increasing trend toward the preparation of electrodeposited nickel matrix nanocomposite coatings using ceramic particles with diameters below 100 nm [3]. This kind of nano nickel composite coating may find application in microtechnology and in thin film coatings. In most of the research papers, researchers have employed commercial oxide powders. Aruna et al. for the first time incorporated solution combustion synthesized (SCS) nanosize powders, such as zirconia [4], alumina [5,6], ceria [7], yttria-doped ceria (YDC) [8], alumina-zirconia [9], and so on, into the Ni matrix during electrodeposition [3–9]. The advantage of this approach is the readiness with which the particles can be prepared by SCS and utilized for preparing the composite coatings. The steps involved in the process are as follows: (i) preparation of a nickel bath like sulfamate bath or Watts bath, (ii) preparation of nanosize oxides by solution combustion method, (iii) dispersion of SCS synthesized particles in the nickel

Concise Encyclopedia of Self-Propagating High-Temperature Synthesis
http://dx.doi.org/10.1016/B978-0-12-804173-4.00044-2

bath, and (iv) coelectrodeposition of particles along with growing Ni layers under optimum conditions [2].

Some of the highlights of the research work carried out on electrodeposited Ni composite coatings containing SCS powders are as follows:

(i) In general, Ni composite coatings electrodeposited at lower corrosion density ($\leq$0.77 A/dm^2) exhibited higher microhardness because of the incorporation of a large number of particles. The highest hardness was observed (750 KHN at 50 gF load) for Ni-monoclinic zirconia without the addition of additives.

(ii) Almost all the coatings exhibited improved corrosion resistance. However, when soft particles like ceria were incorporated, the wear resistance did not improve much.

(iii) In general, it was observed that whenever yttrium oxide-doped oxides, such as YDC and yttria-doped zirconia-alumina (YZA) composite, were employed as the distributive phases, the corrosion resistance of Ni composite coating improved to a greater extent than the Ni composites containing oxide particles without yttria.

(iv) Ni-YZA composite coating exhibited a synergistic combination of improved microhardness and higher corrosion and wear resistance with a lower friction coefficient [9].

The Ni-ZrO_2 coating exhibited excellent corrosion resistance compared to plain nickel coatings by withstanding a continuous salt spray test for 168 h and prolonged 2400 h salt spray test (Fig. 1A). This coating after a surface treatment of fluoroalkane silane also exhibited very high water contact angle of 158 degrees. The Ni composite coatings containing SCS TiO_2-CeO_2 powder (Fig. 1B) and $ZnAl_2O_4$ exhibited lower wear loss and higher corrosion resistance. However, the issue of agglomeration of nanosize particles in the Ni matrix must be overcome.

Fig. 1 (A) Optical microscopy image showing the incorporation of zirconia particles in Ni matrix. (B) FESEM image showing SCS CeO_2-TiO_2 particles.

REFERENCES

[1] Hovestad A, Janssen LJJ. J Appl Electrochem 1995;40:519–27.
[2] Walsh FC, Ponce de Leon C. Trans IMF 2014;92(2):83–98.
[3] Ahmad YH, Mohamed AMA. Int J Electrochem Sci 2014;9:1942–63.
[4] Aruna ST, Rajam KS. Scr Mater 2003;48:507–12.
[5] Aruna ST, Diwakar S, Jain A, Rajam KS. Surf Eng 2005;21(5):209–14.
[6] Aruna ST, William Grips VK, Rajam KS. J Alloys Compd 2009;468:546–52.
[7] Aruna ST, Bindu CN, Ezhil Selvi V, William Grips VK, Rajam KS. Surf Coat Technol 2006;200 (24):6871–80.
[8] Aruna ST, William Grips VK, Rajam KS. J Appl Electrochem 2010;40:2161–9.
[9] Aruna ST, Ezhil Selvi V, William Grips VK, Rajam KS. J Appl Electrochem 2011;41:461–8.

Electrodes

Alexander E. Kudryashov, Yury S. Pogozhev, Evgeny A. Levashov

Cemented carbides on the base of tungsten carbides, such as WC-Co, WC-TiC-Co, and WC-TiC-TaC-Co, are usually used as electrode materials (EMs) for electrospark deposition (ESD) technologies. However, they are characterized by high erosion resistance and do not always ensure required performance of the items. Application of SHS technologies (forced SHS pressing, centrifugal casting, and extrusion) makes it possible to obtain progressive EMs [1] on a tungsten-free basis, which can be divided into three groups.

The first group of EMs includes materials that consist of a refractory compound (carbides, borides, carbonitrides) and a metallic binder composing 10%–60% of the material. This group contains dispersion-hardened metalloceramic materials with the effect of simultaneous dispersive strengthening of the carbide grains and the metallic binder as a result of concentration of lamination (through controlled solid-solution transformations) of the oversaturated solid solutions and formation of excess nanophases (of $Me^{V}C$ and Me^{V} types) inside the carbide grains and within the metallic binder (γ' phases) [2,3]. Such EMs are referred to as CNT—double titanium-niobium carbide (Ti,Nb)C, CTs—double titanium-zirconium carbide (Ti,Zr)C, and CTTa—double titanium-tantalum carbide (Ti,Ta)C. Multicomponent alloys based on Ni-Co-Al-Cr are used as the binder. This group of the EMs also contains alloys based on chrome, molybdenum [4], zirconium borides, silicides and aluminides. EMs based on $M_{1-n}AX_n$ phases, as well as refractory superalloys based on Ni and Co and TiAl and NiAl are successfully used [5].

The second group of EMs comprises dispersion-strengthened electrodes obtained by alloying the reactionary mixture with nanodispersed powders of the refractory compounds: ZrO_2, Al_2O_3, NbC, Si_3N_4, W, WC, WC-Co, and Mo-Al_2O_3 [6]. These are nanomodified materials in systems such as TiC-Ni, TiC-Cr_3C_2-Ni, TiC-NiAl, TiC-Ti_3AlC_2, TiB_2-NiAl, and NiAl-TiN. The nanodispersed component is introduced in amounts that do not exceed 7%, which is sufficient to substantially increase the quantity of crystallization centers within the melt. Such alloying causes a substantial diminution of the synthesis product's grain sizes (up to a factor of 5–7) and an increase in the EMs' erosion capacity. Application of the dispersion-strengthened EMs in ESD technology makes it possible to obtain wear- and heat-resistant coatings, which have homogenous composition and structure. Also, this EM group contains EMs with additives of mineral raw materials, such as datolite concentrate $CaB[SiO_4](OH)$. These EMs have an elevated erosion capacity. The boron and silicon in this case serve as the flux.

Concise Encyclopedia of Self-Propagating High-Temperature Synthesis
http://dx.doi.org/10.1016/B978-0-12-804173-4.00045-4

The third group comprises EMs for the thermoreactive electrospark surface strengthening (TRESS) process [7]. The TRESS combines ESD and the exothermic chemical reaction that occurs within the interelectrode space. The chemical reaction is maintained by a combination of the pulse discharge energy and the heat released because of the chemical reaction. The heat released because of the chemical reaction is comparable to the discharge energy and increases the total energy of the formation process of the coatings (the treatment productivity). Electrodes of this group are formed or sintered (a solid-phase sintering) reactionary powder mixture, for example, of Ti-C-Ni-Al, Ni-Al, Ti-B, Ti-Al, Ni-Al-diamond, and Ti-Al-diamond compositions. Such EMs can be obtained by extrusion forming, drawing the mixture in a metallic shell, extrusion, and rolling.

SHS EMs of all three groups are successfully applied when performing the tasks of electrospark strengthening and arc surfacing of tools and machine parts. The electrospark treatment of expensive large-size tools (stamps, steel, and cast iron forming rolls) and responsible parts that operate under extreme conditions is the most effective.

REFERENCES

[1] Levashov EA, Merzhanov AG, Shtansky DV. Galvanotechnik 2009;9:2102–14.
[2] Levashov EA, Shtansky DV, Lobov AL, Borovinskaya IP. Int J SHS 1993;2(2):165–73.
[3] Levashov EA, Shtansky DV, Lobov AL, Merzhanov AG. Phys Met Metallogr 1994;77(2):118–24.
[4] Levashov EA, Pogozhev YuS, Potanin AYu, Kochetov NA, Kovalev DYu, Shvyndina NV, Sviridova TA. Ceram Int 2014;40:6541–52.
[5] Sanin VN, Ikornikov DM, Andreev DE, Yukhvid VI, Levashov EA, Pogozhev YuS. Int J SHS 2014;23 (4):232–9.
[6] Levashov EA, Vakaev PV, Zamulaeva EI, Kudryashov AE, Pogozhev YuS, Shtansky DV, Voevodin AA, Sanz A. Thin Solid Films 2006;515:1161–5.
[7] Levashov EA, Nikolaev AG, Rogachev AS, Kudryashov AE, Ohyanagi M, Koizumi M, et al. Int. Patent WO 99/18258 of 03.10.1997; Patent of USA No. 6336950 B1 of 08.01.2002; Japanese Patent No. 4020169 of 12.12.2007; European Patent No. 1035231 of 17.07.1998.

Electromagnetic Field Generation in SHS

Karen S. Martirosyan

Experimental and theoretical investigations revealed that an electromagnetic field was generated during gas-solid and solid-solid combustion of metals [1–4]. Electrical and magnetic fields have been detected during oxidation [1,2], nitridation of metal particles [5], and metal sulfidation [6]. It was demonstrated that qualitatively similar temporal voltage signals, but with different amplitudes, are generated by the oxidation of several metals (Ti, Zr, and Hf) that belong to the same group (IVA) of the periodic table and form oxides with the same oxidation state (MeO_2) [3]. Electrical fields up to 2 V and 100 mA formed during the initial stages of oxidation of single metal particles of Zr, Ti, Fe (Fig. 1). These experiments have shown that the generated voltage is significantly higher than the thermoelectric effect at the electrodes, which is on the order of mV. This fact rules out the possibility that thermal emission is the major contributor to the observed voltage. The estimates of the electric current because of the thermal emission made by the Richardson-Dashman equation predict thermal emission currents on the order of 10^{-2} A/cm^2 for the oxidation of Ti at 2000 K. However, it was experimentally shown that the measured current density is on the order of 10 A/cm^2 during the metal oxidation in the micropowder form at about 2000 K and even higher at 10^6 A/cm^2 during Ti single-particle oxidation [7].

It was experimentally shown that the voltage signal obtained during metal oxidation strongly depended on the specific chemical reaction and was mainly due to formation of electrical charge carriers by the chemical reaction. The different rate of diffusion of the charge carriers across the growing oxide shell creates an electric double-charge layer. The formation of positive and negative charge carriers, such as electron holes, oxygen ions,

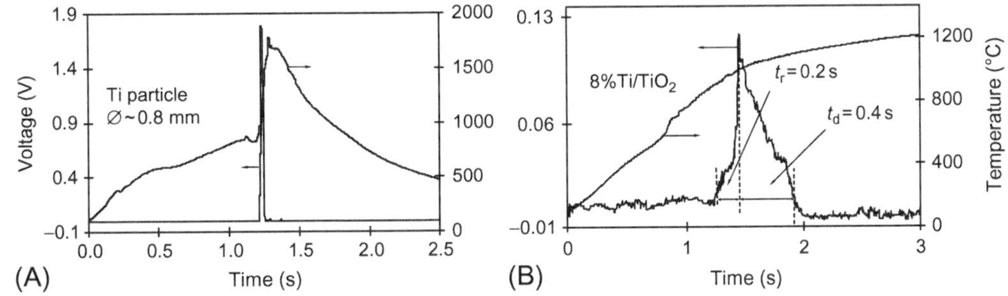

Fig. 1 The temporal voltage and temperature generated during the combustion of (A) a single Ti particle [2] and (B) an 8% mixture of Ti/TiO_2 [7].

Concise Encyclopedia of Self-Propagating High-Temperature Synthesis
http://dx.doi.org/10.1016/B978-0-12-804173-4.00046-6

vacancies, and electrons, on the particle surface is a necessary but insufficient condition for the generation of a voltage by the reaction. The formation of the electrical voltage and its polarity and amplitude depend on the difference in the diffusive velocity of the charge carriers [7].

Several prediction models for the generation of an electrical field formed during the oxidation and sulfidation of metal particles have been developed [3,4,6]. The oxidation model assumes that the electrical field is due to the different rates of diffusion of the positive and negative charge carriers through the growing mixed-ionic-electronic conducting oxide shell. The model predicts that the amplitude and temporal evolution of the voltage depend strongly on the ratio between the thickness of an exterior oxide layer, which usually forms before the reaction starts, and the Debye screening length. Moreover, the sign of the excess electrical charge on the surface depends on the ratio between the diffusion coefficients of the positive and negative charge carriers. The mode and stability of the propagating reaction front also have a strong impact on the generated voltage signal.

The motion of the high-temperature front during combustion synthesis of ferrite materials generates residual magnetization in cylindrical product samples [8]. Qualitatively different magnetic field maps were generated in ferrite samples synthesized by different combustion modes. It was shown that an internal transient magnetic field of up to 5 mT formed in the postcombustion region when its temperature was below the Curie temperature [8]. This result explains why the Earth's field had no impact on the spontaneous magnetization field.

REFERENCES

[1] Nersesyan MD, Claycomb JR, Ming Q, Miller JH, Richardson JT, Luss D. Appl Phys Lett 1999;75:1170.
[2] Martirosyan KS, Filimonov IA, Nersesyan MD, Luss D. J Electrochem Soc 2003;150:J9.
[3] Filimonov IA, Luss D. AIChE J 2004;50(9):2287–96.
[4] Filimonov I, Luss D. AIChE J 2005;51:1521.
[5] Martirosyan KS, Setoodeh M, Luss D. J Appl Phys 2005;98(5):054901–6.
[6] Markov AA, Filimonov IA, Poletaev AV, Vadchenko SG, Martirosyan KS. Int J SHS 2013;22(2):69–76.
[7] Martirosyan KS, Filimonov IA, Luss D. AIChE J 2004;50(1):241–8.
[8] Martirosyan KS, Nawarathna D, Claycomb JR, Miller JH, Luss D. J Phys D Appl Phys 2006;39:3689–94.

Electrothermography

Suren L. Kharatyan

Electrothermography is a method for studying the kinetics of high-temperature solid-phase reactions at strongly isothermal and nonisothermal conditions, including those taking place at SHS of a variety of inorganic materials.

The principle of electrothermography is that the investigated metallic sample in the form of a thin wire or foil is placed into the reaction chamber and heated up by the passage of a direct electrical current in a reactive gas medium (N_2, CH_4, SiH_4, H_2, O_2, etc.). In the case of solid + solid or solid + liquid interactions (e.g., Mo/Si, W/B, Ti/C) the metallic wire is previously cladded by a second solid component (e.g., Si, C, B, Me) and heated in vacuum or in an inert gas medium. For preparation of such samples an in situ CVD technique is generally used. In all these cases, the metallic sample is simultaneously a reagent and a source of high temperature, and sometimes a sensor of temperature, which provides high performance for the executing devices and controllable heating of the sample to elevated temperatures. In experiments, the sample is heated with almost no gradient. That is, there is no temperature distribution along its section and the length because the parameter $\mathrm{Bi} \to 0$ and length of the wire (L) are vastly larger than its diameter (d): $L/d \approx 10^3$.

Usually the sample's temperature is determined by measuring a calibrated photodiode signal.

One of advantages of the electrothermography method is the possibility of measuring in situ the rate of the chemical heat release (dq/dt) and determining the type of kinetic function within the framework of one experiment (one sample) for any given temperature regime of interaction. The heat release rate is determined on the basis of a nonstationary equation of thermal balance as a difference of electrical powers, released on the wire at the first—reactive, $p_1(t)$, and the second—inert, $p_2(t)$, heating (with the same temperature-time history).

$$\text{At the first heating (with chemical heat release)}: \frac{1}{4}\rho c d \frac{dT}{dt} = \frac{dq}{dt} + p_1(t) - h(T) \quad (1)$$

$$\text{Second heating (when reaction is completed}: dq/dt = 0): \frac{1}{4}\rho c d \frac{dT}{dt} = p_2(t) - h(T) \quad (2)$$

Concise Encyclopedia of Self-Propagating High-Temperature Synthesis
http://dx.doi.org/10.1016/B978-0-12-804173-4.00047-8

$$\text{For heat evolution rate}: \frac{dq}{dt}(T, t) = p_2(t) - p_1(t) \tag{3}$$

Eq. (3) is true for both the isothermal and nonisothermal interaction regimes.

Apart from precision in situ measurements of the heat release rate, valuable information on the kinetics and mechanism of the process is received by examining the samples cooled rapidly from high temperatures at various stages of interaction via gravimetry, metallography, SEM, TEM, EDS, XRD, and so on. As the thin samples are characterized by a short time of thermal relaxation, very high rates of quenching are realized (up to 50,000 K/s). For the same reason, the approach used allows the study of nonisothermal high-performance processes with a characteristic duration stated from 0.01 s with millisecond resolution, including those taking place at different zones of the combustion wave of SHS systems.

Three generations of electrothermography apparatus were developed over the past 15 years at the Institute of Chemical Physics National Academy of Science, Republic of Armenia. The current high-speed scanning electrothermography-3 (HS SET-3) setup is a working prototype on which new combinations of functional capabilities and modules are being actively tested and evaluated. It consists of three main parts: (i) reaction chamber made from stainless steel equipped with two optical windows and water jacket for thermostating, (ii) electronic control unit with PC, and (iii) gas control panel.

Custom Microsoft Windows-based software and intuitive graphical user interfaces have been designed for the HS SET-3 model. The latter allows performance of experiments in a variety of regimes, such as constant or variable regimes (preliminary programmed regimes with arbitrary function) of temperature, electroconductivity, electrical power, current, and voltage.

Here are the specifications of the HS SET-3 setup:

• Operating temperature interval:	20–3500°C
• Heating rate of the sample:	$10–10^6$ K/s
• Cooling rate of the sample:	from 10 up to 50,000 K/s
• Measurement frequency:	up to 20 kHz
• Registration time:	from 0.0001 s to 1 h
• Sample length/diameter or thickness (for thin metal wire or foil) :	5 – 10 cm (length)/50 – 300 μm (diameter/thickness)
• Reaction atmosphere:	vacuum or gas (up to 20 atm.)

Many binary systems (Ti-H_2, Zr-H_2, Ti-N_2, Zr-N_2, Ta-N_2, Nb-N_2, Mo-Si, W-Si, Ta-Si, W-C, Mo-C, Ta-C, Nb-C, Ti-C, Zr-C, Ni-Al, etc.) [1–4] were precisely investigated by the electrothermography method. Recently the method was utilized also for studying the kinetics and mechanism of metal reduction from oxide [5].

REFERENCES

[1] Kharatyan SL. Kinetic modelling of combustion wave in SHS systems. Int J SHS 1992;1(4):536–43.
[2] Chatilyan HA, Kharatyan SL, Harutyunyan AB. Diffusion annealing of Mo/$MoSi_2$ couple and silicon diffusivity in Mo_5Si_3 layer. Mater Sci Eng A 2007;459:227–32.
[3] Kharatyan SL, Chatilyan HA, Arakelyan LH. Kinetics of tungsten carbidization under non-isothermal conditions. Mater Res Bull 2008;43(4):897–906.
[4] Kharatyan SL, Chatilyan HA, Aghayan MA, Rodriguez MA. Non-isothermal phenomena in Mo/Si diffusion couple: reaction kinetics and structure formation. Int J SHS 2013;22(1):18–26.
[5] Manukyan KhV, Avetisyan AG, Shuck ChE, Chatilyan HA, Rouvimov S, Kharatyan SL, et al. Nickel oxide reduction by hydrogen: kinetics and structural transformations. J Phys Chem C 2015;119:16131–8.

Elemental Synthesis

Inna P. Borovinskaya

The start in the development of self-propagating high-temperature synthesis (SHS) as a scientific and technical direction is connected with the discovery of "solid-flame" combustion in which initial and final products are solid. The first systems of solid-flame combustion included mixtures of transition metals of Groups IV–V (Ti, Zr, Hf, Nb, Ta) with nonmetals (B, C, Si) that formed refractory compounds (metal borides, carbides, silicides) during combustion. Later, combustion of metals, boron, and silicon was carried out in gaseous (and even liquid) nitrogen, resulting in metal and nonmetal nitrides [1]. Combustion reactions of elements occur with great heat release ($Q = 10^2$–10^3 cal/g), high adiabatic temperatures in the synthesis wave (up 5000°C), remarkable heating rate ($W = 10^3$–10^6 degree/s), and synthesis wave propagation velocity ($U = 0.5$–20 cm/s). In spite of the stiff terms of the synthesis, it is possible to obtain the compounds of the required phase and chemical composition by determining the optimum ratio of initial components and setting heat removal from the burning system. The entire series of nonstoichiometric compounds can be obtained by an ordinary change in a solid reagents ratio. In multiphase systems such as silicides, borides, and carbides of Groups V–VI, a deviation from the optimum ratio of reagents results in the appearance of undesirable intermediate products. Multiphase systems are especially sensitive to organization of heat removal from the burning reactive mass [2]. Even thermally unstable compounds such as transition metal (Ti, Zr, Hf, etc.) hydrides are known to be synthesized by combustion in gaseous hydrogen. In these processes, hydrogen acts as an oxidizer, though in many other reactions it is a fuel agent.

A remarkable group among the processes is synthesis of chalcogenides and phosphides by the reaction of metals with sulfur, selenium, tellurium, and phosphor. In this case, the nonmetal is partially evaporated, but it is necessary to govern the process of its gasification in order to get a high-quality final product.

The elemental SHS was especially developed while working out production methods and implementation of pilot- and industrial-scale technologies of various powder refractory compounds. In the development of powder SHS technologies, great attention is paid to the product purity, which depends on the raw materials' state and synthesis process, including melting and evaporation of initial and intermediate products that can result in significantly incomplete combustion. In addition, contamination of reactive masses by the materials of evaporating boats and supports are controlled. It should be noted that there is practically no external contamination of SHS products because

Concise Encyclopedia of Self-Propagating High-Temperature Synthesis
http://dx.doi.org/10.1016/B978-0-12-804173-4.00048-X

Table 1 Different classes of elemental SHS-materials

Materials	Components: Fuel	Components: Oxidizer	SHS products
Hard alloys	Ti, Cr, Ta, Mo, Zr, Nb, etc.	C, B, N_2	Tungsten-free hard alloys of STIM group
Nitride ceramics	B, Si, Al, Ti, Zr	N_2	Multifunctional structural and tool ceramics, etc.
Oxide ceramics	Cu, Nb, Ti, Zn, Fe, Al	O_2 (air)	Superconducting ceramics, niobates, tantalates, ferrites

of the high reaction rates and the short time the reactive mass stays in the reactor in comparison to furnace elemental synthesis. Moreover, the high temperatures of the synthesis are favorable for the "self-purification" processes (evaporation of volatile admixtures, deoxidation, etc.). Deoxidation and the lack of access to outside oxygen in the hermetically closed SHS reactors make elemental SHS one of the efficient methods for inorganic material production with a low content of harmful admixtures, especially oxygen [2]. In different countries today, there are semiindustrial and industrial production lines based on the elemental synthesis of powders of simple and complex refractory compounds used as raw materials for manufacturing materials and items of high-temperature, fine-grade and technical-grade ceramics, hard alloys, cermets to be used in chemistry, metallurgy, mechanical engineering, electronics, aerospace instrumentation, nuclear and automobile industries, medicine, and so on. Elemental SHS is the most profitable method in comparison with other technologies in terms of yielding products with cheap raw materials. Customers have also noticed its advantages, that is, the improvement in its operational features, which has led to high demand on the market [3]. Elemental SHS is also of key importance in the synthesis variants connected with direct production of items and materials (without the stage of powder preparation), including technologies of SHS compaction (combination of SHS with pressing), SHS gas-statting (SHS at high gas pressures), SHS extrusion, and so on.

This type of elemental synthesis is used to obtain materials of various types and practical applications. Some specific examples are shown in Table 1.

REFERENCES

[1] Merzhanov AG. Self-propagating high-temperature synthesis and powder metallurgy: unity of goals and competition of principles, In: Capus JM, German RM, editors. Particulate materials and processes: advances in powder metallurgy. Proceedings of the 1992 powder metall. World Congress, vol. 9, San Francisco, CA, 21–26 June 1992. Princeton, NJ: Metal Powder Ind. Fed. Publ.; 1992. p. 341–68

[2] Borovinskaya IP. Chemical backgrounds of SHS product technology. In: Maksimov YuM, editor. Self-propagating high-temperature synthesis. Tomsk: Tomsk University; 1991. p. 33–5.

[3] Merzhanov AG, Borovinskaya IP, Prokudina VK. Self-propagating high-temperature synthesis: resources, technologies, economy. Resour Technol Econ 2006;3:30–5.

Emission Phenomena During SHS

Ramil M. Gabbasov, Alexander I. Kirdyashkin, Yury M. Maksimov, Vitaly G. Salamatov

Chemical reaction processes (involving liquid and solid substances [1–2]) are known to be accompanied by acoustic emission during the combustion of gas systems [3].

The experiments showed that the acoustic oscillations were generated during SHS [4,5]. The combustion process was conducted in a sealed reaction chamber in an argon atmosphere at an extreme pressure of 0.5–1.0 atm. Two types of reaction samples were used in the experiments: one type was a cylindrical briquette of the powder systems Ni-Al, Ni-Ti, Ti-B, Zr-B, Mo-B, and FeO-Al with a relative density of 0.3–0.6. The reaction was initiated using an electric coil in the form of a propagating combustion wave. The time of ignition and combustion of the sample and the combustion wave passing through the waveguide were recorded according to the signals of the thermocouples placed near the ends of the sample and waveguide. The second type used Ni-Al, Ni-Ti twisted wires. The systems were ignited by heating the twisted wires to the melting temperature of a low-melt reagent by an electric current. After reaching the critical temperature of twisted wires, a self-accelerating exothermic reaction was initiated between the components. Then the twisted wires were spontaneously divided into two parts, the current in the circuit was interrupted, and the self-sustaining chemical reaction continued to develop as a combustion wave in each part of the twisted wires.

Acoustic emission was recorded in the powder mixture by a piezoceramic sensor through the waveguide (tungsten wire). One end of the wire was embedded into the sample, and the other end was attached to the acoustic sensor. The results demonstrated that the combustion process was initiated near one end of the twisted wire while another end was connected to the acoustic sensor. In this case, the reaction system performed the role of the waveguide. The waveguide (twisted wire) and the acoustic sensor were connected through the corundum clutch, which excluded the galvanic connection between the sample and the sensor.

The experiments showed that the acoustic emission of SHS took place in the mode of low-ordered discrete pulses (Fig. 1A, the repetition frequency from 5 Hz to 50 KHz) and highly ordered oscillations (Fig. 1B, the frequency 0.85–1.10 MHz), whose characteristics depended on the structure and composition of the initial system.

Concise Encyclopedia of Self-Propagating High-Temperature Synthesis
http://dx.doi.org/10.1016/B978-0-12-804173-4.00049-1

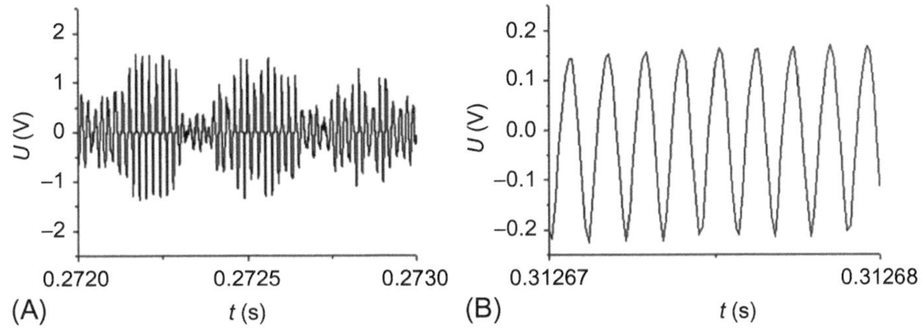

Fig. 1 Signals of acoustic emission during combustion of the Ni-Al powder system: Panel (A) shows the discrete pulses; Panel (B) shows the highly ordered self-oscillations.

The proposed reasons for this effect are the thermal and chemical self-oscillations of the combustion wave, the formation and development of cracks in the sample, and the phase and chemical transformations of the medium.

REFERENCES

[1] Mintz RI, Melekhin VP, Ievlev IYu, Bukhalenkov VV. Solid State Phys 1972;14(5):1582–3.

[2] Zadumkin SN, Khokonov KhB, Shokarov KhB. JETP 1975;68(4):69–77.

[3] Smirnov AN. Ros Khim Zh 2001;45:29–34.

[4] Maksimov YuM, Kirdyashkin AI, Gabbasov RM, Salamatov VG. Combust Explos Shock Waves 2009;45(4):454–60.

[5] Kirdyashkin AI, Gabbasov RM, Maksimov YuM, Salamatov VG. Combust Explos Shock Waves 2013;49(6):454–60.

Emulsion Combustion Synthesis

Osamu Odawara

An emulsion combustion method (ECM) has been proposed by K. Takatori [1] for making fine oxide powders with an emulsion of aqueous solution of precursor compounds of the desired powder product by following certain steps: (i) atomizing the w/o (water in oil) type emulsion in which metal nitrates are dissolved in an aqueous phase, (ii) burning the oil in the atomized emulsion droplet, and (iii) oxidizing the metal salts in the microspheres of aqueous solution. An aqueous solution of the metal nitrates is stirred with kerosene and a small amount of emulsifier to obtain w/o type emulsion. In many cases, the aqueous phase is ~65 vol.% of the emulsion, and the dispersed droplets are about 1–2 μm in diameter. The constituent metal ions in the aqueous droplets are rapidly oxidized at high temperature by combustion of the surrounding kerosene. Most of combustion synthesis is carried out at combustion temperatures higher than 1000 K. Air is introduced into the flame so as to achieve complete combustion. The residence time of the powder at high temperatures is estimated to be less than one-half second.

Hollow alumina fine powders are synthesized by the EMC using w/o type emulsions with different aqueous microsphere sizes [2]. The powders are spherical and hollow with a very thin shell from 10 to 20 nm in thickness when synthesized with large aqueous microspheres from 200 nm to 1 mm. Large particles are obtained from small aqueous microspheres of 6 nm, which cause the adhesion of particles during the firing process with the short distance between microspheres. The emulsion process and the combustion technology are combined so that the reaction fields are of submicron size and the reaction times a fraction of second. The characteristic feature of the synthesized powder by the ECM process is a thinner shell than those obtained by the other processes. The rapid heating of aqueous microspheres by combustion of the oil phase causes immediate supersaturation, precipitation, and oxidation of aluminum nitrate solution at the surface of the aqueous microsphere. The inner substances of the microspheres are consumed in the formation of the dense shell in a short time.

The ECM process is characterized by (1) an isolated small reaction field in which constituent metal ions are mixed homogeneously in the aqueous phase, (2) a short reaction period achieved by the combustion of the thin kerosene film surrounding each aqueous droplet, (3) a continuous fabrication procedure that contributes to lower production costs. The ECM process is classified in chemical synthesis processes such as coprecipitation, sol-gel process, and flame spray pyrolysis (FSP). The ECM process

Concise Encyclopedia of Self-Propagating High-Temperature Synthesis
http://dx.doi.org/10.1016/B978-0-12-804173-4.00050-8

differs from the standard organic solution-fed FSP, because the higher amount of aqueous phase (e.g., 65 vol.%) in the precursor emulsion decreases the flame temperature, favoring the precipitation of precursor in the liquid phase rather than precursor evaporation, and oxidation at the gas phase for particle formation during spray pyrolysis. By comparing the ECM process and the FSP process for synthesis of pure and mixed SiO_2 and ZnO nanoparticles, lower flame temperatures in the ECM process than in the FSP process result in mixed gas and liquid phase reaction, forming ZnO particles with inhomogeneous sizes [3]. The significant differences between ECM and the conventional spray pyrolysis are from the size of the reaction field and the reaction period. An isolated small reaction field of about 1 μm is easily prepared using the emulsion process. The reaction period of the conventional spray pyrolysis is longer than that of the ECM. In the ECM process, each dispersed droplet of emulsion is heated by the combustion of a kerosene film surrounding the droplet, which substantially reduces the reaction period of the ECM.

The ECM process is capable of producing a wide variety of fine powders. It is particularly well suited for making metal oxides for which water-soluble precursors are widely available. The aqueous droplets in the emulsion contain precursor compounds in the ratios needed to produce the desired composition of the powder product [4]. Since the precursors are dissolved and mixed in the aqueous solution, the chemical composition (e.g., ceramic stoichiometry) can be more exactly controlled, and the method is able to provide highly reliable products of unsurpassed compositional homogeneity.

The powders made by the ECM process are inherently spherical solid and/or hollow particles. Spherical solid particles having a narrow size distribution are considered to be ideal for obtaining high packing densities and are desired for making dense ceramic and metal films in the electronics industry. Hollow particles can be attractive as insulating and lightweight filler materials as well as catalyst carriers because of their small size, very thin shell (~10 nm), and high specific area (~50 m^2/g). The ECM process is a promising method for the synthesis of hollow particle of oxides used as fillers (e.g., ZnO) and catalysts (e.g., TiO_2 and ZrO_2) [5].

REFERENCES

[1] Takatori K. R&D Rev Toyota CRDL 1997;32:1–12 [in Japanese].
[2] Takatori K, Tani T, Watanabe N, Kamiya N. J Nanopart Res 1999;1:197–204.
[3] Tani T, Watanabe N, Takatori K. J Nanopart Res 2003;5:39–46.
[4] Tani T, Watanabe N, Takatori K, Pratsinis SE. J Am Ceram Soc 2003;86:898–904.
[5] Chandradass J, Balasubramanian M, Kim KH. Emulsion combustion synthesis. In: Lackner M, editor. Combustion synthesis: novel routes to novel materials. Sharjah: Bentham Science Publishers Ltd.; 2010. p. 25–33

Exothermicity of SHS Reactions

Alexander G. Merzhanov, Yury M. Maksimov

The main peculiarity of combustion is the exothermicity of the process [1]. High combustion temperatures lead to high reaction rates and complete conversion of initial substances in final products. There are necessary and sufficient conditions for propagation of the combustion wave according to the theory of combustion. The necessary conditions are given as follows

$$\beta = \frac{RT_{ad}}{E} \ll 1 \tag{1}$$

$$\gamma = \frac{RT_{ad}^2}{cEQ} \ll 1 \tag{2}$$

where $T_{ad} = T_0 + Q/c$ is the adiabatic temperature of the process, Q is the heat of reaction, E is the activation energy, c is the specific heat at constant pressure, T_0 is the initial temperature of reagents, and R is the gas constant. The parameters of the combustion reaction Q, E, c and the quantity T_0 characterize the initial state of the system. The sufficient condition is determined by the relationship between the heat release in the combustion wave and the heat loss to the environment.

In Eqs. (1), (2), small values of the parameters β and γ determine the strong temperature sensitivity and high exothermicity of the reaction. At the same time, in the combustion wave, the exponential temperature dependence of the heat release rate results from the fact that the material is heated along with a sharp, progressive self-acceleration and the very high temperatures with significant ΔT, where $\Delta T = T_{ad} - T_0$ (Fig. 1).

For the combustion wave propagation, the heat loss should be lower than a certain critical value (3)

$$\eta < \eta_{cr} \tag{3}$$

The value η is determined by the expression (4)

$$\eta = \frac{T_{ad} - T_c}{T_{ad} - T_0} \tag{4}$$

where T_c is the combustion temperature in the presence of heat losses. For the case $\eta_{cr} = \gamma$. For the synthesis of metal borides (IV–VI group), the adiabatic combustion

Concise Encyclopedia of Self-Propagating High-Temperature Synthesis
http://dx.doi.org/10.1016/B978-0-12-804173-4.00051-X

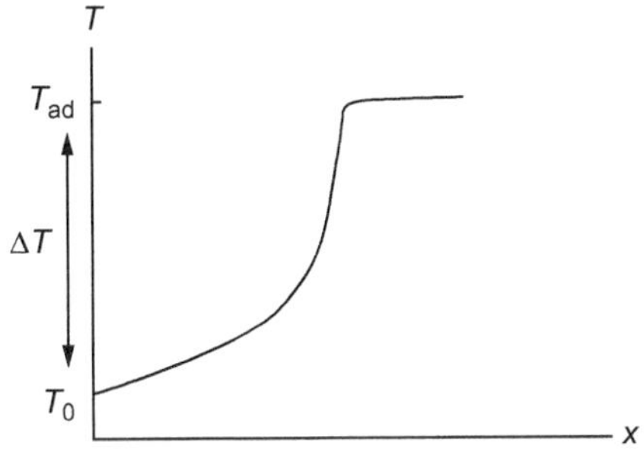

Fig. 1 Typical thermogram of the SHS process in the combustion wave: *T* is the temperature and *x* is the spatial coordinate.

temperatures are in the range of 1400–3500 K. The highest T_{ad} are observed for the reaction with the formation of metal borides of group IV.

Adiabatic temperatures during the combustion reaction of metals in nitrogen are high in the most cases (2500–5000 K). For metal nitrides such as Ti, Zr, Hf, and Al, the combustion temperatures exceed the melting temperatures by several hundred degrees. Dissociation of nitrides can take place because of the very high temperatures of Me-N combustion. The degree of dissociation and the combustion temperature depends on the nitrogen pressure. Comparison of the reaction combustion temperatures during the formation of various compound classes shows that the highest T_{ad} are reached during reactions with the formation of nitrides and that the lowest T_{ad} are reached during reactions with the formation of silicides (1500–3000 K). Carbides and borides are in the intermediate position.

One of the ways to control SHS is in the change in the combustion temperature, which is achieved by preheating the reaction mixture or by dilution of the reagent mixture with a final reaction product. Thus, for example, dilution of the metal and boron mixture leads to a reduction of the combustion temperature by 200–900°C. The T_{ad} calculations allow us to determine the conditions for the reagents in the reaction zone. There is the liquid-phase and the solid-phase combustion. In a liquid-phase mode ($T_c > T_p$ of reagents and reaction products), the chemical reaction develops only in the liquid phase. In a solid-phase mode ($T_c < T_p$ of reagents and products), the chemical reaction is determined by the process of reaction diffusion.

Numerous investigations show that, in most cases, when the IV–VII group metals of the periodic table interact with boron, carbon, nitrogen, silicon, and sulfur, these temperatures are high and reach 2000–4000 K.

REFERENCE

[1] Merzhanov AG. Combustion processes in chemical engineering and metallurgy. Chernogolovka: USSR Academy of Sciences; 1976. p. 5–29.

Experimental Methods

Alexander S. Rogachev

Experimental methods for investigation of the SHS processes and products involve a wide variety of original, often unique, methods, especially developed for this task and many well-known advanced analytical techniques. All experimental methods can be divided in two groups: dynamic (or time-resolved) methods and static methods. Dynamic methods collect data *in situ*, during the process. Here are a few of the dynamic methods [1,2]:

1. Video recording (with macro- or micro-lens, common or high speed, black/white or color) of the process. Recording rate ~100,000 frames per second and optical resolution of a few micron have been achieved [3,4].
2. Thermal vision (IR) recording of the dynamics of temperature field; the maximum rate ~10,000 frames per second with resolution ~10 μm [5].
3. Pyrometry and micropirometry using high-frequency detectors allow time resolution $\sim 10^5$–10^6 measurements per second in one point (spot, minimal size ~10 μm) [6,7].
4. Temperature profile measurements by thermocouples and micro-thermocouples. Taking into account the high temperature of the process, mostly C-type thermocouples (W-Re) are used; minimal thickness of the micro-thermocouples is ~7–10 μm [8].
5. Time-resolved X-ray diffraction or time-resolved synchrotron-radiation diffraction allow monitoring of structural transformations directly during the process. Time resolution achieved for transmitting synchrotron radiation diffraction of thin-film SHS is better than 1 μs per each XRD-pattern, with space resolution ~10 μm [9]; for reflected synchrotron radiation in bulk samples, space resolution ~1 ms [10].
6. Time-resolved transmitting electron microscopy (TEM) was developed recently and applied for study of thin-film SHS [11,12]. An outstanding short time of the TEM picture acquisition of 8 ns in combination with high TEM magnification was obtained; however, this method remained unique and, probably, the most expansive among the dynamic methods of experimental diagnostics of SHS.

Static methods allow investigation of final and intermediate products, as well as initial components *ex situ*. The most commonly used methods for testing SHS products' structure and properties are XRD, SEM + EDS, TEM, macro- and micro-indentation for hardness and microhardness (HRA, HV), bending strength, wear tests, electrical and thermal conductivity, and others. The static methods also can be applied for investigation of phase and structure transformations in the SHS wave; however, for this purpose, the

Concise Encyclopedia of Self-Propagating High-Temperature Synthesis
http://dx.doi.org/10.1016/B978-0-12-804173-4.00052-1

reaction wave must be quenched in order to "freeze" different stages of the transformations in one sample. Supercritical heat losses are used for extinction and quenching of the SHS process by placing relatively thin (or wedge-shaped) samples between cold metal blocks with high heat conductivity [2,13,14]. Cooling rate reaches 10^3–10^4 K/s, which allows freezing of many intermediate microstructure and phases. Samples with the "frozen" SHS wave represent the whole pattern of phase and structure transformations, from initial mixture to final product.

REFERENCES

[1] Merzhanov AG, Mukasyan AS. Solid-flame combustion. Moscow: Torus Press; 2007 [Rus.].
[2] Rogachev AS, Mukasyan AS. Combustion for material synthesis. Boca Raton, London, New York: CRC Press Taylor & Francis Group; 2015. 397 p.
[3] Varma A, Rogachev AS, Mukasyan AS, Hwang S. Complex behavior of self-propagating reaction waves in heterogeneous media. Proc Natl Acad Sci U S A 1998;95:11053–8.
[4] Rogachev AS, Mukasyan AS. Experimental verification of the discrete combustion models for the microheterogeneous systems which formed condensed combustion products. Combust Explos Shock Waves 2015;51(1):53–62.
[5] Rogachev AS, Mukasyan AS, Varma A. Volume combustion modes in heterogeneous reaction systems. J Mater Synth Process 2002;10(1):31–6.
[6] Anselmi-Tamburini U, Maglia F, Spinolo G, Munir ZA. Use of two-color array pyrometry for characterization of combustion synthesis waves. J Mater Res 2000;15(2):572–80.
[7] Garkol DA, Gulyaev PYu, Yevstigneev VV, Mukhachev AB. A new high-speed brightness pyrometry method to investigate self-propagating high-temperature synthesis. Combust Explos Shock Waves 1994;30(1):72–6.
[8] Zenin AA, Merzhanov AG, Nersisyan GA. Thermal wave structure in SHS processes (on the example of the boride synthesis). Combust Explos Shock Waves 1981;16(1):63–71.
[9] Trenkle JC, Koerner LJ, Tate MW, Gruner SM, Weihs TP, Hufnagel TC. Phase transformations during rapid heating of Ni/Al multilayer foils. Appl Phys Lett 2008;93:081903.
[10] Sharafutdinov M, Alexandrov VV, Evdokov O, Naumov D, Pirogov B, Pismenskaya E, et al. The study of Ni+Al self-propagating high-temperature synthesis using synchrotron radiation and a two-dimensional DED-5 detector. J Synchrotron Radiat 2003;10(5):384–6.
[11] Kim J, LaGrande T, Reed B, Taheri M, Armstrong M, King W, et al. Imaging of transient structures using nanosecond in situ TEM. Science 2008;321(5895):1472–5.
[12] Kim J, LaGrande T, Reed B, Knepper R, Weihs T, Browning N, et al. Direct characterization of phase transformations and morphologies in moving reaction zones in Al/Ni nanolaminates using dynamic transition electron microscopy. Acta Mater 2011;59(9):3571–80.
[13] Rogachev AS, Shkiro VM, Chausskaya ID, Shvetsov MV. Gasless combustion in the Ti–C–Ni system. Combust Explos Shock Waves 1988;24(6):720–6.
[14] Varma A, Rogachev AS, Mukasyan AS, Hwang S. Combustion synthesis of advanced materials: principles and applications. In: James W, editor. Advances in chemical engineering, 24. Academic Press; 1998. p. 79–226.

Ferrites

Roman V. Minin, Volya I. Itin

Oxide ceramic materials, such as ferrites, are widely used in radio engineering, radio electronics, computer engineering, and other industrial sectors in developed countries. The application of these materials continuously expands, and the production of ferrite products constantly increases. At the same time, the basic method for the production of ferrite-based materials and products, the volume of which is thousands of tons, is a classic furnace synthesis including the solid-phase sintering of a green mixture consisting of oxide mixtures at high temperatures [1].

In recent decades, crucially new methods have been developed for the production of ferrites, which are based on the SHS proceeding in the mode of filtration combustion in the mixture of metal powders and oxides of other elements in the atmosphere of the reacting gases, such as oxygen or air.

Thus, to obtain composite oxide ferrimagnetics by using SHS, a general chemical formula are used as follows [2] (Fig. 1).

The main reaction for the synthesis of oxide ceramics is the reaction during the combustion of iron in oxygen that is supplied to the reaction zone by filtration from the external environment or that is contained in the green mixture consisting of peroxides, superoxides, or carbonates. In the latter case, the interaction mechanism of reagents combines the features of the gas-free and filtration combustion. The key parameters that determine the degree of the phase formation—that is, the chemical and phase composition of the final product—the structure and electrical and physical properties are the fuel concentration, dispersion of the components, the porosity of the medium, and the pressure of the reacting gas [3].

In addition to the preceding parameters, the electric and magnetic fields, the preliminary thermal impact, and the mechanical activation (also in high-energy devices) are used to control the SHS process.

For the systems with the components possessing magnetic properties, it is reasonable to use a magnetic field that causes a rearrangement of the particles in the reagents and the change in the oxygen exchange. The effect of the constant magnetic field on the combustion, chemical, and phase composition of the final product, as well as the magnetic properties is due to the rearrangement of the reagent particles that are oriented along the field lines and that create the optimal contacts for further development of the reaction to obtain ferrites [4,5].

Concise Encyclopedia of Self-Propagating High-Temperature Synthesis
http://dx.doi.org/10.1016/B978-0-12-804173-4.00053-3

$$\frac{\alpha Me}{\text{Metal fuel}} + \frac{\beta Me'_k O_p}{\text{Solid oxidizer}} + \frac{\gamma Me''_m O_n}{\text{Oxide filler}} \xrightarrow{O_2(gas)} \frac{\delta MeMe'_x Me''_y O_z}{\text{Final product: composite oxide}}$$

Fig. 1 The general route for the composite oxides synthesis.

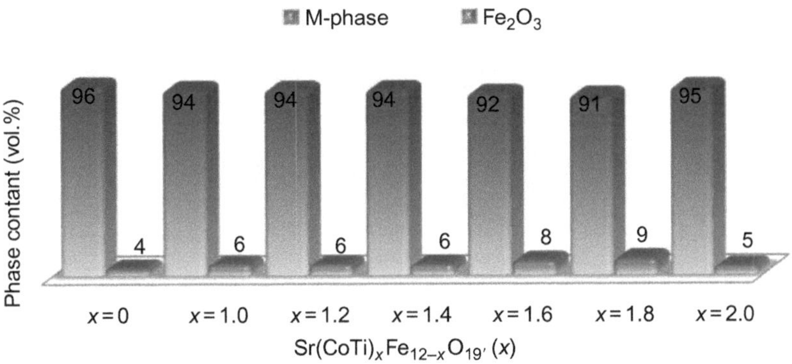

Fig. 2 Phase composition of the ferrites $Sr(CoTi)_xFe_{12-x}O_{19}Sr$ synthesized by SHS with preliminary mechanochemical activation and subsequent ferritization.

The preliminary mechanical activation of the powder mixture expands the number of systems that can react in the combustion mode. To obtain the complex ferrites doped with different elements, including composites, there is a need in the short-term mechanical activation of the initial components in the high-energy planetary ball mill (Fig. 2).

At present, the simple and complex, cubic, and hexagonal ferrites are obtained using the SHS method. At the same time, the laboratory and practical experiments showed that the phase and structure formation during the SHS process was so high that the dispersion of the electrophysical parameters were significantly reduced for the final product. As a result, the industrial production efficiency of magnetic materials and products was significantly increased.

Using the SHS method for the production of ferrites led to a significant increase in the efficiency, the reduction in the synthesis temperature, electricity costs, and duration of the technological cycle and production costs. At the same time, the static and dynamic magnetic properties, including the radio-absorbing capacity, were highly competitive with the analogs synthesized by the traditional ceramic technology.

REFERENCES

[1] Merzhanov AG, Nersisyan MD. Self-propagating high-temperature synthesis of oxide materials. Mendeleev Chem J 1990;35:35–43.
[2] Amosov AP, Borovinskaya IP, Merzhanov AG. Powder technology for the self-propagating high-temperature synthesis of materials. Moscow: Mashinostroenie; 2007. p. 567.

[3] Komarov AV, Nersesyan MD, Avakyan PB, Merzhanov AG. Int J SHS 1993;2(3):239–46.
[4] Merzhanov AG. In: Merzhanov AG, editor. Combustion processes and materials synthesis. Chernogolovka: ISMAN; 1998. p. 511.
[5] Maksimov YuM, Kirdyashkin AI, Smolyakov VK, Itin VI. SHS: theory and practice. Chernogolovka: ISMAN; 2001. p. 215–35.

Ferroalloys

Yury M. Maksimov, Ludmila N. Chukhlomina

Combustion of alloys in nitrogen can differ significantly from combustion of metals. In addition to filtering characteristics, the process of nitriding may be influenced by factors as affinity of the alloy components to nitrogen, polymorphic transformations in a combustion wave, and the ratio between the diffusion coefficient of the alloy and the diffusion coefficient of the reaction product, which results in different combustion mechanisms and structural formation of final products. Therefore, for each alloy grade, there are optimal synthesis conditions that are determined experimentally [1].

COMBUSTION OF FERROVANADIUM IN NITROGEN

Nitrided ferrovanadium is used to create high-speed steel, rails with low-temperature reliability and increased impact toughness, and structural steel for engineering, mining, and oil industries.

Vanadium and iron form a number of α-solid solutions. In alloys with a Fe mass content of 52.3% at a temperature lower than 1225°C, the σ-phase of the equiatomic composition with homogeneity area is formed because of the ordering of the solid solution structure at $T_0 = 20°C$ with a change of Fe from 45% to 65%. The nitriding process makes a significant contribution to after combustion during the combustion of the α-FeV alloy. If the iron concentration is more than 40%, there is a sharp increase in the combustion rate and no aftercombustion. Overnitriding of α-(FeV) leads to the formation of multiphase combustion products. The combustion products include β(V_8N), β(V_3N), γ(V_3N_2), δ(VN) vanadium nitrides, and also α-Fe. In contrast to α-FeV, the combustion products σ-(FeV) are two-phase and contain α-Fe and δ-VN.

Another fact of considerable importance is that nitrided σ-(FeV) is a nonporous material with a density of 6.2–6.5 g/cm^3 (for comparison, ferrovanadium nitrided by the conventional vacuum thermal method has a density of 4.5 g/cm^3).

COMBUSTION OF FERRO-SILICON IN NITROGEN

The silicon-iron alloy forms silicides: Fe_3Si, Fe_5Si_3, FeSi, and $FeSi_2$. The peculiarity of ferro-silicon is fragility and a tendency for spreading. A fine fraction is formed during grinding. Combustion of ferro-silicon in nitrogen is conducted in the filtration mode,

Concise Encyclopedia of Self-Propagating High-Temperature Synthesis
http://dx.doi.org/10.1016/B978-0-12-804173-4.00054-5

since the content of reacting gases in the sample pores is not enough to propagate the combustion front. The reacting gas is supplied to the combustion front by filtration. In filtration combustion, the important factors influencing the process are the nitrogen pressure, the porosity of samples, and the diameter. Experiments have shown that combustion of ferro-silicon in nitrogen occurs in the self-oscillating mode and not in the stationary mode. Self-oscillating combustion of Fe-Si results in a low degree of product nitriding and the chemical and phase heterogeneity. In order to achieve the phase and chemical homogeneity of SHS products, the initial alloy is diluted with a final product of combustion. The additives of ammonium fluoride and magnesium considerably influence the degree of ferro-silicon nitriding.

FERRONIOBIUM NITRIDING IN THE COMBUSTION MODE

During combustion of ferroniobium and also ferro-silicon, nitrogen is supplied to the reaction zone by filtration from the surrounding volume. The combustion rate of ferroniobium in nitrogen increases with an increase in pressure, as the delivery rate of gaseous nitrogen to the reaction also increases. There is a layered combustion mode at high filtration rates of nitrogen; there is a surface combustion mode at the low filtration rate of nitrogen.

According to X-ray phase analysis, the products of ferroniobium nitriding in the combustion mode contained the γ-Nb_4N_3, β-Nb_2N, δ-NbN, $Nb_{4-y}Fe_{2+y}N$, and α-Fe phases. Lower niobium nitrides dominate in the combustion products of ferroniobium at the pressures of 0.2–1 MPa. The increase in pressure leads to the increase in proportion of niobium mononitride, and the composite material Fe-NbN is formed at the pressure above 2 MPa.

COMBUSTION OF FERROTITANIUM IN NITROGEN

Ferrotitanium is used as an alloying material in the production of stainless and heat-resistant steel. Nitrided ferrotitanium is used in manufacturing steel for metal building construction. Steels with carbonitride microalloying (Ti, Al, N) have a higher strength, toughness, cold resistance, and corrosion resistance than steel with the same chemical composition without nitrogen. The surface mode is observed during combustion of ferrotitanium and is connected with the filtration rate of the reacting gas. A characteristic peculiarity of the samples burned in the surface mode is essential macro-inhomogeneity. The degree of nitriding for the external layers is significantly higher than for the internal ones. In order to achieve the phase and chemical homogeneity of SHS samples and the maximum nitriding degree, various additives are introduced in the initial alloy: prenitrided ferrochromium and ammonium chloride. Studies have shown that the phase composition of the combustion products is determined by the basic parameters of synthesis: the pressure of nitrogen, the density of the sample, and the composition of the initial mixture.

COMBUSTION OF FERROCHROMIUM IN NITROGEN

Nitrided Fe-Cr is widely used in modern industry as an alloying material in the manufacturing of almost any steel containing nitrogen. The use of nitrided ferrochromium is most effective for manufacturing of austenitic and austenitic-ferritic stainless steel. According to thermodynamic data, the interaction of ferrochromium with nitrogen in the combustion mode can be realized when the content of chromium in the alloy is higher than 60% and the nitrogen pressure is higher than 6 MPa. In practice, it is necessary to conduct nitriding of ferrochromium with preheating of the initial alloy. Products of ferrochromium SHS-nitriding contain α-iron, mononitride CrN, subnitride Cr_2N, and complex nitride $(Cr, Fe)_2N$ according to the results of XRF.

COMBUSTION OF FERROBORON IN NITROGEN

Nitrided ferroboron FB-20 (20% of boron). Independent combustion of ferroboron in nitrogen in the SHS mode cannot be realized because of the low boron content in the alloy. Adding 5%–10% amorphous boron to the initial ferroboron causes the presence of hexagonal modification and α-Fe in the combustion products of boron nitride. Nitrided ferroboron is used as a catalyst in organic catalysis to produce hydrogen from water and water-organic media.

REFERENCE

[1] Maximov YuM, Chukhlomina LN, Braverman BSh, Smironov LA. Self-propagating high-temperature synthesis of nitrogen-containing alloys. Novosibirsk: Nauka; 2014. 232 p.

Foams

Iñigo Agote, Miguel A. Lagos, Manuel Gutiérrez, Ara Sargsyan

Inorganic porous materials have been widely investigated owing to their outstanding properties, such as large surface area, low density, mechanical stability, and good permeability. In general, porous ceramics and metals possess inherent advantages of high melting points, high corrosion and wear resistance, low thermal mass, and thermal conductivity, thus they find applications in catalyst supports, wastewater treatment, heat insulation, and metal filtration.

There are many different routes to obtaining porous materials, such as liquid processing (casting), powder metallurgy, and combustion synthesis. Many authors have studied the synthesis of porous materials by SHS. For example, Borovinskaya et al. [1] studied the formation of graded porosity TiC specimens. Morancais et al. [2] investigated the synthesis of highly porous SiC ceramic. Merzhanov et al. [3] focused on the influence of the gravity on the synthesis of highly porous NiAl compounds. These are but a few examples of works related to the synthesis of porous materials by SHS. In this section, some interesting examples of foams obtained by SHS are presented.

Porous NiTi shape-memory alloys (SMA) with ideal pore characteristics and high strength for hard tissue implants have been fabricated by combustion synthesis, using forced ignition and thermal explosion modes [4,5]. The desired products, such as B2 (NiTi) and B19 (NiTi), are the predominant phases in the porous NiTi SMA. In addition, the SHS process resulted in the formation of several second phases, such as Ti_2Ni and Ni_4Ti_3. Porous NiTi SMA has a high compressive strength (208 MPa), which can meet the strength demand of hard tissue implants for heavy load-bearing applications, and a low Young's modulus (2.26 GPa) similar to that of natural bones.

$MoSi_2$-Al cermet foams may be obtained under the combustion mode using PTFE as foaming agent [6]. It was established that increasing the amount of Al in the initial mixture lead to a decrease in porosity, while increasing the amount of PTFE resulted in obtaining foams with porosity up to 70%. For higher porosities, the distribution of pores became nonuniform, and the combustion products were inhomogeneous. It was revealed that varying the amount of foaming agent and regulating ambient pressure macro- and micro-pores with a diameter between 0.2 and 1 mm and between 0.01 and 0.05 mm, respectively, may be obtained.

Concise Encyclopedia of Self-Propagating High-Temperature Synthesis
http://dx.doi.org/10.1016/B978-0-12-804173-4.00055-7

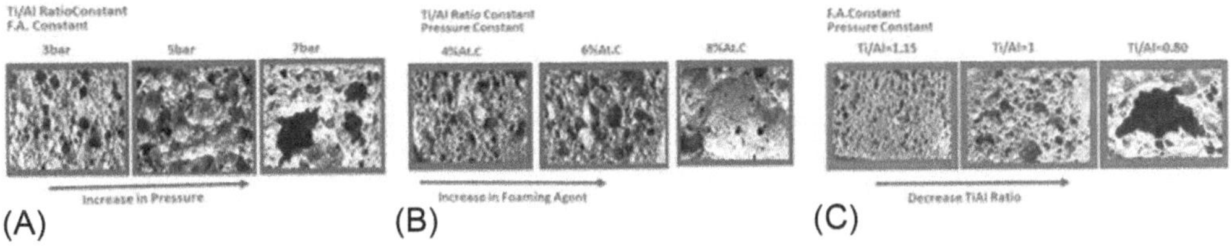

Fig. 1 Increase in the pore size and porosity as the reaction pressure (A), foaming agent increase (B), and Ti/Al ratio (C).

Interesting works have been done in the field of the synthesis of TiAl foams by SHS [7]. Most of the works related to this material have been performed using thermal explosion mode, mainly because of the low exothermicity of the Ti-Al system. Representative examples of the use of chemical activators are found in the work of Kharatyan and Agote [8,9]. In their work, a polymer is used as an activator and as a foaming agent. The use of a polymeric compound has a twofold aim: first, to activate the reaction so that it can be conducted in self-sustaining mode; second, to generate the porosity. Once the Ti-Al system is chemically activated, a cellular structure can be obtained. Three main parameters govern the pore size and degree of porosity: Ti/Al ratio, foaming agent concentration, and synthesis pressure (Fig. 1). The most relevant parameter for obtaining porous structures with controlled porosity was found to be the synthesis pressure. Porosities between 90 and 71 can be obtained, with most of this porosity being open porosity. The obtained pores size is in all the cases smaller than 3 mm.

REFERENCES

[1] Borovinskaya IP, Merzhanov AG, Uvarov VI. Mater Sci Forum 1999;308–311:151–6.
[2] Morancais A, Louvet F, Stanlety D, Bonnety JP. J Eur Ceram Soc 2003;23:1949–56.
[3] Merzhanov AG, Rogachev AS, Rumanov EN, Sanin VN, Sytchev AE, Shcherbakov VA, et al. Cosm Res 2001;39(2):210–23.
[4] Chu CL, Chung CY, Lin PH, Wang SD. Mater Sci Eng A 2004;366:114–9.
[5] Biswas A. Acta Mater 2005;53:1415–25.
[6] Hobosyan MA, Khachatryan HL, Davidov A, Kharatyan SL. Chem Eng J 2011;170:286–91.
[7] Arakawa Y, Kobashi M, Kanetake N. Intermetallics 2013;41:22–7.
[8] Aboyan LS, Nersisyan HH, Kharatyan SL, Saiu R, Orrú R, Cao G, et al. Ceram Int 2001;27(2):163–9.
[9] Agote I, Besné M, Gutiérrez M, Sargsyan A, Kharatyan SL. Int J SHS 2004;13(3):209–18.

Forced SHS Consolidation

Evgeny A. Levashov, Alexander M. Stolin

For obtaining porous free items of required geometric shapes, a variety of methods have been developed; during the SHS combustion wave, the product is consolidated by pressurization. The experiment was based on the following diagram of the process [1,2].

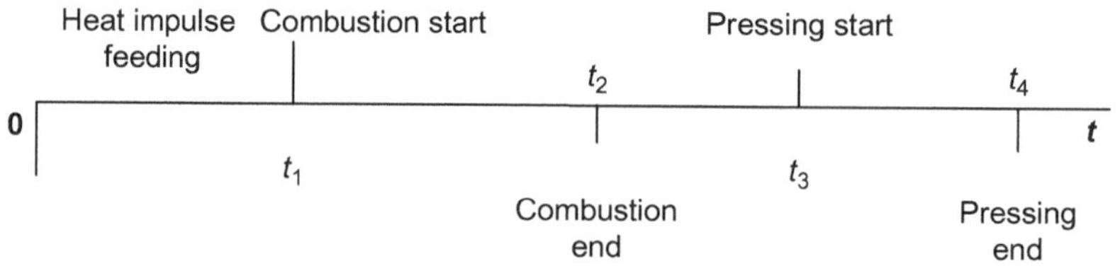

Homogeneity and preservation (without failure) of the dense product billet is conditioned by the inequalities: $t_1 \ll t_2$; $t_2 - t_1 \ll t_p$; t_p is the characteristic time of the sample thermal relaxation (cooling) to the brittle state.

The conditions for obtaining materials and items with minimal residual porosity while preserving their shapes and sizes are the following: $t_d = t_3 - t_2 = (t_{opt})_1$ is the pressing delay time; $t_p = t_4 - t_3 = (t_{opt})_2$ is the time of keeping under pressure; $P = P_{opt}$, where "opt" is the optimal value.

The following table gives the quantitative correlations of the typical times range for the forced SHS consolidation [3].

Initiation time (t_1)	0.3–0.7 s
Combustion time ($t_2 - t_1$)	0.5–15 s
Pressing delay time ($t_3 - t_2$)	1–10 s
Time of pressing and keeping under pressure ($t_4 - t_3$)	2–80 s
Time for unloading the mold	3–120 s
Total process time	35–225 s

The delay time (t_d) is one of the most important characteristics of the process for attaining minimal porosity. The dependence of the residual porosity on t_d is extreme, and the minimum is at a certain optimal value $(t_{opt})_1$. An increase in the porosity at

Concise Encyclopedia of Self-Propagating High-Temperature Synthesis
http://dx.doi.org/10.1016/B978-0-12-804173-4.00056-9

$t_d > (t_{opt})_1$ is connected with the loss of product plasticity when passing to the elastic state. Searching for the optimal parameters (delay and keeping time, pressure) is a specific applied research task [3]. Each of these parameters influences the structure and properties of the dense SHS material. The delay time greatly affects hardness and the strength of the alloys. Forced SHS consolidation is an advanced technology for fabrication of billets, including large ones, from various materials: synthetic hard instrumental materials (STIM), ceramics, and imtermetallides [4,5]. Simplicity of implementation is an advantage of this technology.

After application of external pressure to the combustion products, they consolidate (shrink). The mechanism of consolidating the synthesis products is determined by the final product composition. The products that contain a melt or a plastic phase consolidate the most under loading. The availability of a relatively low-melt metallic binder in the synthesis product composition makes the grains of the refractory compound (carbides, borides, nitrides, etc.) slide much faster, and the melt acts as a lubricant in this case. If a melt does not form during the combustion process and the synthesis products are solid, consolidation is possible because of sliding of the grains and diffusive healing of the pores. While compacting a product in the viscoplastic state, consolidation is performed in several stages: sliding of the grains, plastic deformation or viscous flow because of the intergranular slide, and diffusive healing of the pores.

SHS pressing, SHS extrusion, SHS rolling, and reactive forging are technological varieties. The concept of forced SHS consolidation was developed in work of I.P. Borovinskaya, V.I. Ratnikov, A.N. Pityulin, V.L. Kvanin, E.A. Levashov, and A.M. Stolin, among others. The concepts have been successfully applied for obtaining various items, for example, forming rolls, draw dies, press accessories, stamping plates, disposable cutting plates, targets for magnetron sputtering, and electrodes for electrospark alloying.

REFERENCES

[1] Merzhanov AG. Advanced SHS ceramics: today and tomorrow morning. In: Soga N, Kato A, editors. Ceramics: toward the 21st century: proc. of symp. on ceram. Commemorating the centennial of the ceram. soc. of Japan, 16-18 Oct., 1991, Yokohama, Japan Tokyo: Ceram. Soc. Jap. Publ.; 1991. p. 378–403

[2] Merzhanov AG. Int J SHS 2001;10(2):237–57.

[3] Borovinskaya IP, Pitulin AN. SHS-produced hard alloys. In: Borisov AA, De Luca L, Merzhanov A, editors. Self-propagating high-temperature synthesis of materials. Translated by Scheck YB, combustion science and technology book series, vol. 5. New York: Taylor & Francis; 2002, p. 301–22.

[4] Levashov EA, Pogozhev YuS, et al. In: Costas S, editor. Advances in ceramics—synthesis and characterization, processing and specific application. Croatia: INTECH; 2011. p. 520.

[5] Stolin AM, Bazhin PM. Theor Found Chem Eng 2014;48(6):751–63.

Founders: A.G. Merzhanov, I.P. Borovinskaya, V.M. Shkiro

Inna P. Borovinskaya

A.G. Merzhanov, I.P. Borovinskaya, and V.M. Shkiro discovered "The Phenomenon of the Wave Localization of Solid-State Autoretarding Reactions," referred to as *solid flame*. This discovery started the development and direction of the theory and practice of scientific and technical self-propagating high-temperature synthesis (SHS). This direction connected combustion science with materials science and gave rise to theoretical investigations in the field of solid system combustion that resulted in solid products, and it also gave rise to studies in the theory and practice of combustion processes aimed at inorganic compounds synthesis.

The main author and initiator of SHS and the founder of the new science was A.G. Merzhanov. He also coined the name of the SHS method and new science as *structural macrokinetics*.

Alexander Grigorievich Merzhanov (*Nov. 27, 1931–Jul. 31, 2013*): Academician of the Russian Academy of Sciences, the founder and first director of the Institute of Structural Macrokinetics and Materials Science, RAS.

A.G. Merzhanov is a physicist and outstanding scientist in the field of chemical physics, is a classic researcher in the study of combustion science, and is a follower of Noble Prize winner N.N. Semenov. A.G. Merzhanov started his scientific work at the Institute of Chemical Physics in the USSR Academy of Sciences in 1954. In 1960 he headed the laboratory of ignition and combustion transition to that of detonation and became one of the four first directors of laboratories in the town of Chernogolovka. Merzhanov broadened the theoretical background of combustion and explosion processes in condensed systems, developed the original investigation methods of kinetics of nonisothermal processes, and created new directions in macrokinetics. He expanded experimental diagnostic methods for solid flame, studied the peculiarities of solid flame and filtration combustion, discovered nonsteady modes, and worked out the theory and mathematical models for nonsteady combustion.

In 1987 the Institute of Structural Macrokinetics of the USSR Academy of Sciences was founded by the State Resolution of the Council of Ministers and Presidium of USSR Academy of Sciences (later it was renamed the Institute of Structural Macrokinetics and Materials Science). Merzhanov was the first director of the institute and general director

Concise Encyclopedia of Self-Propagating High-Temperature Synthesis
http://dx.doi.org/10.1016/B978-0-12-804173-4.00057-0

of the Interbranch Scientific and Technical Unit "Termosintez," which united scientific and practical work in the field of SHS in many USSR republics and towns.

Merzhanov continued developing combustion science and also headed the investigation of SHS application for the production of inorganic compounds and materials of different classes and for the development of innovative energy-saving technologies. He created a society of scientists that combined the ability to solve major fundamental problems with the ability to use the results for important practical purposes and prepared 50 Doctor of Science and 150 PhD programs. Merzhanov is the author of more than 800 research papers and 120 patents, the initiator and one of the organizers of the International Symposia on Combustion and Explosion in Russia, and the organizer and permanent chairman of 13 symposia on SHS in different countries of the world. Merzhanov has provided significant help to young people. He founded the Scientific and Educational Center (REC ISMAN) and organized departments and scientific and educational centers in several institutions in the country. He is the founder of the honorary professor at universities in China and India, and academician of the World Academy of Ceramics (WAC).

Inna Petrovna Borovinskaya: Doctor of Chemical Sciences, professor, and specialist in the field of synthesis of compounds based on radioactive elements, including fluorides, uranium, and fluoronitro compounds as components of solid propellants. Since 1962 she has worked at the Branch of the Institute of Chemical Physics, USSR Academy of Sciences, in Chernogolovka.

Borovinskaya is a prominent specialist in the field of combustion chemistry and synthesis of advanced materials and is author and coauthor of more than 250 scientific papers and 100 patents. She is a coauthor of the scientific discovery, "The Phenomenon of the Wave Localization of Solid-State Autoretarding Reactions," which appeared to be the basis for SHS processes. Studying chemistry and the mechanism of combustion of gasless condensed systems and the regularities and mechanism of refractory inorganic compounds formation in the combustion mode, Borovinskaya achieved important fundamental results in physical chemistry and theory of SHS processes; established and described the mechanism of filtration combustion of porous bodies in gas; discovered and studied thoroughly the two-stage mechanism for the conversion of materials (burning and afterburning) and two combustion modes (layer-by-layer and surface); discovered a new autowave process of spin combustion, which is of independent interest in nonlinear dynamics; and formulated the idea of nonequilibrium structure formation of SHS products.

Borovinskaya synthesized dozens of simple and complex nitrides, borides, carbides, chalcogenides, and other compounds by SHS, including nanostructured materials and products. She was the first who obtained a cubic modification of boron nitride by SHS, an unknown cubic phase of tantalum nitride, BN_xC_y compounds, BPN_x, metastable, amorphous nitride phases, materials with original (framework, bridge) structures, to synthesize materials and items of complicated geometry in a single step by combining SHS with physical effects. Borovinskaya took an active part in the development of the pilot-scale SHS technologies of powder, materials, and products production from advanced ceramics, hard alloys, among others.

Borovinskaya is the head of an active team of scientists, engineers, and students. Among her students are doctors and candidates of sciences. She is a member of the Scientific Council of ISMAN, a member of the Scientific Council on Combustion and Explosion, and a member of the editorial board of the (Allerton Press, Inc.), the journal Universities Proceedings Non-ferrous metallurgy (Russian Journal of Non-Ferrous Metals).

Borovinskaya is an honorary professor at Wuhan University of Technology and Harbin Institute of Technology (China). In 2002 she was elected an academician of the World Academy of Ceramics. She delivered lectures on SHS in different countries. She is an active participant of 13 international symposia on SHS and conferences on combustion processes and materials science. She was received government awards, medals, and diplomas at international exhibitions. Borovinskaya takes an active part in the implementation of achievements in industry. Because of her work, SHS technologies are used in various factories in our country and abroad (China, Korea, Spain, etc.).

Valentin Mikhailovich Shkiro (*22.09.1942–08.09.2009*): Doctor of Physical and Mathematical Sciences, senior researcher. He graduated from Tomsk State University in 1964. He has worked at the Institute of Chemical Physics since 1965, first as a trainee researcher (1965–67) and then as a junior researcher.

Shkiro was engaged in the investigation of combustion regularities and mechanism of condensed gasless systems. This subject was also under study when the interaction of metals with nonmetals at the SHS of refractory inorganic compounds was investigated. He was a coauthor of the discovery of "The Phenomenon of the Wave Localization of Solid-State Autoretarding Reactions."

Shkiro's main areas of interest were the systems of metal (titanium, zirconium, tantalum, hafnium) and carbon. These systems were used to study the main combustion regularities, including dependences of combustion rate and chemical and phase composition on various process parameters (initial components ratio, metal particles size, nonmetal structure, temperature, pressure, etc.).

Shkiro managed to establish some peculiarities of the interaction between metals and nonmetals in the combustion mode. Thus at combustion of titanium, the capillary spreading of the metal was described, and at combustion of tantalum-carbon system, the intermediate product and stages of the process were discovered. These results were of particular interest for both the science and the technology of carbide synthesis. He dedicated his last years to the investigation of structure formation of SHS products.

Shkiro was the author of more than 40 publications and over 10 patents and author's certificates on the process of refractory inorganic compounds production. He delivered lectures at many conferences.

Shkiro was awarded medals and diplomas at international exhibitions and the Certificate of Honour of the Russian Academy of Sciences.

Front of the Combustion Wave

Peter M. Krishenik

The wealth of experimental and theoretical studies on combustion in SHS systems clearly shows that the process of flame propagation takes place within a relatively narrow reaction zone. The thermal front of a chemical reaction propagates across a layer of porous material at a finite constant velocity u. As a first approximation, the combustion wave can be divided into two areas: (a) unburned heterogeneous mixture and (b) reaction products, which are separated by a moving surface, the combustion front.

From a strictly mathematical point of view, the thermal front of gasless combustion can exist in infinitely long charges as an asymptotic solution to the system of autonomous differential equations with partial derivatives of the parabolic type. Generally, such equations are used to describe the processes of unsteady heat transfer with nonlinear (typically exponential) temperature dependences and distributed heat sources with restricted power. The behavior of real systems over a limited length differs from that over an infinite length, but this difference is negligible and is related to low reaction rates at low temperatures.

Concise Encyclopedia of Self-Propagating High-Temperature Synthesis
http://dx.doi.org/10.1016/B978-0-12-804173-4.00059-4

Functionally Graded Materials

Evgeny A. Levashov, Irena Gotman

Materials with variable volume compositions and structures inherently have unique functional properties that are not typical of homogeneous materials and that are called functionally gradient (or graded) materials (FGMs). Synthetic gradient materials of SIGMA grade with symmetric (SIGMA-1 with a hard and wear-resistant layer based on TiC or TiB_2 and a ductile strong core based on Ni) and asymmetric profiles of the metallic binder distribution across the billet (SIGMA-2 with a high concentration of the binder based on Ni on the one side and a low concentration of the binder on the other side) [1] have been developed. Gradient materials differ from laminated composites by the absence of abrupt transitions between areas with different properties. In gradient alloys, the binder content, the grain size, as well as the chemical composition of the grains, and the binder change monotonously.

SIGMA materials obtained using forced SHS compaction have an elevated impact strength, which led to their use as armor plates, stamp tools, cathode targets for ion-plasma sputtering, and bed plates for diamond synthesis. Table 1 exemplifies the mechanical properties of the SIGMA-2 alloy (TiC-Ni system with nickel concentration gradient from 100% to 25%) in comparison with standard cemented carbides WC-Co and with STIM grade alloys. The SIGMA alloy fracture diagrams have two characteristic areas of the brittle and ductile failure. The fracture work for the gradient alloy is 20 times higher than corresponding homogeneous alloy, and its impact strength attains 1.3 kg m/cm^2.

The gradient material principle was used when obtaining, through forced SHS compaction, functionally gradient targets with a ceramic working layer, for example, based on TiB_2-TiN or Ti_5Si_3-TiN, and with a TiB_2-50%Cu sublayer [2], the latter having high thermal and electrical conductivity. Gradient materials are naturally formed through deposition by welding, as well as through centrifugal SHS casting in tubes.

Synthesis of various lightweight FGMs by Reactive Forging (pressure-assisted thermal explosion-based method) was reported [3]. Initial layered compositions of three different FGMs prior to processing are given in Table 2. The reagent powder particle size was <20 μm for TiC-Ni, B_4C, and Ti, <5 μm for Al and <0.01 μm for TiO_2.

Compacts up to 46 mm in diameter prepared by compaction of the corresponding stacked layers were placed between the rams of a testing machine and were preheated from 850°C to 1100°C (above T_m of the lowest melting phase). A relatively low pressure

Concise Encyclopedia of Self-Propagating High-Temperature Synthesis
http://dx.doi.org/10.1016/B978-0-12-804173-4.00060-0

Table 1 Mechanical properties of SIGMA and other standard hard alloys

Alloy grade	Impact strength (J/cm^2)	Hardness (HRA)
SIGMA-2	1.3	90.0
VK20 (WC-20%Co)	0.48	84.0
VK8 (WC-8%Co)	0.35	87.5
T15K6 (WC-15%TiC-6%Co)	0.08	90.0
T30K4 (WC-30%TiC-4%Co)	0.07	92.0
STIM-2 (TiC-20%Ni)	0.15	90.0
STIM-2A (TiC-20%NiMo)	0.12	87.0

Table 2 Initial layer compositions of the powder compacts for FGMs synthesis

FGM	#1	#2	#3
1st (outer) layer	(70Ti-30C)-37(TiC coated with 15Ni) (vol.%)	($3TiO_2$-7Al)-50 B_4C	(Ti-3Al)-50 vol.% B_4C
2nd layer	(70Ti-30C)-15(Ti-Ni eut.) (vol.%)	($3TiO_2$-7Al)-20 B_4C	(Ti-3Al)-20 vol.% B_4C
3rd layer	Ti6Al4V alloy plate	$3TiO_2$-7Al	Ti-3Al
4th layer		Ti6Al4V alloy plate	Ti6Al4V alloy plate

Table 3 Selected properties of FGMs synthesized from powder compacts shown in Table 2

Type of FGM	1	2	3
SHS products in the 1st and 2nd layer	TiC, Ti_2Ni	B_4C, TiAl	B_4C, $TiAl_3$
Vol.% of TiC or B_4C in the outer layer	91	53	57
Average density of FGM (g/cm^3)	<5.0	<5.0	<3.5

(50 MPa) was applied to ensure the rapid heating and squeezing of molten Al or Ti-Ni eutectics into the pores between hard ceramic particles—that is, by the short distance infiltration method [4]. Nearly fully dense FGM materials were obtained with compositions and densities shown in Table 3. All FGMs exhibited a gradual change in microstructure and hardness with a smooth transition between the layers. The outer layers of all the three FGM types contained high volume fractions of carbides and possessed correspondingly high hardness and wear resistance. The low density of the synthesized FGMs and the high hardness of their outer layers make them attractive candidates for light body armor.

REFERENCES

[1] Pityulin AN, Bogatov YuV, Rogachev AS. Int J SHS 1992;1(1):111–8.
[2] Levashov EA, Larikhin DV, Shtansky DV, Rogachev AS, Grigoryan HE, Moore JJ. Int J SHS 2002;11 (1):99–117.
[3] Gutmanas EY, Horvitz D, Gotman I. SHS processing of functionally graded carbide based composites. In: Vincenzini P, editor. CIMTEC 2002—10th Int. ceramic congress—part B, vol. 2, Techna Srl; 2003. p. 295–306.
[4] Gutmanas EY, Gotman I. J Eur Ceram Soc 1999;19:2381–93.

Fundamentals: Experiment

Alexander S. Rogachev

Basic experimental investigation of SHS processes among other studies includes

- Measuring main global phenomenological characteristics of the process, such as combustion wave propagating velocity, maximum combustion temperature, temperature profile, etc.
- Determining of dependences of the combustion characteristics on various experimental parameters, e.g., dependence of the propagating velocity on initial temperature of the sample, fuel-to-oxidizer ratio, porosity, reactants particle size, and so on [1]
- Using the obtained dependencies, evaluation of effective kinetic parameters, especially energy of activation of the process
- Revealing sequences and mechanisms of microstructure and crystal (atomic) structure transformations
- Studying phase and chemical composition, microstructure, and properties of final SHS-products

Investigation of a variety of physical, chemical, and structural transformations, which take place during SHS, requires a comprehensive set of advanced *experimental methods* [2]. The use of a particular method depends primarily on the studied structural level. Thus the macrostructure can be examined by direct observation with a relatively small magnification and macro-photo and video recordings. Transformations of the microstructure can be investigated by the methods of optical and scanning electron microscopy (SEM) and transmission electron microscopy (TEM), mainly using quenched samples. The evolution of the crystal structure and phase composition can be studied by diffraction of X-rays and neutrons and/or electrons, sometimes in situ. For SHS in these three structural levels, which are commonly recognized in materials science, it is necessary to add the thermal structure of the combustion wave. Such thermal structure can be examined through macro- and micro-video recording, as well as by micro-thermocouples and pyrometric measurements. The chemical composition and distribution of the elements are typically determined by energy-dispersive X-ray spectroscopy (EDS).

Depending on the characteristic time of investigated stages of structural transformation, dynamic or static methods are used. Fast processes and short-lived structures are studied by dynamic methods, such as high-speed video recording, time-resolved X-ray diffraction (TRXRD), high-speed optical spectroscopy, and so on. The slower

Concise Encyclopedia of Self-Propagating High-Temperature Synthesis
http://dx.doi.org/10.1016/B978-0-12-804173-4.00061-2

processes and structures that may be quenched in the products can be studied by static methods (XRD, SEM, TEM, EDS, etc.), which are more accurate. Finally, experimental approaches should be selected depending on whether the combustion wave is studied as a whole (on the macro level) or the physicochemical transformations are investigated at the microscopic level. In the former case, it is necessary to arrange the combustion front in a heterogeneous medium, whereas in the latter case, the reaction cell or a transformation of the individual particle of this medium can be studied separately (referred to as experimental modeling of the reaction cell).

REFERENCES

[1] Merzhanov AG. Self-propagating high-temperature synthesis. In: Kolotyrkin YaM, editor. Physical chemistry: current problems. Moscow: Khimiya; 1983. p. 8–44 [in Russian].

[2] Rogachev AS, Mukasyan AS. Combustion for material synthesis. Boca Raton, London, New York: CRC Press Taylor & Francis Group; 2015. 397 p.

Fundamentals: Theory

Alexander S. Rogachev

According to definition of SHS given by A.G. Merzhanov, SHS is a kind of combustion process that yields valuable solid compounds (materials) [1]. Therefore a comprehensive theory of this process must combine the theory of combustion and the theory of product structure formation. Theory of SHS must also include two general modes of the process: the volume combustion mode and the combustion wave propagation (autowave) mode. The theory of the volume combustion mode is based on the N.N. Semenov theory of thermal explosion [2], while the combustion wave propagation theory for SHS processes originated from the classical theory of flame propagation in premixed gases developed by Ya.B. Zel'dovich [3]. These theories were adapted for combustion and thermal explosions modes of SHS, taking into account specific characteristics of heat and mass transfer in the SHS systems.

The combustion theory, developed for the gas flame, can be extended to the heterogeneous condensed systems, called "quasi-homogeneous models." The basic concept of this approach is the principle of thermal homogeneity, which implies that for heat transfer processes, the heterogeneous reactant media can be represented as a homogeneous one with certain averaged thermophysical characteristics. It was also assumed that all features of heterogeneous reactions can be taken into account through the reaction kinetics function. The principle of thermal homogeneity for heterogeneous media was introduced in several papers in the following manner: regardless of the particle size, the condensed mixture may be treated as a thermally homogeneous medium if the rate of heat release in this medium, controlled by diffusion of reagents, is much slower than the rate of heat transfer between the particles [4,5]. The chemical heterogeneity is accounted for only in a specific form of kinetic functions that determine the rate of the reaction heat release in a virtually homogeneous medium. Thus the main postulate of the quasi-homogeneous models can be formulated as following: in the steady state combustion mode, the local heat release from a chemical reaction can never occur faster than the rate of dissipation of the heat along the reacting medium. The quasi-homogeneous approach is widely used also in the infiltration combustion theories [6].

However, the experimental study of the SHS waves using the high-speed micro video recording shown indicated that in many cases the assumption of the thermal homogeneity of the medium cannot be applied to the real processes of gasless combustion. It was

Concise Encyclopedia of Self-Propagating High-Temperature Synthesis
http://dx.doi.org/10.1016/B978-0-12-804173-4.00062-4

found that the reaction can be transmitted along the chain of particles and that combustion of each particle is accompanied by a short, bright flash, followed by a relatively long period of the combustion front stagnation, followed by the rapid reaction of the neighboring particle, and so on [7,8]. Several theoretical papers proposed the mathematical models that account for the thermal microheterogeneity of the reactive media [9–11]. These theories are called "micro-heterogeneous models" of SHS.

Theoretical description of the products' structure formation in SHS systems is based on the consideration of *heterogeneous reactions* and *grain growth* processes. A reactive diffusion model was considered as the main mechanism of the heterogeneous SHS reaction for a long time; thus the mathematical formalism of this model was well developed. However, numerous experimental results revealed many specific processes, such as capillary spreading, reactive dissolution-precipitation, and others that determine microstructure of the products. A vapor-liquid-solid mechanism plays a leading role in the products' structure formation in solid-gas systems. For the SHS systems with reduction stage (thermite type), phase separation in products are of the most importance for products' macrostructure and microstructure. Development of the theory of all these structure-forming processes represents tasks for the near future.

REFERENCES

[1] Merzhanov AG. Self-propagating high-temperature synthesis: twenty years of research and findings. Combustion and plasma synthesis of high-temperature materials. New York, NY: VCH Publishers Inc.; 1990. p. 1–53.

[2] Semenov NN. Thermal theory of combustion and explosions. Usp Fiz Nauk 1940;23(3):251–92 [in Russian].

[3] Zel'dovich YaB, Frank-Kamenetsky DM. The theory of thermal flame propagation. Zh Fiz Khim 1938;12(1):100–5 [in Russian].

[4] Khaikin BI. Theory of combustion waves in heterogeneous condensed media. In: Merzhanov AG, editor. Combustion processes in chemical technology and metallurgy. Chernogolovka: Izd. Inst. Chem. Phys.; 1975. p. 227–44 [in Russian].

[5] Hardt AP, Holsinger RW. Propagation of gasless reactions in solids: 2. Experimental study of exothermic intermetallic reaction rates. Combust Flame 1973;21(1):91–7.

[6] Aldushin AP, Merzhanov AG. Theory of infiltration combustion: general principles and state of the art. In: Matros YuSh, editor. Propagation of heat waves in heterogeneous media. Novosibirsk: Nauka; 1988. p. 9–52 [in Russian].

[7] Merzhanov G, Rogachev AS. Structural macrokinetics of SHS processes. Pure Appl Chem 1992;64 (7):941–53.

[8] Varma A, Rogachev AS, Mukasyan AS, Hwang S. Complex behavior of self-propagating reaction waves in heterogeneous media. Proc Natl Acad Sci USA 1998;95:11053–8.

[9] Rogachev AS, Merzhanov AG. On the theory relay-race combustion wave propagation in the heterogeneous media. Dokl Ross Akad Nauk 1999;365(6):788–91.

[10] Smolyakov VK. "Roughness" of the gasless combustion front. Combust Explos Shock Waves 2001;37 (3):274–84.

[11] Rogachev AS. Microheterogeneous mechanism of gasless combustion. Combust Explos Shock Waves 2003;39(2):150–8.

Gas Transport Technology

Victoria V. Kurbatkina, Evgeny A. Levashov

This technology is intended for coating deposition with thicknesses up to 150 μm, based on borides, nitrides, silicides, intermetallides, carbides, and other compounds included in an SHS charge mixture [1,2]. A gas transport additive—that is, halogen (G_n), for example, iodine or its compounds—is introduced into the reaction mixture of the components A and B. Within the heat penetration zone, volatile halogenides AG_m and BG_m, are formed, after which they are transported as gases to the substrate, where they are chemically absorbed; and at the combustion temperature, dissociation of the halogenides and the chemical reaction of A and B take place and an AB-containing coating is formed. This technology combines the refractory compound formation process with chemical transport reactions. Owing to the gas transport reactions, the coatings can be deposited on items of any shape and composition.

The coating deposition scheme is shown in Fig. 1. Because of the main SHS reaction between A and B, the combustion front propagates with speed U_r. In addition, the following reactions take place in the preheating zone:

$$\mathrm{A} + \mathrm{B} \rightarrow \mathrm{AB} + \mathrm{Q}$$

$$\mathrm{A} + m/n \times \mathrm{G}_n \rightarrow \mathrm{AG}_m$$

$$\mathrm{B} + m/n \times \mathrm{G}_n \rightarrow \mathrm{BG}_m$$

At low temperatures, gaseous compounds are formed, and at high temperatures, these compounds dissociate. The AB coating on the surface of the part is formed at high temperatures within the post combustion zone:

$$\mathrm{AG}_m \rightarrow \mathrm{A} + m/n \times \mathrm{G}_n$$

$$\mathrm{BG}_m \rightarrow \mathrm{B} + m/n \times \mathrm{G}_n$$

$$\mathrm{A} + \mathrm{B} \rightarrow \mathrm{AB}$$

However, the diffusion length of the volatile halogenide is of the same order of magnitude as the width of the warmed zone and amounts to ~0.1 mm. A and B reagents formed on the surface of the part (after the dissociation of halogenides) diffuse into the depth. As a result, a wide diffusion zone is formed at the coating-and-substrate interface. The adhesion of the gas-transported SHS coatings to the substrate is high. The

Concise Encyclopedia of Self-Propagating High-Temperature Synthesis
http://dx.doi.org/10.1016/B978-0-12-804173-4.00063-6

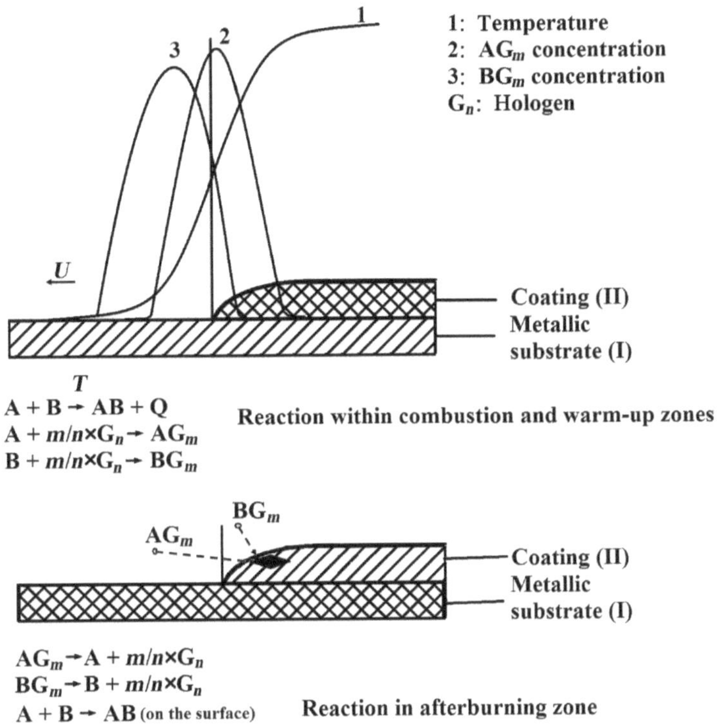

Fig. 1 The scheme of a gas transport SHS coating deposition.

characteristic times of the combustion process, for preheating the substrate, and for cooling the burned charge are the main technological process parameters. The mentioned parameters are determined using thermophysical properties of the reaction mixture, which depend on the heat transfer conditions, thermal conductivity, and sizes of the parts being coated. By regulating the mixture composition and stages of chemical reactions in the combustion wave, it is possible to obtain, within one process, multicomponent coatings (borides, silicides, intermetallides, etc.) with a thickness from 5 to 150 μm. The coating growth rate is at the level of chemical gas-phase deposition rates, which is much faster than diffusive and galvanic deposition. The gas transport SHS technology is effective for coating depositions on steel, hard alloys, ceramics, and graphite.

REFERENCES

[1] Merzhanov AG. Advanced SHS ceramics: today and tomorrow morning. In: Soga N, Kato A, editors. Ceramics: toward the 21st century: proc. of symp. on ceram. commemorating the centennial of the Ceram. Soc. of Japan, 16–18 Oct., 1991, Yokohama, Japan. Tokyo: Ceram. Soc. Jap. Publ.; 1991. p. 378–403

[2] Merzhanov AG. Int J SHS 2001;10(2):237–57.

Grain Growth

Alexander S. Rogachev

Solid products of SHS, primarily nucleates during chemical reaction in the combustion front, are in the form of very small crystalline clusters or grains [1]. After the reaction finishes, a primary product usually consists of submicrometer grains and some residual melt. Temperature of the primary product is close to the maximum combustion temperature ~1500–3000 K. These conditions are favorable for the fast growth of the grains. For example, dynamics of the carbide grains growth behind the quenched combustion front in the Ti-C-Ni system is shown in Fig. 1. Because of the presence of a nickel binder, which practically does not react with the titanium carbide, the carbide grains are clearly distinguished from each other. After removing the nickel binder (in hydrochloric acid), the morphology and size of the grains become clearly visible. It can be seen that the rapid grain growth—increasing the size tens of times—is accompanied by an even more rapid

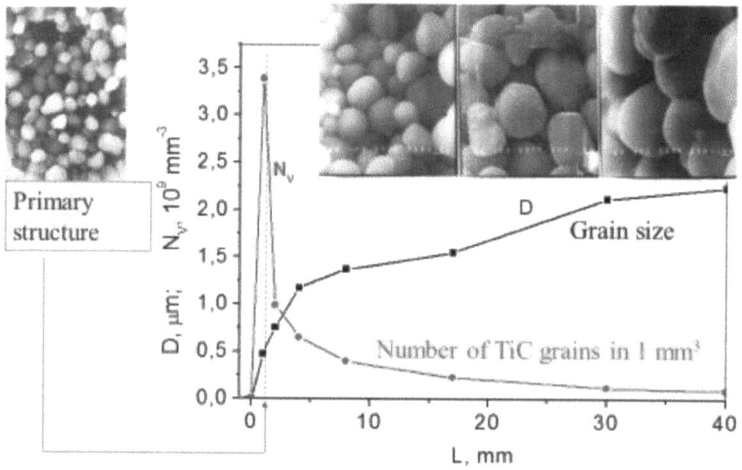

Fig. 1 Dynamics of TiC grain growth behind the combustion front in the Ti-C-Ni system. The point $L = 0$ corresponds to the position of quenched combustion front (reaction zone); increasing L indicates movement toward the final product.

Concise Encyclopedia of Self-Propagating High-Temperature Synthesis
http://dx.doi.org/10.1016/B978-0-12-804173-4.00065-X

decrease of the number of grains. In other words, the grains become larger because of collective recrystallization or coalescence.

The foregoing features of the secondary structure formation are also typical of other SHS systems in which grain growth of the refractory product takes place in a metal matrix with a lower melting point. The pattern becomes more complicated in the systems of filtration combustion (e.g., the metal-nitrogen system), because of variable concentrations of gas reactant(s) across the sample and the possibility of gas-phase reactions. In the thermite-type SHS systems, the products are completely molten, and grain nucleation and growth are governed by the cooling rate.

The driving force of the grain growth is decreasing of the specific surface energy of the grains: the larger mean grain size, the smaller specific surface per unit volume. Two main mechanisms are probably responsible for the growth. First, Ostwald ripening is based on dissolution of the smaller grains into the matrix with simultaneous precipitation of the same product on the surface of larger grains. This situation occurs because the solubility of the solid phase is higher in the vicinity of the surface with larger curvature, and therefore small grains with large curvatures may dissolve in the same matrix, where large grains grow through the precipitation of the solid product on their surface. The second mechanism is fusion (sintering) of the grains. Kinetics and detailed mechanisms of the grain growth in SHS processes have not been studied adequately at this time.

The fast grain growth in SHS products is a primary difficulty for the production of nanocrystalline powders and materials. To solve this problem, the combustion temperature must be decreased, and the ripening of grains must be extinguished. For example, *solution combustion synthesis* provides conditions for freezing of the initial particle sizes, as they appear in the combustion zone; therefore this method currently serves as the main technology for synthesis of the submicrometer and nanosize products by combustion.

REFERENCE

[1] Rogachev AS, Mukasyan AS. Combustion for material synthesis. Boca Raton, London, New York: CRC Press Taylor & Francis Group; 2015. 397 p.

Hard-Facing by Welding

Vladimir I. Yukhvid, Vladimir N. Sanin

The SHS method allowed obtaining a wide range of hard-alloy coatings based on metal carbides and borides, intermetallics, stainless steel, and so on, on the surfaces of different samples made from steel, titanium, and aluminum. Combustion of high-temperature mixtures of oxides with aluminum and nonmetals under gas pressure resulted in the formation of a cast composite coating that is uniformly distributed on and strongly adherent to the metal base [1]. The uniform distribution was influenced by the initial mixture mass, combustion temperature, gas pressure, the value of overloading, and so on. The minimum thickness of SHS coatings was 2–3 mm, while the maximum depended on the mass ratio of the base and the green mixture layer; it was also restricted by the sample melting.

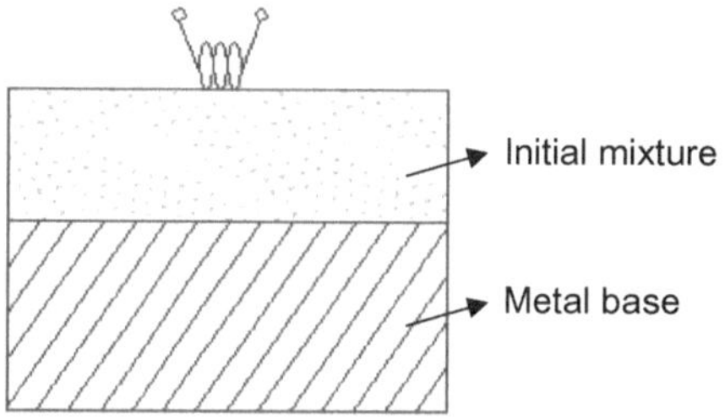

Scheme of SHS hard-facing

Samples with hard-facing

The SHS method allowed obtaining a wide range of hard-alloy coatings based, for example, on metal carbides and borides, intermetallics, and stainless steel on the surfaces of different samples made from steel, titanium, and aluminum [2]. Analysis of the samples by metallography, X-ray diffraction, and electronic microscopy methods showed three zones: a coating, a transition zone, and a metal base. The cast coating contains the target elements formed during SHS and the material of the base. This material is obvious and appears in the coating because of melting of the base surface, which is in contact with the high-temperature melt, and intermixing with the synthesis products. As a rule, the elements contained in the cast coating are uniformly distributed by its height, which proves the intensive convective mixing.

Three connected processes occur in the system of combustion product melt-base after combustion:

1. Motion of metal drops in the oxide medium because of gravity and migration of the drops from the melt onto the base surface

Concise Encyclopedia of Self-Propagating High-Temperature Synthesis
http://dx.doi.org/10.1016/B978-0-12-804173-4.00066-1

2. Spreading of the metal drops on the base surface and a uniform layer formation
3. Heat exchange between the "hot" two-phase melt and the cold metal phase when the surface layer of the base melts and the "weld" joint connecting the cast coating and the base forms [3]

Accordingly, the mathematical model of the process of hard-facing on steel bases [3] was developed and analytical evaluation of the following SHS hard-facing parameters was made:

- The minimum combustion temperature of the initial mixture that is necessary for melting the surface layer of the base (T_{c^*}) is

$$T_{c^*} = T_0 + (T_m - T_0)[(1+\gamma)/\gamma] \quad (1)$$

  where T_0 is the initial temperature of the base; T_m is the melting point of the base material; $\gamma = \sqrt{(\lambda_1\rho_1c_1/\lambda_2\rho_2c_2)}$, $\lambda_1\rho_1c_1$ and $\lambda_2\rho_2c_2$ are the thermal activities of the melt and the base; λ_1 and λ_2 are the thermal conductivity coefficients of the melt and the base; ρ_1 and ρ_2 are the densities of the base and the melt; and c_1 and c_2 are their thermal capacities.
- The coating thickness (Δ), which is the sum of the thickness of the combustion metal product layer (Δ_1) and that of the base melt layer (Δ_2), where $\Delta_1 = \eta_\phi(M_c/\rho_2 S)$, $\eta_\phi = A\left[\rho_1 c_1 d_1^2(ng)/\alpha\nu_0\right]\exp(-E/RT_c)[1-\exp[-E(T_c-T_s)/RT_c^2]]$, $\Delta_2 = h[(T_c - T_m)/\{T_m - [T_0 - (L/c)]\}]$, η_{ph} is completeness degree of phase separation, M_c is calculated mass of the metal phase of the combustion products, S is specific surface area of the base, A is the dimensionless constant, d_1 is the metal drop diameter, $n \geq 1$, g is gravity acceleration, α is heat transfer coefficient, ν_0 is preexponential factor of kinematic viscosity, E is energy of activation of viscosity, T_c is combustion temperature, and T_s is the oxide phase crystallization temperature.

Dynamics of SHS hard-facing was investigated by computer modeling [4].

REFERENCES

[1] Kachin AR, Yukhvid VI, Vishnyakova GA. Problemy tehnologicheskogo goreniya. Inst Chem Phys vol. 2. Chernogolovka [in Russian]; 1981, p. 22–25.
[2] Levashov EA, Rogachev AS, Kurbatkina VV, Maksimov YuM, Yukhvid VI. Perspekyivnye materialy I tehnologii SVS; Moscow: MISiS Publishing; 2011. p. 1–378.
[3] Yukhvid VI. Izv Akad Nauk SSSR Met 1988;6:130–35.
[4] Gordopolova IS, Ivleva TP, Shkadimskii KG, Yukhvid VI. Int J SHS 1999;8(2):137–51.

Heterogeneous Reactions

Alexander S. Mukasyan

Heterogeneous and homogeneous chemical reactions are classes of reactions that are distinguished based on the physical state of the substances present. In particular, heterogeneous chemical reactions consist of reactants that are components of two or more phases (solid and gas, solid and liquid, two immiscible liquids) or where one or more reactants undergo chemical change at an interface. Thus you can also consider catalytic and noncatalytic heterogeneous reactions. The former involves a small amount of a special substance, the catalyst, which though not consumed in the reaction, leads to a significant increase in the reaction rate. Combustion synthesis reactions are noncatalytic heterogeneous reactions (HR).

Noncatalytic HR includes several classes of systems: (A) solid reactants, fluid (liquid, gas) products; (B) solid reactants, fluid and solid products; (C) solid and fluid reactants, solid products; (D) solid and fluid reactants, fluid products; (E) solid and fluid reactants, fluid and solid products; (F) fluid reactants, solid products; among others. Based on thermodynamic analysis, you can determine to which group of HR the CS system belongs. Adiabatic combustion temperature (T_{ad}) for the thermite, $MoO_3 + 2Al$ mixture, at 10 atm of argon pressure, is ~4600 K, and the equilibrium products (Mo and Al_2O_3) are in liquid state, which represents the (A) class of HR systems. The T_{ad} for Ni-Al (1:1 mol) system equals 1911 K. This temperature is above the melting point (m.p.) of its precursors and equal to the m.p. of its product. However, only part (~60%) of the product can melt in reaction front. Thus this system belongs to the (B) class of HR. Combustion of boron in nitrogen ($P = 100$ atm, $T_{ad} = 2400$ K) leads to the formation of a solid product (BN). Thus this reaction belongs to the (C) type of HR. Combustion of coal in oxygen with formation of gaseous carbon oxides represents the (D) type of HR systems. The reaction mechanisms for each of these classes of heterogeneous noncatalytic reaction are different. An attempt to classify these mechanisms for CS processes was made in Ref. [1].

Fig. 1 schematically represents the microstructural models, which were developed to describe the reaction in different types of noncatalytic HR systems. The first microstructural models were developed independently [2,3]. For these models, the elementary reaction cell consists of alternating lamellae of two reactants (A and B), which diffuse through a product layer (C) (Fig. 1A).

For the next two types of theoretical models, the elementary reaction cell involves a spherical particle of one reactant surrounded by a melt of the other reactant. In the

Concise Encyclopedia of Self-Propagating High-Temperature Synthesis
http://dx.doi.org/10.1016/B978-0-12-804173-4.00067-3

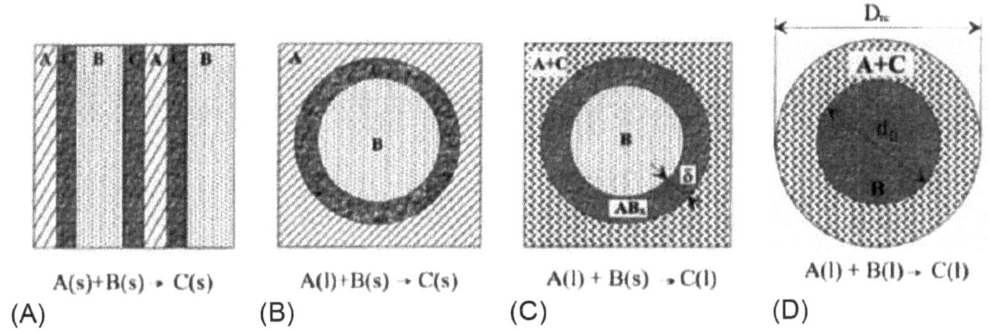

Fig. 1 Geometry of the reaction cells considered in the theoretical models.

first case, the product layers (C) grow on the surface of the more refractory particles (B) because of the diffusion of atoms from the melt phase (A) through the product layer (see Fig. 1B). At a given temperature, the concentrations at the interphase boundaries are determined from the phase diagram of the system [4]. The other model (see Fig. 1C) assumes that upon melting of reactant A, a layer of initial product forms on the solid reactant surface. The reaction proceeds by diffusion of reactant B through this layer, which is assumed to remain at a constant thickness during the reaction [5]. The final product crystallizes (C) in the volume of the melt after saturation. Based on this model, a kinetic expression for the diffusion-controlled rate has been developed [6]. For the case where both reactants melt in the preheating zone and the liquid product forms in the reaction zone, a simple combustion model using the reaction cell geometry presented in Fig. 1D was developed [7]. After melting, the interdiffusion of the precursors and the formation of a liquid product occur simultaneously. Numerical and analytical solutions were obtained for both kinetic- and diffusion-controlled reactions. The detailed analysis of these and other models is given in Ref. [8].

REFERENCES

[1] Varma A, Rogachev AS, Mukasyan AS, Hwang S. Combustion synthesis of advanced materials: principles and applications. In: Wei J, editor. Advances in chemical engineering, vol. 24. New York, NY: Academic Press; 1998. p. 79–226.
[2] Aldushin AP, Martem'yanova TM, Merzhanov AG, Khaikin BI, Shkadinskii KG. Combust Explos Shock Waves 1972;8:159.
[3] Hardt AP, Phung PV. Combust Flame 1973;21:77.
[4] Nekrasov EA, Smolyakov VK, Maksimov YM. Combust Explos Shock Waves 1981;17:513.
[5] Aleksandrov V, Korchagin M. Combust Explos Shock Waves 1988;23:557.
[6] Kanury AM. Metall Trans A 1992;23A:2349.
[7] Okolovich EV, Merzhanov AG, Khaikin BI, Shkadinskii KG. Combust Explos Shock Waves 1977;13:264.
[8] Rogachev AS, Mukasyan AS. Combustion for materials synthesis. New York, NY: CRC Press, Taylor & Francis; 2015. p. 340.

High Gas Pressure SHS of Ceramics

Inna P. Borovinskaya

Ceramics based on transition metals of Group IV–VI, boron, aluminum and silicon nitrides, and carbides have a specific place among the materials characterized by high-operation properties.

Porous and dense ceramics are usually obtained by sintering in furnaces and hot pressing. Gasostatic methods and a new approach of Spark Plasma Sintering are also used. As a rule, these methods use composites and compounds synthesized a priori. The most advantageous method is a direct one-stage process of simultaneous synthesis of a compound of a required composition and formation of its structure and geometry. One such methods is SHS gas-statting (a combination of SHS with high gas pressures) [1,2]. This method is based on self-propagating, high-temperature synthesis at high reactive or inert gas pressures. Special SHS gasostats were designed in ISMAN. They allow synthesizing and forming materials and items simultaneously at gas pressures of up to 800 MPa and temperatures from 1000°C to ≥3000°C arising because of the heat of exothermic interaction. Within the direct synthesis of materials and items by SHS with high gas pressures, some factors have a particular importance: a possibility to control the material porosity and form its macrostructure and microstructure, to obtain final products of a preset chemical and phase composition, and even predetermined properties. The most efficient ways to control the processes of SHS gas-statting and synthesize high-quality products are use of the volume effect of the reactions, activation of the material volume shrinkage by changing the combustion mode (from surface to layer-by-layer one), formation of initial and intermediate product melts at high temperatures, exclusion of final product dissociation at high gas pressures and temperatures, use of the effect of the "nonuniqueness of structure and properties" of materials with the same chemical and phase composition depending on SHS parameters, and so on. The combination of SHS with high gas pressures allows synthesizing a wide range of low-porous, high-porous (5%–60% porosity), and dense (porosity ≤1%) materials based on transition metal nitrides, silicon, boron, and aluminum nitrides and their composites with carbides, borides, oxides, and other inorganic compounds. Moreover, materials with preset electrophysical and chemical characteristics as well as net-shape ones have been obtained. One of the first examples was SHS gas-statting of titanium nitride items, including corrosion-resistant crucibles and electrodes for electrolysis of salt melts. The material chemical composition was close to stoichiometric TiN composition ($TiN_{0.98}$). The items were shaped as crucibles, cylinders,

Concise Encyclopedia of Self-Propagating High-Temperature Synthesis
http://dx.doi.org/10.1016/B978-0-12-804173-4.00068-5

and bricks and didn't require further machining. The SHS process was carried out at filtration combustion of a porous billet made from titanium diluted by titanium nitride of 60% porosity at a gaseous nitrogen pressure of 100 MPa. An analogous approach was used to synthesize composite materials based on zirconium nitride with the addition of (ZrN + ZrO_2); (ZrB_2 + SiC), and so on of 1%–10% porosity.

An original method of preliminary layer-by-layer pressing of small batches of bidispersed initial powders with the following addition of the layers was developed to obtain materials with regulated porosity. In this case, structuring of powders occurred in the volume of a long layer-by-layer pressed billet. Long items of TiB_2 and TiAl with a developed ordered porous structure with spherical and nonspherical large and fine pores were synthesized by this method using the billets of structured powders of titanium mixtures with boron or aluminum in inert gas (argon) at 1500–2200°C [3].

The method of SHS gas-statting is actively used to synthesize low-porous and nonporous materials and items based on aluminum, silicon, and boron nitrides and their composites with metal and nonmetal carbides, borides, and oxides. While working with silicon, it was necessary to exclude the combustion product dissociation in the Si-N system and silicon particle melting in the warming-up zone to select optimum terms of gaseous nitrogen flow to the combustion zone, and so on. Short times of the SHS process and high-temperature gradients are very important here. When compact silicon nitride is synthesized, the warming-up rates of the initial mixture are 10^2–10^8°C/s and the gas pressure gradient between the combustion front and external environment can achieve 1000 atm. These peculiarities stimulate mass and heat transfer, provide crystal formation and growth, form particle morphology, and influence the material physicochemical properties. For example, the SHS material based on silicon nitride and consisting of needlelike β-Si_3N_4 particles with the length-diameter ratio of ~100 and rather high porosity (about 40%) is characterized by high-temperature (T=1400°C) ultimate strength 120–150 MPa. It is comparable to the values of conventional silicon-nitride materials with lower porosity. Another example is development of insulating sleeves for furnaces of oriented crystallization for production of aircraft engine blades. The material with rather low mechanical strength (σ_{bend}=30 MPa, σ_{compr}=50 MPa) demonstrates attractive properties of an electrical insulator ($U_{break\text{-}down}$ ~ 1 kW/mm) at $T \leq 2000$°C and $P = 10^{-1}$ atm in the medium of metal (Ni, Al, Fe) vapors. According to industrial tests, boron nitride insulators obtained by the direct SHS method are twice as stable as their industrial analogs sintered from BN and BNC powders.

The specific features of SHS at high gas pressures are the most obvious in the production of materials and items from the composites of silicon, aluminum, boron nitrides with carbides, borides, oxides [4,5]. When the previously mentioned approaches of porosity, structure, and properties regulation are used, it is possible to synthesize materials with unique characteristics and operating at temperatures up to 3500°C (composites of BN+oxides [SiO_2, ZrO_2], materials dispersion-strengthened by particles of refractory compounds with Si_3N_4 and SiC_xN_{1-x} nanostructured matrixes, and those with melt AlN matrixes and melt TiB_2 grains or bridging structures). These materials demonstrate high strength characteristics at temperatures of up to 2000°C, have a great variety of electrophysical properties at the same chemical and phase composition (insulator-conductor), low friction coefficients (lower than that of bronze), and high corrosion resistance. The detailed description of the terms of nitride-based material SHS can be found in the articles 0135 "Nitrides and Nitride Ceramics," "Black Ceramics," and "Non-Metal Ceramics."

Therefore the method of SHS gas-statting allows combining synthesis with sintering and obtaining porous and dense material and items in one step. This method is not as widely used as other SHS technologies. Some work in this area was carried out by Japanese scientists headed by Miyamoto [6]. This variation implies that the initial reagents are loaded into a pyrox glass ampule, which is sealed hermetically. The ampules are put into a graphite crucible surrounded by an igniting mixture and then placed into the gasostat (a chamber with high-pressure argon). During the igniting mixture combustion (T_c~3000°C), the ampule's contents warm up from 1600°C to 1700°C, and when the SHS reaction occurs, the ampule walls soften, and the contents are subjected to isostatic combustion at P~100 MPa. The mixture of Ti + C + 40 wt% Ni was chosen to synthesize TiC-Ni as well as layered mixtures with different Ni content. The final product in this case was functionally graded material TiC-Ni. Another functionally graded material is a compound of $MoSi_2$ (resistant to oxidation) with TiAl (extra-strong intermetallics). A more complicated functionally graded material of ($MoSi_2$-SiC)-TiAl composition is characterized by high toughness. SHS gas-statting is actively studied by Polish scientists, mainly for synthesis of ceramics with gaseous nitrogen. Silicon nitride-based ceramics and intermetallic composites are the main subjects of their investigation [7,8].

REFERENCES

[1] Borovinskaya IP, Loryan VE, Mukasyan AS. Gas-stattic technology of ceramic materials. In: Technology, equipment, materials 1988;1:16–20.

[2] Merzhanov AG. The chemistry of self-propagating high-temperature synthesis. J Mater Chem 2004;14 (12):1779–86.

[3] Ponomarev MA, Loryan VE, Merzhanov AG. Structure formation peculiarities in the mixtures of titanium spherical particles with boron at densification. Fizicheskaya mezomekhanika 2012;15(5):85–90.

[4] Borovinskaya IP, Loryan VE, Zakorzhevsky VV. Combustion synthesis of nitrides for development of ceramic materials of new generation. In: Gromov AA, Chukhlomina LN, editors. Nitride ceramics: combustion synthesis, properties, and applications. Weinheim: Wiley-VCH Verlag GmbH & Co. KGaA; 2014 [chapter 1].

[5] Mukasyan AS. Combustion synthesis of boron nitride ceramics: fundamentals and applications. In: - Gromov AA, Chukhlomina LN, editors. Nitride ceramics: combustion synthesis, properties, and applications. Weinheim: Wiley-VCH Verlag GmbH & Co. KGaA; 2014 [chapter 2].

[6] Miyamoto Y. SHS sintering at high pressure Khachoyan AV, Trans. In: Koidzumi K, editor. Khimiya sinteza sghiganiem. Moscow: Mir; 1998. p. 128–43.

[7] Skibska M, Lojkowski W, Morawski A. The effect of high nitrogen pressure on combustion synthesis of silicon nitride, In: Proc. 7th CIMTEC-world ceramics congress, Montecatini Terme, Italy; June, 1990. p. 24–30.

[8] Misiolek W, German RM. Reactive sintering and reactive hot isostatic compaction of aluminide matrix composites. Mat Sci Eng A 1991;A144(1):1–10.

High-Speed Micro Video Recording

Joshua M. Pauls, Alexander S. Mukasyan

High-speed micro video recording (HSMVR) is used in combustion synthesis to quantitatively investigate the degree of heterogeneity in the combustion front. These observations permit the classification of combustion mode as either microheterogeneous or quasihomogeneous [1,2]. Advancements in CMOS technology have dramatically increased the imaging rate in digital cameras and allowed HSMVR to become ubiquitous and indispensable throughout the many fields of combustion science [2–5].

Many initial applications [2,5] defined high speed as being any frame rate >1000 fps; however, they made only qualitative use of this rich information. As HSMVR has progressed from film cameras [5] to digital cameras [6,7], methods have simultaneously developed to analyze features such as instantaneous velocity (u_{inst}), hesitation time (t_h), and roughness, all of which characterize the combustion front [1,4,7].

The cameras used include National Instruments' Photonfocus [1], Photron's Fastcam [8], and, most commonly, Vision Research's Phantom [6,7]. When coupled with a long-distance microscope lens, such as Infinity's K2 series [4,6], HSMVR can obtain spatial resolution as fine as 1 μm [4,7] and temporal resolutions as fast as 10 μs [7] (100,000 fps). The most recent published results include data obtained by cameras with a spatial resolution of 10 μm and frame rates of 50,000 fps [7,8].

Limitations in this technique are determined by the exposure time and amount of available illumination, although in this regard, the luminescence found in combustion systems is quite beneficial. An optimum range must also be selected with regard to the number of pixels captured and the frame rate. A compromise of imaging an area of ~1 mm^2 at a frame rate of 50,000 fps is well within the capabilities of the available recording devices [7].

The typical instrumental setup is similar to that shown in Fig. 1, although the camera may be positioned horizontally as well as vertically.

During video analysis, an intensity threshold [4] is selected that defines the combustion front location (Fig. 2). The combustion front may then be analyzed on a frame-to-frame basis for features of heterogeneous combustion such as t_h and u_{inst} [1,9].

The advances in digital recording technology over the past 15 years have made HSMVR accessible to researchers around the world.

Concise Encyclopedia of Self-Propagating High-Temperature Synthesis
http://dx.doi.org/10.1016/B978-0-12-804173-4.00069-7

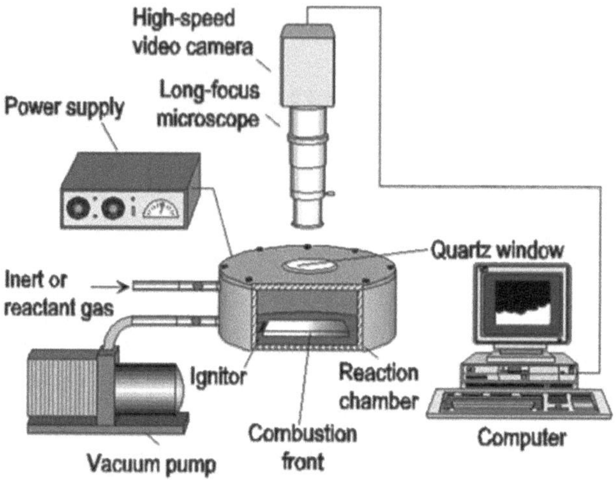

Fig. 1 Typical experimental setup for HSMVR.

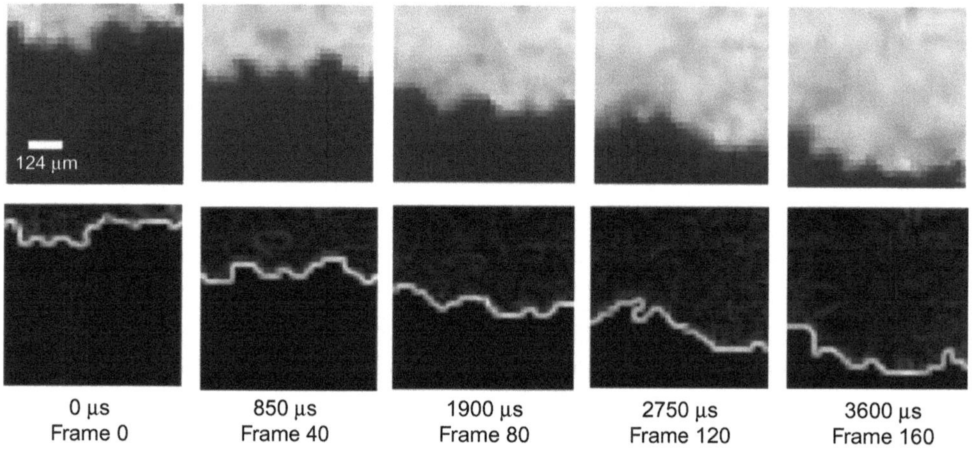

Fig. 2 Selected frames from HSMVR analysis of combustion of nickel and aluminum.

REFERENCES

[1] Rogachev AS, Kochetov NA. Fiz Goren Vzryva 2006;42(4):61–70.
[2] Kochetov NA, Rogachev AS, Merzhanov AG. Phys Chem 2002;389(1):65–7.
[3] Sick V. Proc Combust Inst 2013;34:3509–30.
[4] Mukasyan AS, Rogachev AS, Varma A. Chem Eng Sci 1999;54:3357–67.
[5] Gaffet E. Adv Sci Technol 2006;45:979–88.
[6] Shafirovich E, et al. Combust Flame 2008;152:262–71.
[7] Rogachev AS, Mukasyan AS. Fiz Goren Vzryva 2015;51(1):53–62.
[8] Apperson SJ, et al. J Propul Power 2009;25(5):1086–91.
[9] Mukasyan AS, Rogachev AS, Varma A. AIChE J 1999;45(12):2580–5.

High-Temperature Superconducting Materials

Vladimir B. Balashov

The Y-123 and Bi-2223 high-temperature superconducting materials (HTSC) were obtained by the SHS method in the 1980s [1]. SHS of $YBa_2Cu_3O_{7-\delta}$ ceramics was conducted in argon by using the solid oxidizers $NaClO_4$, J_2O_5, and BaO_2.

$$0.5Y_2S_3 + 2BaO_2 + 3Cu + 2NaClO_4 \rightarrow YBa_2Cu_3O_{7-\delta} + 2NaCl + 1.5SO_2\uparrow + O_2\uparrow \quad (1)$$

$$0.5Y_2S_3 + 2BaO_2 + 3CuS + 3.5NaClO_4 \rightarrow YBa_2Cu_3O_{7-\delta} + 4.5SO_2\uparrow + 3.5NaCl + O_2\uparrow \quad (2)$$

For sodium perchlorate used as an oxidant, the maximum combustion temperature was limited by the boiling temperature of the NaCl(T=1465°C). The synthesis products of Y-123 HTSC were disks (cylinders) of 10–50 mm in diameter. The relative density ρ=0.5–0.7 was connected to the partial gasification of reaction products. In order to increase the density of the samples and stabilize the superconducting phase, additional annealing was applied at T=915°C, followed by a decrease in the temperature at a rate of 100°C/h. The transition temperature measured for different samples was equal to T_c=90–93 K according to the reactions (1) and (2), and the value $\triangle T_c$=0.5 K was observed for the narrow transition range.

The samples obtained by the SHS were the ceramic materials considered to be of an unordered system of fine (5–20 μm) crystals in weak point contact with each other [2]. The Josephson Effect was manifested in the points of the poor contact with the small values of the critical current during the transition to the superconducting state. This information was explained by the fact that the thin surface layer (3–5 nm) of crystallites differed from the phase 1-2-3. Therefore, for $T \leq T_c$, the superconducting ceramics were slightly coupled and represented a "Josephson granular medium," the properties of which depended on the parameters of weak intercrystallite couplings, and the surface resistance in the microwave range (9.4 GHz) was comparable to the resistance of copper.

Concise Encyclopedia of Self-Propagating High-Temperature Synthesis
http://dx.doi.org/10.1016/B978-0-12-804173-4.00149-6

SHS of bismuth ceramics (reaction 3) was conducted using iodides, taking into account their unique properties (BaJ_2, $T_{melting}=711°C$, CaJ_2, $T_{melting}=779°C$).

$$2Bi + 2BaJ_2 + 2CaJ_2 + 3Cu + 4J_2O_5 \rightarrow Bi_2Ba_2Ca_2Cu_3O_{10} + J_2\uparrow + O_2\uparrow \quad (3)$$

Both iodides dissolve in acetone and ethyl alcohol, which is important during mixing with powder reagents such as an alloy of bismuth and copper. The alloy consisted of 2 mol of bismuth and 3 mol of copper (69 wt.% of bismuth, $T_{melt}=820$–$850°C$). The use of the calcium and barium iodides was logical from a kinetic point of view because the melting point of the oxides was much higher:

$$T_{melt}(BaO) = 1920°C \quad T_{melt}(CaO) = 2580°C$$

The calcium and barium iodides were readily oxidized by the reaction (3) and then treated as the reagents for copper and bismuth in the atomic state.

SHS of Bi-2223 HTSC was conducted under an argon pressure of 9–10 MPa with a twofold excess of oxygen ($4J_2O_5$) in excess of the stoichiometry according to the reaction (3) [3,4]. The Bi-2223 high-temperature superconductor obtained was sintered, fine-milled (crystallite size of 1–5 μm) and porous $\rho_{rel}=0.56$. After cooling and extraction, the condensate of iodine was washed from the synthesized samples in an alcoholic solution of potassium iodide. The measured critical transition temperature was $T_c=112$ K.

REFERENCES

[1] Didenko AN, Balashov VB, et al. Study of the properties for high-temperature superconducting ceramics with a narrow temperature transition. In: Proceedings. Problems of high-temperature superconductivity. Part 1, Sverdlovsk; 1987.
[2] Balashov VB, Maksimov YuM, et al. Study of high-temperature superconducting ceramics obtained by the HTSC method. Information report on the sixth all-union school-seminar, "Theory and Practice of SHS Processes" Chernogolovka, 1989.
[3] Balashov VB, Maximov YuM, Perov GA, Merzhanov AG. Method for obtaining of superconducting materials. Patent (RU 1810027); 1997.
[4] Nersesyan MD, Borovinskaya IP, Merzhanov AG, Maksimov YuM, Balashov VB. Method for producing materials in the combustion mode. Patent (RU 1824023); 1997.

Hydrides and Deuterides of Transition Metals

Seda K. Dolukhanyan

The interest in hydrogen and metal hydrides is explained by global problems, that is, the environmental issues and the exhaustion of fossil fuel reserves. The hydrides of transition metals possess unique physical and chemical properties. The application range of metal hydrides is very broad: in the hydrogen energetic, the hydrides are ecologically pure hydrogen accumulators and are potential components of fuels (hydrogen storage); in the nuclear energetic, they are used in biological protection against neutron beams.

The SHS technique appeared to be an excellent method for fabrication of hydrides. In 1975 in the Armenian A.B. Nalbandyan Institute of Chemical Physics of NAS laboratory, high-temperature synthesis of the SHS processes in the Me-H and Me-D were predicted and accomplished for the first time. Furthermore, this direction became a priority of the research in this laboratory [1–3]. Initially, it was assumed that the Me-H_2 system did not satisfy the basic principles of SHS. However, preliminary thermodynamic analysis showed that the interaction of metals with H_2 takes place with a considerable release of heat. As a result of a systematic investigation of combustion processes of metals in hydrogen and deuterium, more than 40 binary hydrides and deuterides were synthesized: $ScH_{2.4}$, $ScD_{0.73}$, $YH_{2.2–2.9}$, $YD_{2.1}$, $PrH_{2–2.8}$, $PrD_{2.4}$, $NdH_{2.2–2.6}$, $NdD_{2.36}$, $SmH_{2.24–3}$, $SmD_{2.38}$, $GdH_{2–2.8}$, $GdD_{2.3}$, $TbH_{2–2.8}$, YbD_2, $DyH_{2.26–2.8}$, DyD_2, $HoH_{2.26–2.8}$, $HoD_{2.12}$, TiH_2, TiD_2, ZrH_2, $ZrD_{1.96}$, HfH_2, $HfD_{1.93}$, $VH_{0.71}$, $NbH_{0.9–1.85}$, $TaH_{0.55}$, and so on.

Fundamentals of SHS during combustion of metals in H_2 were defined. The factors that control the regime and velocity of combustion wave propagation, temperature, and completeness of reaction were elucidated. A two-step combustion mechanism for hydride formation in SHS wave was demonstrated: (i) a solid solution of hydrogen in metal formed in the combustion front, and (ii) a stoichiometric hydride formed because of the solid solution saturation by hydrogen in the postcombustion zone [1–5]. The combustion of metals in hydrogen proceeds at a relatively low temperature (250–1000°C). It has been shown that the completeness of the reaction does not depend on hydrogen pressure, dispersion of initial powder, and sample density. These characteristics differentiated the Me-H systems from the other SHS systems (Me-N, Me-B, Me-C, and so on).

Concise Encyclopedia of Self-Propagating High-Temperature Synthesis
http://dx.doi.org/10.1016/B978-0-12-804173-4.00070-3

These investigations resulted in development of "hydrides' material science" and the development of a series of prospective, highly productive, ecologically safe technological SHS processes for hydrides production, which currently allow synthesis of a large assortment of cheap and high-quality hydrides. The technological processes of TiH_2, ZrH_2, and HfH_2 hydrides production were worked out, and more than 20 tons of hydrides were produced in the experimental factory at Yerevan (Armenia). The SHS method has essential advantages in comparison with the traditional types of hydride production, that is, high productivity, high quality of hydrides, low energy consumption, ecologically safety, and so on. Various laborious operations, demanded at hydrides synthesis by traditional techniques, such as activation of metals, deep purification of H_2, use of fine powders, are excluded. It is worth noting that in addition to powders, in the synthesis of hydrides of TiH_2, ZrH_2, and HfH_2 in SHS mode, the shavings, sponge, and waste appearing at metal machining can be used as precursors.

Another novel direction was also elaborated: the synthesis of hydrides of transition metals in the accelerated electron beam, that is, thermal-radiation synthesis (TRS) [6]. The systematic investigation of TRS processes in Me-H systems was performed jointly with the physicists of the Yerevan Physics Institute on the Linear Electron Accelerator. The influence of the accelerated electron beam on the formation of binary hydrides was studied. More than 20 different binary hydrides were synthesized, among them a super-stoichiometric $HfH_{2.2-2.4}(H/Me > 2)$. TRS and SHS reactions are analogous in many respects. A main difference between them is the mode of exothermal reaction: frontally spreading in SHS and volume combustion synthesis (such as a thermal explosion) in TRS. Some unexpected effects of an electron beam on the process of hydrides synthesis were observed. After switching of the electron beam and cooling the sample to room temperature, a sharp increase in temperature (between 420°C and 620°C) was registered at the point when hydrogen filled the camera. This jump showed the interaction of irradiated metal with hydrogen at room temperature. The analyses confirmed the hydrides formation in these reactions. We call this phenomenon "cold synthesis."

REFERENCES

[1] Dolukhanyan SK, Nersesyan MD, Nalbandyan AB, Borovinskaya IP, Merzhanov AG. Dokl ANSSSR 1976;231(3):675–8.
[2] Dollukhanyan SK, Nersesian MD, Borovinskaya IP, Merzhanov AG. Author's certificate 552293. Bull. of Inventions; 1977. p. 12.
[3] Dolukhanyan SK. J Alloys Compd 1997;253–254:10–2.
[4] Dolukhanyan SD. SHS of binary and complex hydrides. In: Borisov AA, De Luca L, Merzhanov AG, editors. Self-propagating high-temperature synthesis of materials. NY: Taylor and Francis; 2002. p. 219–37.
[5] Dolukhanyan SK, Aleksanyan AG, et al. Int J SHS 2010;19(2):85–93.
[6] Dolukhanyan SK, Shekhtman VSh, et al. J Alloys Compd 2002;330–332:551–8.

Hydridonitrides and Carbohydrides of Transition Metals

Seda K. Dolukhanyan

The refractory compounds (carbides and nitrides) of IV group metals interact with hydrogen and form ternary carbohydrides and hydridonitrides. The study of SHS in Me-C-H and Me-N-H three-component systems is instigated by the fact that ternary hydrides are more heat-stable than binary hydrides ($T_{dissociation} = 800°C–1000°C$ vs. 450°C–800°C). High-temperature hydrides have potential for use as biological protection material in nuclear energetic environments.

Analysis of regularities of combustion in Ti(Zr)-C-H [1] and Ti(Zr or Hf)-N-H [2] systems showed that the combustion modes are defined by the competition between carbon, nitrogen, or hydrogen for interaction with the metal. The critical conditions for transition between combustion modes were stated as were the parameters required for the formation of hydride product (hydrogen pressure, dispersion of initial powder, and sample density). For example, in the Ti-C-H system, the transition between combustion modes occur neatly:

$$Ti + (1-x)C + H_2 \rightarrow TiH_2 + (1-x)C, \quad \text{at } x < 0.4; \quad T_c = 800°–1000°C \quad (1)$$

$$Ti + (1-x)C + H_2 \rightarrow TiC_{1-x} + H_2 \rightarrow TiC_{1-x}H_y, \quad \text{at } x > 0.4; \quad T_c = 1800°C \quad (2)$$

Carbohydrides of titanium, zirconium, and hafnium were successfully synthesized. It was found that different temperatures of combustion initiation with otherwise identical conditions influenced the systems to proceed via one of two parallel reactions. For low-temperature initiation (800°C–1000°C), a $Ti + H_2$ reaction was dominant, whereas for high-temperature initiation (1800°C), a Ti + C reaction was dominant. This observation was the first experimental confirmation of theoretically predicted similarity of the steady combustion mode in systems with parallel reactions.

The combustion of metal in a gaseous (N_2-H_2) environment was also investigated. These experiments were conducted with a wide range of gas pressure $P(H_2)/P(N_2)$ ratios

Concise Encyclopedia of Self-Propagating High-Temperature Synthesis
http://dx.doi.org/10.1016/B978-0-12-804173-4.00071-5

for $P_{tot} = P(H_2) + P(N_2) < 100$ atm. Depending on $P(H_2)/P(N_2)$, combustion may proceed by one of the following reaction routes:

$$Zr + N_2 + H_2 \rightarrow ZrH_2 + N_2 \qquad P(H_2)/P(N_2) > 2 \qquad (3)$$

$$Zr + N_2 + H_2 \rightarrow ZrN + H_2 \qquad P(H_2)/P(N_2) \leq 0.3 \qquad (4)$$

$$Zr + N_2 + H_2 \rightarrow ZrN_x + H_2 \rightarrow ZrN_xH_y \qquad 1 \leq P(H_2)/P(N_2) \leq 1.5 \qquad (5)$$

There are some interesting features of reaction (5), in which high hydrogen content results in the single-phase ternary hydridonitride [2–4]. In this case, combustion is sustained by a reaction with nitrogen. Analysis of quenched samples showed that only solid solutions of nitrogen in Zr formed in the combustion front. Further nitridation behind the combustion front was suppressed by hydrogenation, which became predominant in the postcombustion zone leading to the formation of the hydridonitride. Combustion in the Ti-N-H, Hf-N-H, and Y-N-H systems was found to proceed in a similar manner.

The interactions of solid solutions of nitrogen and titanium, zirconium, or hafnium with hydrogen in the SHS mode were also investigated [5,6]. Combustion proceeded in the low-temperature regime (500°C–600°C), and the combustion products were single-phase hydridonitrides of titanium, zirconium, and hafnium. SHS in systems that contain two metals, along with carbon, nitrogen, and hydrogen (Zr-Nb-C-H, Zr-Nb-N-H, Zr-Nb-C-N, Zr-Nb-C-N-H, Ti-V-C-H, Ti-V-Cr-C-H, Ti-Nb-Cr-C-H, Ti-Nb-W-C-H, and so on) allowed production of a series of complex bi- and tri-metal single-phase carbohydrides and hydridonitrides with cubic and hexagonal structures.

REFERENCES

[1] Martirosyan NA, Dolukhanyan SK, Merzhanov AG. Combust Explos Shock Waves 1983;19(5):569–71.

[2] Dolukhanyan SK, Alexanian AG, Seyranyan GB, Aghajanyan NN, Nalbandyan AB. Dokl AN SSSR 1984;276(1):131–40.

[3] Dolukhanyan SK, Alexanian AG, Hakobian HG. Interaction. Int J Hydrogen Energy 1995;20(5):391–5.

[4] Dolukhanyan SK, Alexanyan AG, Nalbandyan AB, Merzhanov AG. Combust Explos Shock Waves 1985;21(3):337–41.

[5] Aleksanian AG, Dolukhanyan SK. Int J Hydrogen Energy 1996;21(11/12):955–9.

[6] Aghajanyan NN, Dolukhanyan SK, Ter-Galstyan OP. Int J Hydrogen Energy 2011;6:1306–8.

Immobilization of Radioactive Wastes

Tatiana V. Barinova, Inna P. Borovinskaya

Self-propagating high-temperature synthesis (SHS) technology for immobilization of high-level radioactive wastes (HLWs) is based on combustion-assisted embedding of HLWs into chemically inert analogs of natural minerals used as preserving matrices of high chemical and radiation resistance [1]. Green composition used for the purpose normally contains (a) Ti, Al, and Zr powders as reducing agents; (b) oxides such as Fe_2O_3, MoO_3, CuO, V_2O_5, BaO_2, and $Ca(NO_3)_2$; and (c) modifying additives that favor formation of desired mineral phases. HLWs are introduced to the preceding green mixture in the form of calcinates, oxides, or sorbents of the radionuclides extracted from liquid HLWs. Combustion products are obtained as cermet materials formed by two interpenetrating phases: a mineral-like ceramic phase with embedded radionuclides and a metallic phase originated upon reduction of the oxides. The technique of forced SHS compaction [2,3] can be used to decrease the porosity of preserving matrices.

Preserving matrices, with embedded HLWs imitators, were fabricated on the basis of perovskite $CaTiO_3$ and $BaZrO_3$, zirconolite $CaZrTi_2O_7$, titanate zirconate pyrochlore $Y_2(Ti_{1-x},Zr_x)_2O_7$, yttrium aluminum garnet $Y_3Al_5O_{12}$, and pollucite $CsAlSi_2O_6$. Also synthesized were the matrix materials [4] based on titanate pyrochlore $Y_2Ti_2O_7$ containing 10% of embedded $^{238}UO_2$, $^{237}NpO_2$, $^{239}PuO_2$, and $^{241}Am_2O_3$. The $Y_2Ti_2O_7$ matrices with 20 wt% of embedded α-active ^{232}Th were fabricated in conditions for pilot-scale production. Synthesized preserving matrices exhibited high radiation and chemical resistance to the action of imitators and actinoides.

For reprocessing radioactive graphite rods containing nuclide ^{14}C (half-life 5730 years) along with inclusions of nuclear fuel and its decay products, it is suggested [5] to bind the nuclide ^{14}C into the stable carbide $Ti^{14}C$. The SHS immobilization takes advantage of the low yield of secondary radioactive products. In the case of SHS processing, the evaporation of ^{14}C is on the order of three magnitudes weaker than in the case of direct incineration. Highly dense Al_2O_3/TiC matrices were prepared by forced SHS compaction. Mineral-like compounds with perovskite, ilmenite, and garnet structure were designed for immobilization of nuclear fuel and its decay products. Cesium was immobilized [6] in the form of $Cs_2Ti_6O_{13}$.

A pilot-scale production unit for SHS compaction of blocks up to 1.75 kg in their weight was designed. Synthesis was carried out directly in metallic containers designed

Concise Encyclopedia of Self-Propagating High-Temperature Synthesis
http://dx.doi.org/10.1016/B978-0-12-804173-4.00072-7

for subsequent burial. Thus prepared TiC/Al_2O_3 materials with embedded radioactive graphite showed low open porosity (<1%) and high compression strength ($\sigma_c \leq 1$ GPa).

Although the studies on immobilization of nuclear wastes are currently conducted in many countries, the problem of convenient and reliable preserving matrices is yet to be resolved.

REFERENCES

[1] Merzhanov AG, Borovinskaya IP, Makhonin NS, Zakorzhevsky VV, et al. The method of treatment of radioactive waste of variable composition. Russ. Patent 2 065 216; 1996.
[2] Glagovskii EM, Kuprin AV, Pelevin AP, et al. HLW immobilization to stable mineral-like materials by SHS mode. At Energy 1999;87(1):57–61.
[3] Borovinskaya IP, Barinova TV, Ratnikov VI, Zakorzhevsky VV, Ignat'eva TI. Consolidation of radioactive wastes into mineral-like materials by the SHS method. Int J SHS 1998;7(1):129–35.
[4] Vinokurov SE, Kulyako YuM, Perevalov SA, Myasoedov BE. Immobilization of actinides in pyrochlore-type matrices produced by self-propagating high-temperature synthesis. C R Acad Sci Ser C 2007;10:1128–30.
[5] Merzhanov AG, Borovinskaya IP, Makhonin NS, Zakorzhevsky VV, et al. The method of processing solid highly active graphite-containing wastes. Russ. Patent 2 055 220; 1994.
[6] Kobyakov VP, Barinova TV, Borovinskaya IP, Sichinava MA. Phase relations in the TiO_2–$CsNO_3$ system between 550 and 1140 K. Inorg Mater 2011;47(3):290–5.

Impregnated SCS

Alexander S. Mukasyan

Solution combustion synthesis (SCS) is a phenomenon that involves self-sustained chemical reactions in homogeneous aqueous solutions of different oxidizers (e.g., metal nitrates) and fuels (e.g., urea, glycine, hydrazides). It is important that this process allow synthesis of a variety of nanomaterials, including oxides, metals, alloys, and so on. There are many modifications of the SCS process [1,2]. The most advanced ones relate to the process called "Impregnated SCS" [3,4]. Two fundamentally different approaches exist: (i) reactive solution is impregnated into an inert porous media; (ii) reactive solution is impregnated into chemically active media. In the former case, the reaction is initiated in the heterogeneous inert support/solution media, which leads to formation of the desired product in the porous skeleton of the support media. In the latter case, a process typically involves three steps. First, the reactive aqueous solution is impregnated into the carrier, typically a porous cellulose matrix. Second, the media is dried to remove the unbound water. Finally, the reaction is locally initiated at one end of the media, and a combustion wave propagates along the media converting the reactive solution to the desired product and consuming the cellulose matrix as CO and CO_2 species.

Impregnated SCS with inert support is used widely to prepare different types of supported catalysts [5–7]. For example, uniform coating of Pd-substituted ceria on cordierite monolith (Fig. 1) has been achieved by a single-step impregnated SCS method. These simple and inexpensive processes for preparation of the supported catalysts hold great promise for automobile exhaust remediation. A set of effective metals (Cu, Ni, CuNiFe alloys) supported (Al_2O_3, Ce_2O_3) catalysts for hydrogen production from alcohols was also synthesized by this approach [7,8].

The impregnation active layer combustion (IALC) approach is valuable for the synthesis of metal oxides or metals, especially if the corresponding metal nitrate-fuel system has low exothermicity. In this case, additional heat formed through burning of (e.g., cellulose in air) makes the following oxide formation reaction self-sustained. For example, this method was used for fabrication of $SrRuO_3$ perovskites for fuel cell applications [8]. It was shown that reactions between strontium/ruthenium nitrites and glycine were not sufficiently exothermic for combustion under normal conditions. Also, the pure cellulose paper employed in the synthesis did not exhibit a controllable low-temperature smoldering combustion wave. Thus neither exothermic reaction can occur by itself in a low-temperature mode. Conversely, the impregnated system did exhibit the

Concise Encyclopedia of Self-Propagating High-Temperature Synthesis
http://dx.doi.org/10.1016/B978-0-12-804173-4.00073-9

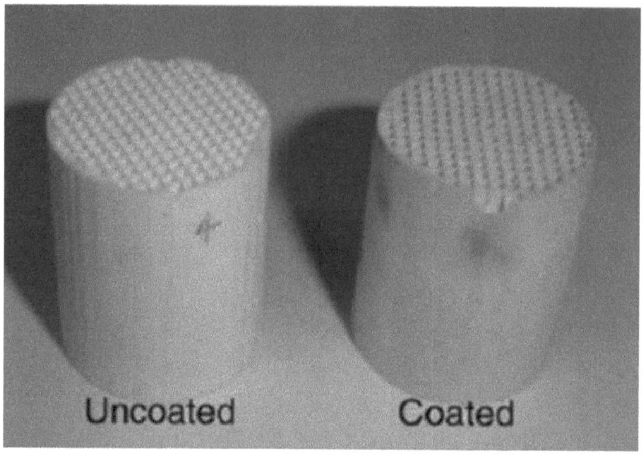

Fig. 1 Uncoated and coated by SCS impregnation by Pd-substituted ceria Cordierite monolith [5].

desired smoldering mode resulting from heat exchange between the burning cellulose and nitrate reduction reactions. As a result, low-temperature synthesis of a nanoscale product was possible with the IALC process [8–10], including $Li_4Ti_5O_{12}$, 7CuO/3ZnO/1ZrO, $SrRuO_3$, and many others. It is important that the theoretical models were developed to describe the ILCS process [11,12]. Based on this approach, a continuous technology for synthesis of nanopowders by SCS with a yield of 0.5–2 kg/h (Fig. 2) has been suggested [13]. This is an important first step toward scaling up SCS.

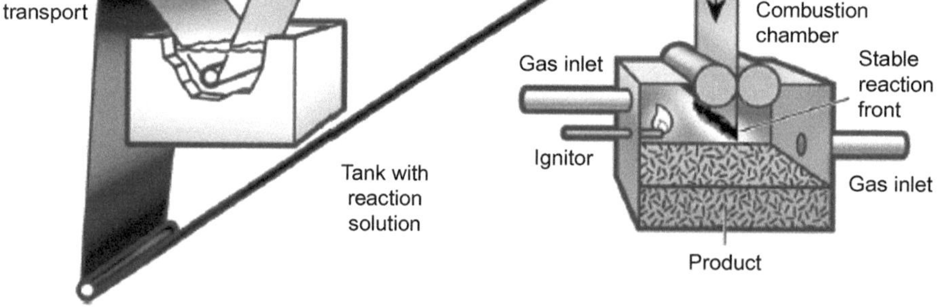

Fig. 2 General scheme for continuous synthesis of nanopowders by the impregnated active layer combustion (IALC) method.

REFERENCES

[1] Aruna ST, Mukasyan AS. Curr Opin Solid State Mater Sci 2008;12:44.
[2] Mukasyan AS, Epstein P, Dinka P. Proc Combust Inst 2007;31(2):1789.
[3] Dinka P, Mukasyan AS. J Phys Chem B 2005;109:21627.
[4] Mukasyan AS, Dinka P. Int J Self-Propag High-Temp Synth 2007;16:23.
[5] Sharma S, Hegde MS. Catal Lett 2006;112:69.
[6] Cross A, Roslyakov S, Manukyan KV, Rouvimov S, Rogachev AS, et al. J Phys Chem C 2014;118:26191.
[7] Cross A, Kumar A, Wolf EE, Mukasyan AS. Ind Eng Chem Res 2012;51:12004.
[8] Lan AD, Mukasyan AS. Ind Eng Chem Res 2008;47:8989.
[9] Yuan T, Wang K, Cai R, Ran R, Shao Z. J Alloys Compd 2009;477:665.
[10] Kumar A, Mukasyan AS, Wolf EE. Appl Catal A Gen 2010;372:175.
[11] Kumar A, Mukasyan AS, Wolf EE. Ind Eng Chem Res 2010;49:11001.
[12] Lennon EM, Tanzy MC, Volpert VA, Mukasyan AS, Bayliss A. Chem Eng J 2011;174:333.
[13] Mukasyan AS, Dinka P. Apparatus for synthesizing nanopowder, WO2007019332-A1; 2008.

Industrial Ferroalloys

Yury M. Maksimov

Modern production of steel is based on alloying with nitrogen-containing ferroalloys, the production of which is highly energy-intensive in the steel industry. Energy consumption in furnaces during production of nitrogen-containing alloying elements ranges from 5000 to 10,000 kW hour per ton, and the alloying element has low nitrogen content and low density. SHS technology significantly decreases the energy consumption that reaches only 0.1 kW hour per ton during production of alloying elements. In addition, the alloys obtained by SHS technology contain 1.5–3 times more nitrogen than analogs, which allows us to reduce the use of alloying elements by several times for the manufacture of steel.

The first experimental site designed for nitriding of ferroalloys by SHS with productivity of 100 ton of the alloy per year was built in 1979 in Tomsk [1]. The initial ferroalloy was ground in a jaw crusher and a pulse chopper; then it was placed in the SHS reactor and burned under nitrogen pressure.

These enterprises apply the nitriding of ferroalloys using the SHS technology: ferrochromium, ferromolybdenum, ferrovanadium, ferrosilicon, and so on. Nitrided ferroalloys of vanadium, silicon, and chromium are widely used in metallurgy. Ferrovanadium nitrided by SHS is used for manufacturing high-speed steel, which reduces the energy consumption by 10 times and decreases the loss of raw material by 5 times. Vanadium-containing SHS alloys are also used for manufacturing structural steel in oil refining and machine building industries, for constructing bridges and rails with elevated low-temperature reliability, and so on [2].

Ferrosilicon nitrided by the SHS technology is used at OAO MMK (Magnitogorsk) for the production of refractories, which allows us to increase the resistance of chutes and trays by 2.5 times, stabilize the output of products from blast furnaces, and reduce significantly the consumption of these masses [3].

Nitrided ferrochrome and chromium are added to the composition of most alloyed steels. These steels have high strength and corrosion resistance and are used in the electrical industry. Adding an alloy nitrided by SHS to steel enhances the absorption of nitrogen by steel up to 90% or 95%, reduces the duration and loss of chromium, and improves the working conditions of steelworkers.

Concise Encyclopedia of Self-Propagating High-Temperature Synthesis
http://dx.doi.org/10.1016/B978-0-12-804173-4.00075-2

Fig. 1 SHS reactors.

Fig. 2 Nitrided alloying elements.

The industrial technology of nitrogen-containing alloying chromium SHS alloys is used at the NTPF "Etalon." The high concentration of nitrogen and the maximum density of alloying elements allow us to produce nitrogen-containing alloys with wide nomenclature [4] (Figs. 1 and 2).

REFERENCES

[1] Maksimov YuM, Ziyatdinov MKh, Vasilyev AP, Merzhanov AG. Experimental-industrial technology for nitrided alloys, Chernogolovka: Offprint BICP AS USSR; 1985. 32 p.

[2] Panfilova LM, Smirnov LA, Filippenkov AA, et al. The use of complex V-containing alloying elements obtained by the SHS method for steel production for various purposes. Problems of production and the use of steels with vanadium, Ekaterinburg: Proceedings of Int Sci-Tech Sem; 2007. p. 127–43.

[3] Maksimov YuM, Ziyatdinov MKh, Kolmakov AD, et al. Metal composition and production process, United States, Patent 4,623,402, Foreign application priority data: Jan. 25, 1980 [SU] USSR 2,865,652, PCT filed: Dec. 25, 1980, date of patent: Nov. 18, 1986.

[4] Ziyatdinov MKh, Shatohin IM. Self-propagating high-temperature synthesis of nitrided ferrochrome. Steel 2009;99:48–53.

Industrialization

Inna P. Borovinskaya

There are three types of SHS production: pilot-scale, semiindustrial, and industrial-scale. A pilot-scale facility produces small batches of products on customer request. The production often uses manual labor. The major task is checking out new technical solutions and potential SHS technologies. Semiindustrial production yields serial products for customers. Usually, such production is completely or partially automated. The major task is industrial-level, which involves checking a new process alongside the elements of business.

Industrial-scale production yields a large output of serial products for realization on the market [1].

The first pilot-scale and semiindustrial facilities appeared in Russia for obtaining trial lots of powders of different refractory inorganic compounds: simple and complicated metal and nonmetal carbides, borides, nitrides, silicides, oxides, chalcogenides, intermetallics, and composites thereof, and later they were developed to direct synthesis (combined with compaction, gas-statting, casting, extrusion, etc.) of consolidated nonporous and porous materials and items (including net-shape products) from refractory compounds and their composites (also those with binding elements, fillers, etc.). The obtained materials and items extended the range of high-temperature ceramics, cermets, hard, and wear-resistant alloys [2]. Then semiindustrial production appeared in different countries: in Japan, TiNi, highly pure TiAl powders, and SiC(C) composites; [3] in Kazakhstan, Furnon refractories based on complex oxides; in the United States, high-quality TiB_2/Al_2O.

Most impressive in China is SHS production of ceramic-lined steel pipes ($d=$ 20–800 mm). The pipes are used for transportation of coal, coke, coal cinder, mineral powder, cement mortar, molten aluminum, oil-water mixture, and so on [1].

Based on semiindustrial production, some independent SHS facilities were developed. Some of them use automated technological lines. One of the facilities reoriented to SHS technology was the Kirovakan plant's (Armenia) manufacturing of high-temperature heating elements based on molybdenum disilicide. The large-tonnage SHS production of nitrided ferroalloys and alloying additives was organized in Magnitogorsk (Ural). This technology involves SHS reactors of 150 L capacity with one-time loading of 500 kg and total output of 2400 tons a year. The combustion heat is used for the space heating. One of the most impressive achievements of the industrial application of SHS technologies is the organization of completely automated SHS production of ceramic powders (in Spain and Russia) [4]. Promising examples of SHS technologies

Concise Encyclopedia of Self-Propagating High-Temperature Synthesis
http://dx.doi.org/10.1016/B978-0-12-804173-4.00076-4

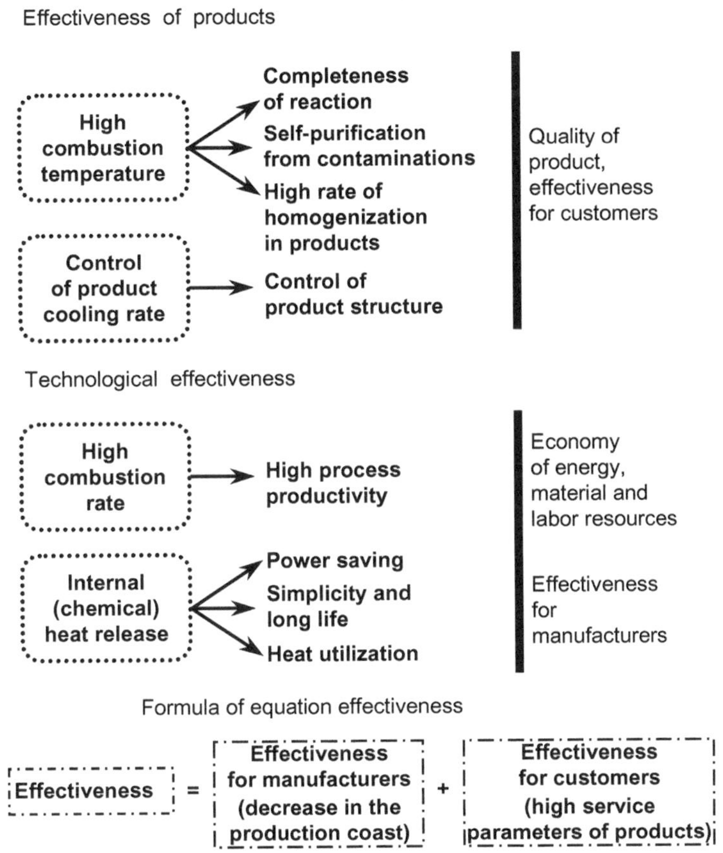

Fig. 1 Physical background of the effectiveness of SHS.

in the organization of complex metallurgical facilities is Russian engineering of special equipment of a plant specializing in production of fine hard-alloyed tungsten carbide powders with a production output of 500 tons/year (with a possible increase up to 2000 tons/year) [5] and Chinese manufacturing of pipes with corrosion-resistant internal coating with a total output of 10,000 tons/year and a setup line of 25 km.

Fig. 1 demonstrates the effectiveness of SHS technologies and products [1,6].

REFERENCES

[1] Merzhanov AG. SHS on the pathway to industrialization. Chernogolovka: ISMAN; 2001.
[2] Merzhanov AG. Fundamentals, achievements, and perspectives for development of solid-flame combustion. Rus chem bull 1997;46(1):1–27.
[3] Miyamoto Y. State of the art in R&D of SHS materials in the world. Int J SHS 1999;8(3):375–84.
[4] Merzhanov AG, Borovinskaya IP, Prokudina VK. Self-propagating high-temperature synthesis: resources, technologies, economy. Resour Technol Econ 2006;4:26–32.
[5] Merzhanov AG, Borovinskaya IP, Prokudina VK, Nikulina NA. Efficiency of the SHS powders and their production method. Int J SHS 1994;3(4):353–70.
[6] Pampuch R. Advanced HT ceramic materials via solid combustion. J Eur Ceram Soc 1999;19 (13–14):2395–404.

Infrared Thermal Analysis

Christopher E. Shuck

The infrared (IR) portion of the electromagnetic spectrum was not known until 1800 when it was discovered accidentally by Sir William Herschel while he was testing the transmission of heat through different types of glass. Thermal vision systems use IR radiation to determine surface temperature. All materials constantly emit IR radiation, which is exponentially coupled to temperature in the following way:

$$W = \varepsilon \sigma T^4$$

Because this relationship is known, the temperature can be determined if the emissivity (ε) is known. For true blackbodies, ε is always 1, but for real systems, this is not the case. Mathematically, the relationship between blackbody radiation and ε is

$$\varepsilon = \frac{W}{W_{BB}}$$

where ε defines the amount of radiation that is produced by the material and a blackbody source at the same temperature. In real materials, three processes can occur, some of the produced radiation may be absorbed (α), some may be reflected (ρ), and some may be transmitted (τ). The sum of these fractions must be unity:

$$\alpha + \rho + \tau = 1$$

For opaque materials, there is no transmittance, simplifying the relationship. According to Kirchhoff's law [1], the spectral emissivity and absorbance of a body are equal at a specified temperature and wavelength. This simplifies the relationship to

$$\varepsilon + \rho = 1$$

Because it is easier to measure the reflectance of a material, this is often completed, and then related to the emissivity. Using these experimentally measured values, the temperature of any process can be monitored.

Use of IR allows for in situ monitoring of the combustion synthesis process. Information on the surface temperature of the entire sample can be monitored over the entire experiment (Fig. 1). This, coupled with advancements on high-speed data acquisition and detectors, allows for insight into the CS process that cannot be accomplished by any other technique.

Concise Encyclopedia of Self-Propagating High-Temperature Synthesis
http://dx.doi.org/10.1016/B978-0-12-804173-4.00078-8

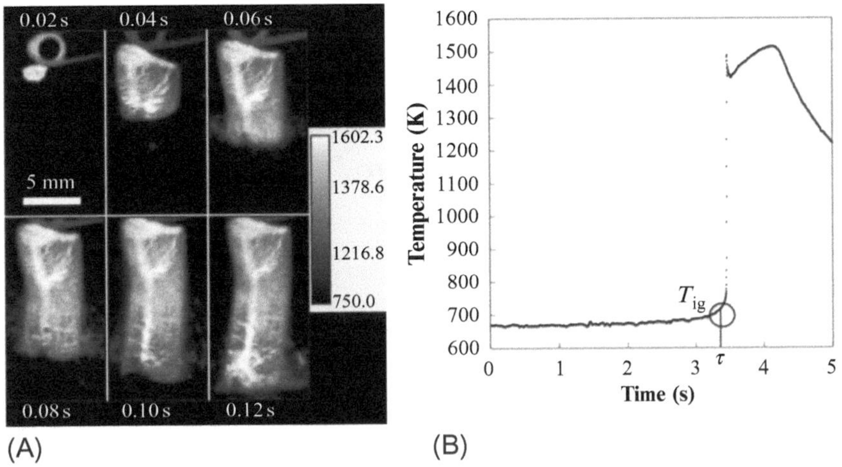

Fig. 1 Sample data that can be attained using a high-speed IR camera, including (A) temperature over the entire sample or (B) at a single point.

State-of-the-art high-speed IR cameras can have frame rates exceeding 15,000 Hz, with spatial resolution of 2 μm; with these specifications, it is possible to intimately examine the combustion process. In many systems, these specifications allow many points to be examined within the combustion front itself, giving unprecedented insight into the front evolution with time.

These techniques have been used in many different experimental studies, including SCS [2], SHS [3], and ILCS [4]. Because each captured frame stores information on the entire surface temperature, time-temperature profiles can be generated from any region on the sample. This makes it possible to monitor ignition of the sample, reaction propagation velocity, time-temperature history at any point throughout the reaction, and information on the mechanism in more complicated reactions. The existence of IR transparent glass, for which sapphire is most commonly used, allows for the temperature profile to be observed even within pressurized chambers.

REFERENCES

[1] Kirchhoff G. Ann Phys Chem 1860, 275–301.
[2] Manukyan KV, et al. J Phys Chem C 2013, 24417–27.
[3] Shuck CE, et al. Combust Flame 2016, 487–93.
[4] Kumar A, et al. Appl Catal A Gen 2010, 175–83.

International Journal of Self-Propagating High-Temperature Synthesis

Yury B. Scheck, Alexander E. Sytschev

This is an international journal embracing a wide scope of topics concerning SHS science and technology and adjacent areas. The journal was founded in 1991 by Professor A.G. Merzhanov and since then has been published quarterly (four issues a year) by Allerton Press Inc., New York c/o International Academic Publishing Company NAUKA/INTERPERIODICA, Moscow, and since 2007 distributed by Springer Science + Business Media (http://link.springer.com/journal/12003); ISSN 1061-3862 (print version); ISSN 1934-788X (electronic version).

Concise Encyclopedia of Self-Propagating High-Temperature Synthesis
http://dx.doi.org/10.1016/B978-0-12-804173-4.00079-X

The journal publishes reviews, original papers, and brief communications on various aspects of SHS such as exothermic processes, macroscopic kinetics, combustion theory, new methods of research, scientific instruments, kinetics and thermodynamics of SHS processes, mechanism of SHS reactions, mathematical modeling of SHS processes, structural macrokinetics of SHS processes, nonlinear dynamics of SHS processes, chemistry/physics of SHS processes, combustion synthesis, powder metallurgy of SHS products, SHS-produced materials, applications of SHS products in R&D engineering, and fundamentals of industrial-scale SHS production.

To date, the edition is abstracted and/or indexed in Academic OneFile, ChemWeb, EI-Compendex, Expanded Academic, Google Scholar, INSPEC, OCLC, Referativnyi Zhurnal (VINITI), SCImago, SCOPUS, and Summon by ProQuest.

The issues of 2007–15 are available online at http://link.springer.com/journal/12003.

Editorial Board contacts:

Email: rogachev@ism.ac.ru; scheck@ism.ac.ru; journal@ism.ac.ru

Telephone: +7(496) 524-6256; +7(496) 524-6232; +7(496) 524-6284

International Symposia on SHS

Yury B. Scheck

International Symposia on SHS have been held biannually since 1991. These international meetings were organized at the initiative of Professor Alexander Merzhanov and headed by him until SHS-2011. This is the main meeting for the scientific and technological SHS community, where all new ideas, results, and products can be presented to the best specialists in the field.

Symposia goals are oriented around the following:

- Reviewing the state of the art in SHS science and technology worldwide with regard to materials design, SHS processing, and applications
- Sharing new ideas, new approaches, and SHS processes to enable more effective penetration of SHS into various fields of manufacture
- Serving as a platform for initiating discussions on innovative SHS methodologies and techniques
- Involving young scientists and engineers and promoting sustained development in the field of SHS, both in fundamental and applied research
- Fostering international collaboration among the worldwide community of scientists, engineers, and technologists active in the field

Symposia topics normally cover a wide range of scientific disciplines such as

Combustion theory and modeling
Structural macrokinetics of SHS processes
SHS chemistry and physics
SHS of advanced and smart materials: metals, ceramics, intermetallics, cermets, etc.
Unconventional SHS reactions and processes
New experimental techniques and applications
Industrialization of SHS processes and products
Biocompatible materials and coatings
Sintering and consolidation
Aerospace applications

According to tradition, every Symposium must be organized in a new place.

Following is a list of previous SHS Symposia in a chronological sequence:

I SHS Symposium (SHS-1991), Alma-Ata, Republic of Kazakhstan
II SHS Symposium (SHS-1993), Honolulu, Hawaii, United States

Concise Encyclopedia of Self-Propagating High-Temperature Synthesis
http://dx.doi.org/10.1016/B978-0-12-804173-4.00080-6

III SHS Symposium (SHS-1995), Wuhan, China
IV SHS Symposium (SHS-1997), Toledo, Spain
V SHS Symposium (SHS-1999), Moscow, Russia
VI SHS Symposium (SHS-2002), Haifa, Israel
VII SHS Symposium (SHS-2003), Cracow, Poland
VIII SHS Symposium (SHS-2005), Cagliari, Italy
IX SHS Symposium (SHS-2007), Dijon, France
X SHS Symposium (SHS-2009), Tsakhkadzor, Republic of Armenia
XI SHS Symposium (SHS-2011), Attica, Greece
XII SHS Symposium (SHS-2013), South Padre Island, Texas, United States
XIII SHS Symposium (SHS-2015), Antalya, Turkey

The next, XIV SHS Symposium (SHS-2017) will be held in Tbilisi, Georgia.

ISMAN

Inna P. Borovinskaya, Mikhail I. Alymov

The Institute of Structural Macrokinetics and Materials Science of the Russian Academy of Sciences (ISMAN) was founded in 1987. The foundation of the institute is connected with the scientific discovery of "The Phenomenon of the Wave Localization of Solid-State Autoretarding Reactions" (solid flame phenomenon) made by the scientists of the Institute of Chemical Physics, USSR Academy of Sciences (its branch in Chernogolovka), A.G. Merzhanov, I.P. Borovinskaya, and V.M. Shkiro in 1967. This discovery marked the start of a new scientific-and-technical direction—self-propagating high-temperature synthesis (SHS) [1]. A.G. Merzhanov was appointed first director of ISMAN and general director of the All-Union Interbranch Scientific-and-technical Unit "Termosintez" headed by the institute. The task of the unit was to unite the efforts of the scientists and specialists working in this field in order to develop SHS from fundamental research to industrial implementation. At that time at ISMAN, there was an excellent team of highly experienced scientists (V.V. Barzykin, B.I. Khaikin, E.A. Grigoryn, A.S. Shteinberg, K.G. Shkadinsky, A.P. Aldushin, A.S. Stolin, I.P. Borovinskaya, V.A. Veretennikov, V.I. Ponomarev, V.K. Prokudina, and others) and young active specialists working in different directions of investigation: combustion theory and practice, mathematics, chemistry of synthesis and analytical chemistry, physical methods of analysis, and so on (V.M. Shkiro, Yu.M. Maksimov, A.S. Rogachev, A.S. Mukasyan, A.N. Pityulin, S.M. Mamyan, V.E. Loryan, V.A. Shcherbakov, M.D. Nersesyan, A.P. Kachin, and others). This team was working actively with a great number of scientific and practical problems of combustion and explosion processes, such as mechanism of high-temperature chemical reactions, regularities of heat release and transfer, behavior of combustion front propagation and combustion product structure, dynamics of combustion product structure, and composition development. The members of the team worked out diagnostic methods of high-temperature, fast-occurring processes, and mathematical modeling of combustion processes (for the common case, non-one-dimensional and nonstandard) and synthesized chemical compounds and materials, developed the technology of refractory compound powders (carbides, borides, nitrides, silicides, etc., and composite materials thereof), coatings, and items. Semiindustrial production lines were made, marketing was carried out, and commercialization was started. Several variants of SHS technologies were developed. The main ones were connected with the production of powders (Powder SHS) and direct one-stage

Concise Encyclopedia of Self-Propagating High-Temperature Synthesis
http://dx.doi.org/10.1016/B978-0-12-804173-4.00081-8

obtaining of materials and items (including near-net shape technology). Among them are SHS sintering (SHS gas-statting), forced SHS compaction, SHS casting (SHS metallurgy), SHS extrusion, SHS welding, and gas-transport SHS. The original equipment was designed at ISMAN to study mechanism of SHS processes, including computer-assisted installations (powder SHS reactors, molds for force compaction, SHS gasostats withstanding gas pressure up to 8000 atm., centrifugal installations to synthesize cast materials, SHS-extrusion devices, etc.). An installation of dynamic X-ray analysis, which allowed observing all the intermediate phases formed during the synthesis, was worked out. A universal SHS installation consisting of three reactors different in their capacity was made for the technological study and powder production.

Many scientific and technological projects of ISMAN that started with the SHS method became very efficient [2,3] and were patented in many countries all over the world, including the United States, Japan, the United Kingdom, France, and so on.

The scientific and practical work was carried out with the scientists of many republics of the former USSR: Armenia, Georgia, Kazakhstan, Ukraine, Belorussia, Uzbekistan, and Azerbaijan. Some mutual workshops and conferences were held.

In 1980 ISMAN headed by A.G. Merzhanov started the international activity in the SHS field with the scientists from different SHS Centers, institutes, and universities: J.A. Puszynski with his colleagues from the University of California (Davis, Z. Munur), Alfred University (Alfred, R. Spriggs), University of Notre Dame (Notre Dame, A. Warma), Northwestern University (Evanstone, B. Matkowsky), Washington State University (Pullman, S. Wojcicki), Ryukoky University seta (Otsu, M. Kouzumi), Institute of Technology (Tokyo, O. Odawara), Osaka University (Osaka, Y. Miyamoto), Wuhan University of Technology (Wuhan, R. Yuan), Academy of Mining and Metallurgy (Krakov, R. Pampuch), Indian Institute of Science (Bangalor, K. Patil), and so on. Later some universities in France (Dijon, Poitiers, Paris) and Italy (Turin) joined this activity. Since 1991 the International SHS Symposia discussing scientific and industrial achievements in SHS have been held every 2 years. Fourteen symposia have already been held in different countries of the world. The *International Journal of Self-Propagating High-Temperature Synthesis* with the editorial board in ISMAN is issued. The International SHS Association was founded within the World Academy of Ceramics [4].

ISMAN works in the area of SHS science and technology, which found its application in almost all areas of natural sciences (chemistry, physics, physical chemistry, chemical physics, combustion and explosion chemistry and physics, nature study, etc.). It has founded the backgrounds of a new branch of knowledge—structural macrokinetics—a boundary between combustion science and material science [5], and has developed new alternative resource- and energy-efficient production technologies. ISMAN could easily join such actual directions of modern science and engineering as development of nanosystems and materials, environmental resources management, advanced weapons, military and special equipment, biocompatible materials, chemical energy problems, raw material processing, and so on [3].

Current directions of ISMAN research include

- Investigations in the field of general and structural macrokinetics of the combustion and explosion processes
- SHS
- New materials synthesis and modification at high-dynamic pressure conditions
- Study and use of the combustion and combustion and explosion processes
- Study and development of the new highly effective technologies to produce the modern construction, functional and instrumental materials and coatings; material science studies in general

ISMAN now is a well-known leader in the fundamental searching and applied investigations in the field of combustion and explosion processes (including SHS) and their use of new materials production. Over many years, the institute developed the theory of the heterogeneous compound filtration burning and nonisothermal chain processes. A possibility of the combustion processes regulations by chemical methods was demonstrated and realized in practice. A possibility of gasless compound detonation was theoretically based and experimentally proved. New technologies were developed for the unique superhard and high-temperature materials production, including nanodispersed and bulk nanocrystalline materials.

Today there are two general fields in ISMAN scientific activities. First is the "traditional" SHS with the use of powder, where the main results were obtained 70–80 years ago. There remain some interesting fundamental problems (connected with the combustion theory and material synthesis), but the main results are ready for industrial adaptation. In this field, ISMAN possesses the patents, equipment, production forms, and records, presenting now a noticeable intellectual property. The second direction is connected with the new SHS methods and variants, developed in the middle of 1990s, such as a "solution combustion," multilayer systems burning, and so on. Other new studies are connected with use of the nanosize reagents and/or nanostructured products formation. These new processes have demonstrated high practical value but their mechanisms remain unrecognized. ISMAN has great experience in gasless and filtration combustion and therefore can be a leader of the fundamental investigations in these directions.

REFERENCES

[1] Merzhanov AG. The chemistry of self-propagating high-temperature synthesis. J Mater Chem 2004;14 (12):1779–86.
[2] Merzhanov AG. History and new developments in SHS. Adv Synth Process Compos Adv Ceram 1995;56:3–25 [special issue].
[3] Crider JF. Self-propagating high temperature synthesis—a Soviet method for producing ceramic materials. Ceram Eng Sci Proc 1982;3(9–10):519–28.
[4] Merzhanov AG, Borovinskaya IP. Historical retrospective of SHS: an autoreview. Int J SHS 2008;17 (4):242–65.
[5] Merzhanov AG, Rogachev AS. Structural macrokinetics of SHS processes. Pure Appl Chem 1992;64 (7):941–53.

Joining of Refractory and Dissimilar Materials

Alexander S. Mukasyan

ASTM C71 defines refractories as "nonmetallic materials having those chemical and physical properties that make them applicable for structures, or as components of systems, that are exposed to environments above 1000°F (811 K; 538°C)" [1]. Refractory materials are chemically stable at high temperatures and typically have low coefficients of thermal expansion (CTE). In turn, dissimilar materials are materials that have large differences in their structural and physical properties (i.e., melting point, thermal conductivity, coefficient of thermal expansion). They need to be joined for fabrication of layered composites that consist of joined ceramic-metal parts, for example, a hard ceramic front face and a soft metal back [2]. In general, refractory and dissimilar materials can be welded using either fusion or nonfusion processes [3,4]. For fusion welding, the high energy density processes (electron and laser beams, plasma) are more successful, as they are better able to cause melting. Resistance welding (RW) is another high-energy density option, with virtually no limit to providing powder to reach melting conditions. Nonfusion approaches include friction and explosion welding, as well as solid-state diffusion process.

Joining by SHS or combustion joining is a high-energy density method for welding the refractory and dissimilar materials [5,6]. The most widely used approach for combustion joining of refractory and dissimilar materials is reactive resistance welding (RRW). A detailed discussion of the SHS-welding apparatus can be found elsewhere [5,7,8]. Briefly, referring to Fig. 1, a layer of reactive mixture (1) is contained between two disks of the refractory materials (2) (e.g., C-C composite) that are to be joined. The stack (1, 2) is held in place between two electrodes (3), which are connected to a direct current (dc) power supply (4). The dc current is used to preheat the stack and uniformly initiate the reaction in the reactive layer. The electrodes are also part of the pneumatic system (5), which applies a load to the stack. All operational parameters, such as initial pressure and final pressure, applied current, delay time between ignition and final pressure application, and duration of Joule heating are defined by a programmable logic controller (6) system [7,8].

It has been demonstrated that a variety of materials can be joined by the combustion-based reactive resistive welding method (Table 1).

Concise Encyclopedia of Self-Propagating High-Temperature Synthesis
http://dx.doi.org/10.1016/B978-0-12-804173-4.00084-3

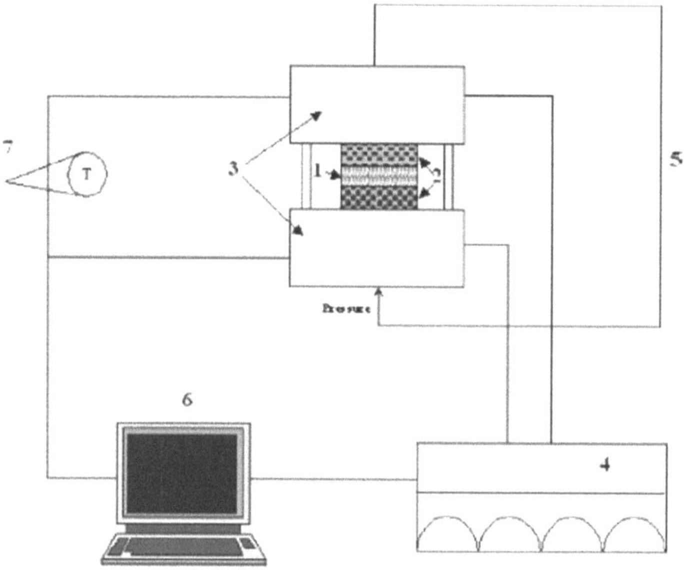

Fig. 1 Schematics of the SHS-joining apparatus.

Table 1 Materials joined with the combustion-based RRW method

Material(s) joined	Reactive media	References
W and Mo W/Mo and graphite Graphite		[5]
WC-8Co and steel	Ti + Ni	[9]
C-C composite	Ti + C	[7,10]
C-C composite	Ti, Mechanically activated Ni + Al	[11]
SiC-Al	Ti, Ti + C	[8]

Typical examples of cross sections for a SiC/Al-alloy joint surface are presented in Fig. 2 [8]. It can be seen that SiC/AA-5083 joint interface is approximately 15 μm in thickness. A close-up in Fig. 2B shows an intermediate phase in between SiC and AA-5083 with thickness in the range 300–600 nm. Thus the developed method with application of an external load provides formation of a nanoscale joint layer, which is a unique feature in the field of joining of bulk dissimilar materials. It was demonstrated that in all cases, it is more important that the mechanical properties of the joint layer are higher as compared to the joined materials.

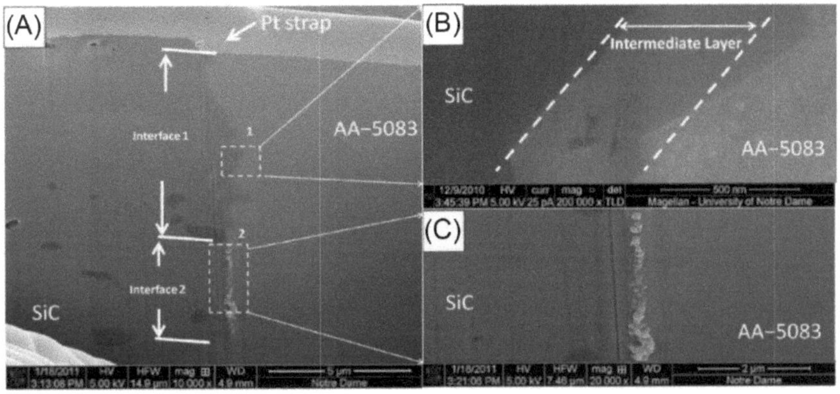

Fig. 2 Typical FIB cross-section of a SiC/AA-5083 joint interface (A) and magnified region of the joint interface 1 (B) and interface 2 (C).

REFERENCES

[1] ASTM volume 15.01 refractories; activated carbon, advanced ceramics.
[2] McCauley JW, Crowson A, Gooch Jr WA, Rajendran AM, Bless SJ, Logan KV, et al. Ceramic armor materials by design, ceramic transactions. 134. Ohio: The American Ceramic Society; 2001.
[3] Messler RW. Joining of materials and structures. Amsterdam: Elsevier; 2004.
[4] Fernie JA, Drew RAL, Knowles KM. Int Mater Rev 2009;5:283–331.
[5] Shcherbakov VA, Steinberg AS. J Self-Propag High-Temp Synth 1993;2:357–61.
[6] Messler RW, Zurbuchen MA, Orling TT. Weld J 1995;74:37–41.
[7] White J, Mukasyan AS, La Forest ML, Simpson AH. Rev Sci Instrum 2007;78:015105.
[8] Lin Ya-C, McGinn PJ, Mukasyan AS. J Eur Ceram Soc 2012;32(14):3809–18.
[9] Shcherbakov VA. SHS Key Eng Mater 2002;217:215.
[10] White J, Simpson AH, Steinberg AS, Mukasyan AS. JMR 2008;23(1):160–9.
[11] Lin Ya-C, Nepapushev AN, Rogachev AS, McGinn PJ, Mukasyan AS. Ceram Int 2013;39:7499–505.

Large-Size Items

Inna P. Borovinskaya, Evgeny A. Levashov

Large-size billets and items obtained by forced SHS consolidation are notable for their high homogeneity and uniform distribution of strength and hardness within the billet volume. Economic estimates have demonstrated that the larger the items are, the more efficient SHS is than other technologies, and if the diameter of the items exceeds 300 mm, SHS technology has no comparable counterparts and is unrivaled [1,2]. In the technology of obtaining large-size items, the design of the reaction mold, where combustion along with formation of structure and shape occurs is important. The press accessories perform several functions: withstand high temperatures; provide for withdrawal of the emitted gases; ensure thermal isolation of hot combusted products in order to increase the thermal relaxation time; and effectively transfer the pressure for deforming the hot products to the porous free state. That is why designing the press accessories optimally is an important and an independent technological task. In this connection, an experience of the Institute of Structural Macrokinetics and Materials Science, RAS deserves attention. Figs. 1–4 give photos of items that illustrate capabilities of the technology for obtaining large-size items from ceramics, hard alloys, high-temperature invar compositions, scale-resistant intermetallides (e.g., the ones based on NiAl, Ni_3Al, TiAl, TiNi, and Nb_3Al),

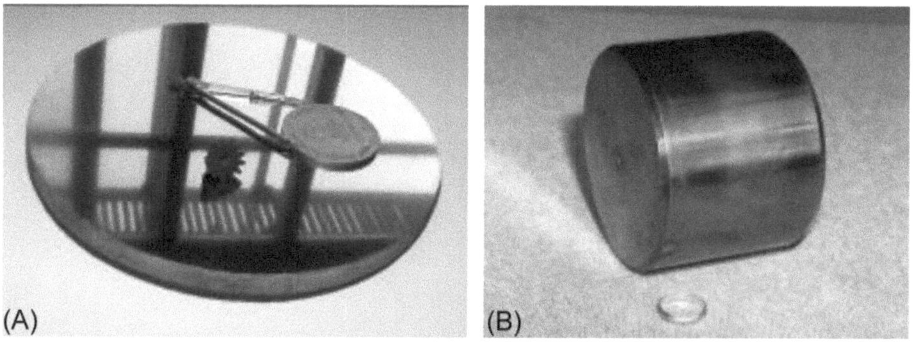

Fig. 1 A billet with a diameter of 360 mm, a height of 20 mm, and a weight of 10 kg (A). A roll for gold production made of TiC-Ni-Mo alloy: the diameter is 115 mm, the height is 87 mm, and the weight is 4.7 kg (B).

Concise Encyclopedia of Self-Propagating High-Temperature Synthesis
http://dx.doi.org/10.1016/B978-0-12-804173-4.00085-5

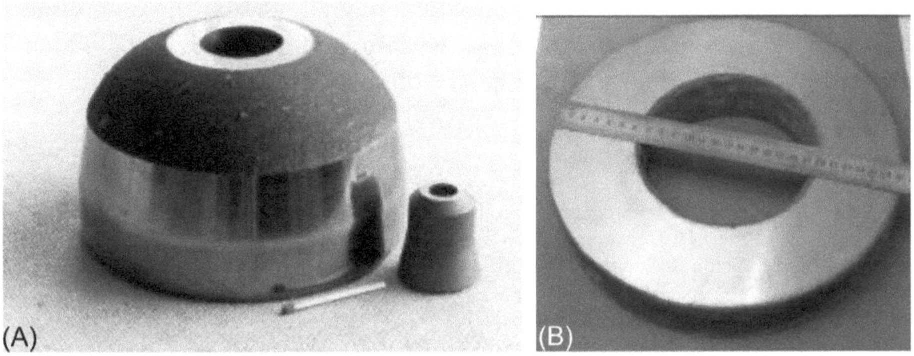

Fig. 2 A TiC-Ni-Mo alloy tube piercing plug: the diameter is 123 mm, the height is 75 mm, and the weight is 4 kg (A). A cupping die billet made of STIM-4 alloy with an outer diameter of 260 mm, an inner one of 135 mm, and a height of 50 mm (B).

Fig. 3 Hard alloy rings sized at Ø218 × Ø140 × 8 mm for a bead mill manufactured by the Drais Company. They operate in pairs by rotating relative to each other with a rate of 820 rpm and serve for filtering hard particles sized at 0.3 mm while preparing paints (A). An industrial batch of STIM-4 alloy rolls for a hot continuous copper formation in a Southwire company's installation (B).

Fig. 4 A TiC-Ni-Mo alloy bed plate with a diameter of 75 mm, height of 20 mm, and weight of 0.85 kg out of accessories for synthesizing artificial diamonds (A). Protective plates made of STIM-4 alloy (B). 48Ti-48Al-2Nb-2Cr alloy billets for turbine blades (C).

functionally gradient materials, diamond-containing alloys, and the ones dispersedly reinforced with nanoparticles. Dr. V.L. Kvanin and Mrs. N.T. Balikhina have contributed much to developing the sphere of large-size items.

REFERENCES

[1] Merzhanov AG. Advanced SHS ceramics: today and tomorrow morning, In: Soga N, Kato A, editors. Ceramics: toward the 21st Century. Proceedings of the symposium on ceramic commemorating the centennial of the Ceramic Society of Japan, 16–18 Oct., 1991, Yokohama, JapanTokyo: Ceramic Society of Japan; 1991. p. 378–403.
[2] Merzhanov AG. Int J SHS 2001;10(2):237–57.

Master Alloys

Iñigo Agote, Miguel A. Lagos, Manuel Gutiérrez

Master alloys are formed by base metals such as aluminum, titanium, and iron combined with a relatively high percentage of one or more other elements or compounds. The master alloy is a semifinished product and is typically manufactured for use as a raw material by the metal industry. Use of master alloys is considered as one of the simplest methods to easily add alloying elements or ceramic particles to molten metals. Because of the precise chemical composition and the homogenous distribution of the elements, master alloys are widely used in the metal industry.

SHS provides an excellent route to produce the master alloys for different applications with minimum energy consumption. The major advantages of the SHS route are the reduced probability of chemical segregation and the fact that the high temperatures that occur during the synthesis reaction create a unique interface between the ceramic particle and the metal matrix, which leads to better adhesion of the reinforcement to the metal matrix and eventually to better properties of the final metal matrix composites.

Different examples of master alloys are synthesized by SHS. In most of them, the exothermic reaction is produced by the formation of a carbide or boride compound surrounded by a metallic matrix. Depending on the exothermicity of the compound being synthesized and the amount of excess metallic phase, the best system stoichiometry will be selected. Typically, in order to maintain the reaction in a self-sustaining mode and obtain a full conversion of the reactants, the amount of the final metallic phase in the master alloy should not exceed 60%.

Heavily loaded compounds (master alloy) synthesized by SHS have obtained ample approval because of their cutting and wear-resistant properties [1]. Refractory metal carbides and borides such as titanium, tungsten or zirconium carbides, and borides [2] are the most common ceramic elements used for these applications. Fe-(TiMo)C master alloys have been obtained by SHS to be used as a ceramic reinforcement source for iron-based casting processes. Complex carbide (TiMo)C surrounded by a metallic matrix (Fe) has been synthesized according to Eq. (1) using different Fe/(TiMo)C ratios.

$$\mathrm{Ti + Mo + C + Fe \rightarrow Fe(TiMo)C + Q(Heat)} \tag{1}$$

The formation of the complex Ti-Mo carbide Fe master alloy substantially improves the wettability of the ceramic particles in the final molten Fe alloy. These master alloys are used to reinforce steels and alloys, such as the austenite of the Hadfield type manganese

Concise Encyclopedia of Self-Propagating High-Temperature Synthesis
http://dx.doi.org/10.1016/B978-0-12-804173-4.00086-7

steels. The polygonal titanium molybdenum carbides, added to ductile and tough austenite grains, cause a small reduction of some mechanical properties (toughness and ductility). However, the impact energy values of the new materials and their ductility are sufficiently high for the working conditions of the typical components made in Hadfield steel (railway crossings, crushing jaws, hammers), so there is no risk of breaking in components made with reinforced steel. Also, the yield strength of the reinforced alloy is higher and thus the stiffness and fatigue resistance are superior, improving the typical properties of the Hadfield steel [3].

Another example of master alloys obtained by SHS is the synthesis of Ti-based master alloys, especially Ti-TiB. This master compound can be subsequently used in an investment casting process to obtain TiB-reinforced net-shape titanium matrix composites. Ti-TiB master alloy can be synthesized to vary the amount of the Ti metallic matrix compound according to the final application needs and the most appropriate synthesis conditions of the mixture (Fig. 1). Note that a minimum TiB content is required to allow a self-propagating synthesis mode. TiB contents below 60 wt% will not be sufficient for a self-propagating reaction mode.

The master compound presents a homogeneous distribution of TiB in the titanium matrix. Needle-shaped TiB reinforcements with diameters of around 2–5 μm and lengths of around 5–20 μm are obtained. By the addition of the master compound, produced by SHS, in situ TiB reinforced titanium matrix composites can be obtained by investment casting. During melting, this master compound is directly and completely dissolved. Throughout the solidification, boron reacts with titanium, and thus in situ TiB

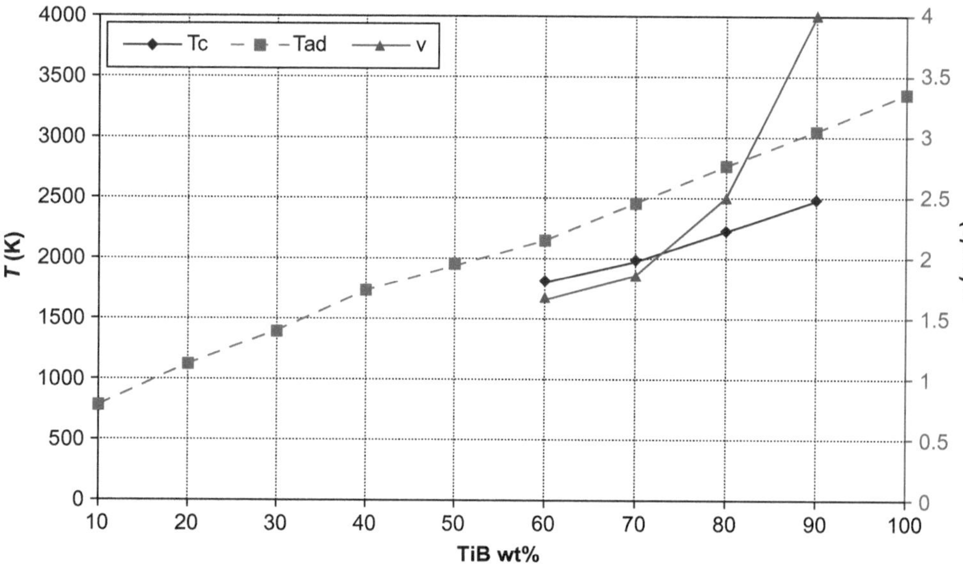

Fig. 1 Influence of the TiB content on the temperature and velocity of the combustion front propagation.

Fig. 2 TiC-Al master alloy SEM image.

reinforcements are again formed [4]. The TiB precipitation has a refining effect that decreases the alpha plates in the CPTi (chemically pure titanium), but in the Ti6Al4V alloy, TiB precipitation decreases both the alpha and the beta grain sizes. Improvements in tensile strength of around 10% compared with the Ti6Al4V have been achieved while maintaining similar ductility values to that of unreinforced alloy [5].

Aluminum-based master alloys have also been obtained by SHS. These maser alloys are added to molten aluminum alloys and can be used either for grain refining or for the alloy reinforcement. Al-Ti-B master alloys are widely used in the aluminum industry as grain refiners to control the microstructure of the aluminum alloys. In situ SHS is an excellent alternative to conventional processes for the creation of these master alloys [6]. Alumimun matrix also has been used in the formation of TiC-Al master alloy, which has been applied to reinforce aluminum alloys (Fig. 2).

The production of master alloys by SHS is an excellent alternative to fabricate metallic matrix compounds containing high amounts of other alloying or reinforcing elements and compounds. Precise chemical composition together with the simple addition of these alloys into the metal melt and their good dissolution and wettability make the SHS route a very attractive alternative to conventional processes to synthesize master alloys for further alloying and reinforcing of metallic melts.

REFERENCES

[1] Tinklepaugh JR, Crandall WB. Cermets. New York, NY: Reinhold, Chapman & Hall; 1960.
[2] Steen WM. Future developments of metals and ceramics. London: Institute of Materials; 1992. p. 261–262.
[3] Goñi J, Coleto I, Agote M, García de Cortazar y, Erauzkin E. Procedure for the obtention of titanium based composites by foundry and components thus obtained. PCT/ES03/00596; 2005 Jun 16.
[4] Agote I, Gutiérrez M, Sargsyan A, Erauskin I. Int J SHS 2010;19:17–22.
[5] Agote I, Garcia de Cortazar M, Silveira E, Egizabal P, Coleto J, Le Petitcorps Y. J Mater 2008;60:40–6.
[6] Vicario I, Poulon-Quintin A, Lagos MA, Silvain JF. Metals 2015;5:1387–96.

Materials and Coatings for High-Temperature Applications

Dmitry V. Shtansky

Materials for high temperature applications include refractory, oxidation-resistant, and heat-resistant materials. Refractory materials can withstand high temperatures without degrading and consequent loss of mechanical properties. Oxides (alumina, silica, magnesia, zirconia, etc.), carbides of refractory metals, carbon, boron nitride, and various composite materials including cermets are attributed to refractory materials. Materials that have the ability to protect various products against high temperature are called heat-resistant materials. Oxidation-resistant materials are those that can resist surface chemical degradation in gaseous environments at high temperatures.

Since SHS occurs at high temperatures, combustion synthesis with simultaneous densification is a promising method for the production of refractory materials. SHS products for high-temperature applications can be fabricated in various forms: powders, bulk materials, and coatings. A complete review of the earlier results in the field of SHS was compiled by Merzhanov [1], and the more recent achievements in SHS of refractory and hard materials were recently reviewed by Liu [2]. Combustion synthesis of refractories was first performed by Merzhanov and coworkers in 1984 from alkali chromates. Magnesia and chromite-magnesia refractory materials were further fabricated for linings of high-temperature units. In addition, SHS was employed for the fabrication of tungsten-free hard alloys based on Ti and Zr carbides and borides with metallic (Ni, Mo, Cu, Fe) binders known as STIM [3]. Many other refractory metallic alloys, intermetallics, dichalcogenides, metal-ceramic, and composite materials are fabricated via SHS, for instance: NiAl, Ni_3Al, Ni_3Al/Al_2O_3, TiC-(Ni,Fe), TiC-TiB_2/Ni, Ti(C, N)/TiB_2, WSi_2-SiC, (Ti,Si)C-BN, and AlN-TiN-BN-SiC. Oxidation studies of Al_2O_3-ZrB_2 and Al_2O_3-TiB_2 composites obtained by means of the SHS process demonstrated their high stability against oxidation up to 900°C with no significant changes in mechanical properties. AlN-SiC-TiB_2 ceramics fabricated by SHS were shown to withstand 16 h at 1100°C with almost no weight gain. Cyclic and isothermal oxidation tests of SHS-fabricated Fe-Al-Si alloys performed at 800°C in air confirmed their excellent oxidation resistance. SHS was successfully utilized to restore gas turbine components by means of simultaneous synthesis and joining of a NiCrAlY layer to a super alloy substrate.

Concise Encyclopedia of Self-Propagating High-Temperature Synthesis
http://dx.doi.org/10.1016/B978-0-12-804173-4.00087-9

SHS was also employed to fabricate ultrafine WC, Al_2O_3, TiB_2, SiC, ZrC, and $MoSi_2$ powders. SHS powder materials are commonly used as precursors for coating deposition. For instance, ZrC–ZrB_2 ceramic composite and ZrB_2–ZrC/Ni cermet coatings were deposited by atmospheric plasma spraying using SHS-fabricated ZrC–ZrB_2 and ZrB_2–ZrC/Ni composite powders. Cermet FeCr-TiC coatings deposited by high-velocity oxygen fuel spraying of SHS powders were reported to withstand abrasive wear resistance and high-temperature oxidation.

It is well known that MAX-phases $M_{n+1}AX_n$ ($n=1$–3) are easily machinable, resistant to thermal shock and high-temperature oxidation, and thermally and electrically conductive. SHS was successfully used to fabricate MAX phases in the ternary systems Ti_3AlC_2, Ti_2AlC, and Cr_2AlC. MAX phase-based $Ti_{2-x}Cr_xAlC$ (with $x=0$, 0.5, 1.5, and 2) targets were employed for deposition of coatings resistant to corrosion and high-temperature oxidation [4].

In the field of surface engineering, several technologies that combine the SHS reaction and coating deposition were elaborated. The SHS process was combined with other methods such as centrifugation, microwave and induction heating, laser cladding, sol-gel, concentrated solar energy treatment, thermal spraying, atmospheric plasma spraying, and so on. Various metallic, cermet, and composite coatings were developed and deposited, for instance NiAl, Ni_3Al, NiAl/Ni-Al-Ti, Al_2O_3-TiC-TiB_2 and Al_2O_3-TiN-TiB_2, and TiB_2-TiC-Al_2O_3-SiC. FeNiCr-TiC coatings fabricated using SHS showed excellent high-temperature and wear and oxidation resistance at 1000°C.

Multicomponent coatings based on refractory transition metal carbides, borides, and nitrides have attracted significant attention as protective coatings operated at high temperatures. The possibilities of various deposition technologies can be extended through SHS targets. Hard tribological coatings, including those for high-temperature applications, were obtained by magnetron sputtering of SHS-fabricated TiBC, TiBN, TiCrB, TiSiB, TiSiN, TiAlSiB, TiZrCO, TiSiC, TiCrSiCN, and TiAlSiCN targets [5]. More recently, the features of combustion reaction and structure formation in CrAlSiB and MoSiB materials designed for high-temperature applications and coating deposition were developed and studied. Coatings for high-temperature applications can also be produced through nonvacuum methods, for instance by electrospark surface modification using electrodes fabricated by SHS method.

REFERENCES

[1] Merzhanov AG. Ceram Int 1995;21:371–9.
[2] Liu G, Li J, Chen K. Int J Refract Met Hard Mater 2013;39:90–102.
[3] Tavadze GF, Shteinberg AS. Production of advanced materials by method of self-propagating high-temperature synthesis. Berlin-Heidelberg, Germany: Springer; 2013.
[4] Shtansky DV, Kiryukhantsev-Korneev PhV, Sheveyko AN, Mavrin BN, Rojas C, Fernandez A, Levashov EA. Surf Coat Technol 2009;203:3595–609.
[5] Shtansky DV, Kiryukhantsev-Korneev PhV, Bashkova IA, Sheveiko AN, Levashov EA. Int J Refract Met Hard Mater 2010;28:32–9.

Max Phases

Yury S. Pogozhev, Artem Yu. Potanin, Evgeny A. Levashov

$M_{1-n}AX_n$ phases are a class of refractory oxygen-free compounds in which M is a transition metal, A is mainly an element of subgroup IIIA or IVA of the Periodic table, and X is carbon or nitrogen. These materials have a unique combination of metal and ceramic properties: low density, high thermal and electrical conductivity, corrosion and scale resistances, and high melting temperature; and they are stable enough at temperatures up to 2000°C [1]. Mechanical treatment can be easily performed because of their layered structure (Fig. 1).

Ti_3AlC_2 $M_{1-n}AX_n$ phase was first synthesized by Jeitschko and Nowotny [2] in 1967. At that time, only its structure was studied at which point the research was interrupted. Later Ti_3AlC_2 was obtained through sintering pressed powder mixtures of Ti, TiAl, Al_4C_3, and C at 1300°C for 20 h within a hydrogen atmosphere [3]. Hot pressing, hot isostatic pressing, spark plasma sintering, and SHS [4–7] methods are used in order to obtain such ternary compounds.

Most research in this area is dedicated to $M_{1-n}AX_n$ phases in Ti-Al-C (T_3AlC_2, Ti_2AlC) and Ti-Si-C (Ti_3SiC_2) systems. The main problem obtaining such compounds through SHS consists in presence not only $M_{1-n}AX_n$ phase but also admixture phases, for example, TiC, Ti_xAl_y, Cr_2Al, Cr_7C_3, Ti_5Si_3, which have adverse effects on the properties. The fact that the chemical and structural transformations occur in stages is the main cause of the phase composition heterogeneity. Thermodynamically stable compounds are formed at intermediate stages. In addition, local breaches of stoichiometry are possible because of a partial evaporation of Al within the SHS wave.

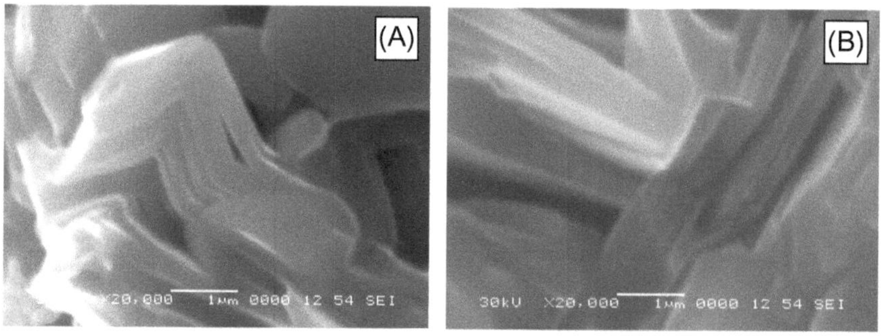

Fig. 1 The microstructure of compact SHS materials Ti_2AlC (A) and Cr_2AlC (B).

Concise Encyclopedia of Self-Propagating High-Temperature Synthesis
http://dx.doi.org/10.1016/B978-0-12-804173-4.00088-0

In the Ti-Al-C system during SHS, $M_{1-n}AX_n$ phases are formed at the stage of the secondary structure formation when the initial TiC_x grains interact with the Ti-Al [7] melt. At the same time, if the reaction mixture is warmed up slowly, intermetallide Ti_xAl_y compounds are formed first and then TiC_x grains are crystallized from the melt oversaturated with carbon. The subsequent interaction between TiC_x and Ti_xAl_y leads to the formation of ternary $Ti_{x+1}AlC_x$ [8] compounds. At the same time, like in cases with the SHS process, $M_{1-n}AX_n$ phases can form when TiC_x interacts with the Ti-Al melt.

$M_{1-n}AX_n$ phases can be obtained not only through synthesis from the elements. In order to obtain a ternary Ti_3SiC_2 compound in combustion mode, Ti powder and SiC [9] whiskers can be used. In this case, the end product contains some amount of intermediate TiC. Experimental SHS under microgravity conditions [9] made the concentration of the ternary Ti_3SiC_2 phase increase.

It is recommended to apply mechanical activation [10] of the reaction mixture in order to increase conversion within the combustion wave, along with structural and phase homogeneity of the $M_{1-n}AX_n$ phases. This approach makes it possible to increase the reactive capacity and decrease the mixture heterogeneity scale. Applying MA in the Ti-Cr-Al-C system made it possible to obtain a 100% $M_{1-n}AX_n$ phase in the end product.

$M_{1-n}AX_n$ phases are promising structural ceramics intended to obtain items for high-temperature purposes for the benefit of aircraft, rocket, and space machinery [1]. Composite SHS target cathodes based on $M_{1-n}AX_n$ phases are successfully applied in ion-plasma sputtering to obtain frictional coatings resistant to high-temperature oxidation.

REFERENCES

[1] Tzenov NV, Barsoum MW. J Am Ceram Soc 2000;83(4):825–32.
[2] Jeitschko W, Nowotny H. Die Monatschrift für Chemie 1967;98:329–37.
[3] Pietzka MA, Schuster JC. J Phase Equilib 1994;15(4):392–400.
[4] Barsoum MW, Ali M, El-Raghy T. Metall Mater Trans A Phys Metall Mater Sci 2000;31(7):1857–65.
[5] Hashimoto S, Takeuchi M, Inoue K, et al. Mater Lett 2008;62:1480–3.
[6] Zhou WB, Mei BC, Zhu JQ, et al. Mater Lett 2005;59:131–4.
[7] Rogachev AS, Gachon JC, Grigoryan HE, et al. J Mater Sci 2005;40:2689–91.
[8] Levashov EA, Pogozhev YuS, Shtansky DV, Petrzhik MI. Russ J Non-Ferrous Met 2009;50 (2):151–60.
[9] Grigoryan AE, Rogachev AS, Sychev AE, Levashov EA. Ref Ind Ceram 1999;40(11–12):484–8.
[10] Levashov EA, Kurbatkina VV, Patsera EI, Pogozhev YuS, Rupasov SI, Rogachev AS. Russ J Non-Ferrous Met 2010;51(5):403–33.

Measurements of Combustion Velocity

Alexander S. Rogachev, Alexander S. Mukasyan

The most informative contemporary method for measurement of the propagating velocity of SHS wave (combustion velocity) is frame-by-frame processing of video records. Usually, the combustion front looks like a sharp boundary between a black field (unreacted part of the sample) and a bright field (products), as shown in Fig. 1. A level of brightness must be selected as a threshold, or marked, to indicate the position of the combustion front (front boundary). For this region, a straight line parallel to the direction of propagation is put over the image (Fig. 1), and an intersection of this line with the front boundary gives us the local coordinate of the combustion front. Applying many straight lines and averaging many coordinates, you can calculate the mean position of the front.

After processing a large number of video frames, dependence of the front coordinate on time can be plotted, and the slope of this dependence gives the propagating velocity [1]. Fig. 2 presents examples of such dependences.

Most of the SHS systems prepared from common powders exhibit stochastic oscillations of propagating velocity in the scale of the reactants' particles (10–100 μm), as shown in Fig. 2B. However, in the scale of the sample (~1 cm), the average slope of the coordinate-time dependences are constant. This average slope is usually accepted as the combustion velocity for considering mixture. Strictly uniform propagation with the constant velocity at all scales was observed in the multilayer reactive nanofilms (Fig. 2B).

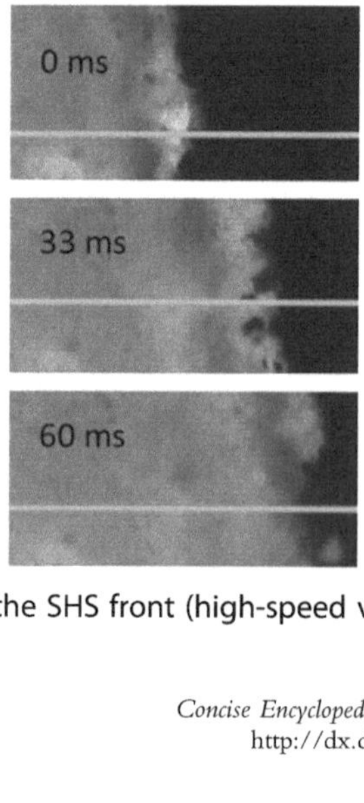

Fig. 1 Examples of video frames of the SHS front (high-speed video).

Concise Encyclopedia of Self-Propagating High-Temperature Synthesis
http://dx.doi.org/10.1016/B978-0-12-804173-4.00089-2

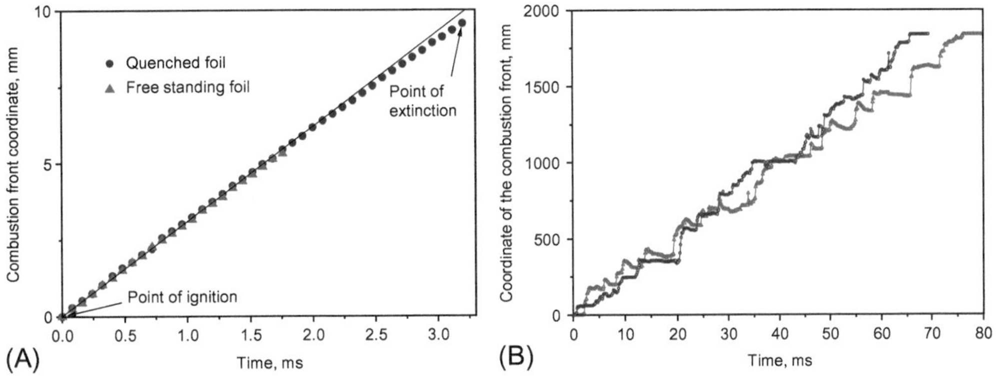

Fig. 2 Front coordinate—time dependence for SHS wave propagation in the Ni/Al multilayer nanofoins (A) [2] and Ti/Si powder samples (B).

REFERENCES

[1] Rogachev AS, Mukasyan AS. Combustion for material synthesis. Boca Raton, London, New York: CRC Press Taylor & Francis Group; 2015, p. 397.

[2] Rogachev AS, Vadchenko SG, Baras F, Politano O, Rouvimov S, Sachkova NV, et al. Combustion in reactive multilayer Ni/Al nanofoils: experiments and molecular dynamic simulation. Combust Flame 2016;166:158–69.

Mechanically Assisted SHS

Victoria V. Kurbatkina, Evgeny A. Levashov

Mechanochemical synthesis (MS) under severe mechanical activation (MA) of the reaction mixture is applied for obtaining various compounds, including ones obtained with use of SHS. Mechanochemistry comprises three main areas: MS, mechanical alloying, and MA. MS is applied to obtain chemical compounds such as intermetallides, sulfides, carbides, borides, and silicides. Mechanical alloying is used to obtain powder composites and alloys as an alternative to the traditional methods of fusion and sputtering, sintering, and grinding. In order to perform MS, an energy barrier sufficient for initiating the chemical reaction must be cleared. Accumulation of the mechanical energy by the reaction mixture is considered MA in SHS technologies. At this stage, the structural transformation of the reaction mixture occurs, the components are ground, and severe plastic deformation occurs, and as a result, composite particles with layered structure are formed, the scale of heterogeneity is reduced, and the reactive surface increases. As a result of MA, an energy reserve is created, which is then realized during SHS. An advantage of MA in SHS technologies is connected with the rate of chemical reactions that prevent annihilation of the energy accumulated by the charge mixture. Increasing the charge mixture's reactivity is the main idea of the SHS MA. Technologically, it is extremely important to search for an optimal "subcritical" reserve of the system's internal energy; if it is exceeded, a spontaneous mechanochemical reaction begins, including one in combustion mode. Fig. 1 outlines the dependence of conversion during MS of the reaction mixture depending on the mechanical treatment time. Point K corresponds to a certain critical time of MA; if it is exceeded, the chemical reaction occurs directly in the mill drums. The effectiveness of MA in the SHS process is determined first by the amount of energy transferred to the charge mixture, which depends on the type and capacity of the equipment used, as well as on composition and properties of activated reagents.

Numerous research reports on MA SHS show that the structural changes include a more uniform distribution of the components (which is important for multicomponent mixtures), growing reactive surface, increasing density of the metallic and nonmetallic reagent structure defects. As a result of the particle deformation, the concentration of three-dimensional and surface defects grows, the internal energy of the system increases; in turn, the heat released and heat release rate also increase, and the interaction starting temperature is reduced.

Concise Encyclopedia of Self-Propagating High-Temperature Synthesis
http://dx.doi.org/10.1016/B978-0-12-804173-4.00090-9

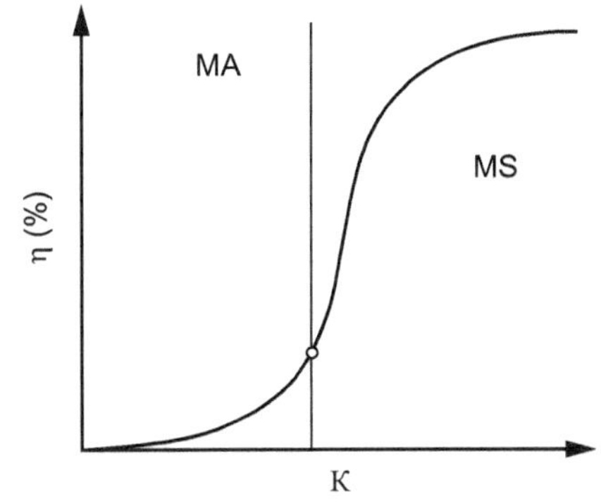

Fig. 1 Conversion level changes by the mechanical activation (MA) time. K is an optimal energy reserve.

Eighty to ninety percent of the MA energy assimilated by the material is consumed to create a new reactive surface, which is not always accompanied with an increase in the specific surface. Simultaneous increase in specific and contact surfaces occurs in the systems consisting of brittle reagents or when at least one of the reagents (carbon black, amorphous boron, etc.) is highly dispersed (the scale of its specific surface is decisive), as in the case of Ti-C, Ti-B, Mo-B, and Cr-B compounds [1–3]. For systems that consist of ductile metal-metal-type components (Ni-Al, Ti-Al, Ti-Ni, Nb-Al, Cu-Al, etc.) [4–7], during MA the specific surface often reduces because of agglomeration of the particles because as a result of the components' deformation flow, multilayer agglomerates with a large reactive surface of the different layers are formed. To assess the efficiency of using MA for SHS processes, it is necessary to determine the contribution of MA to the amount of heat evolution and the change of energy needed to activate the SHS reaction to form the desired product.

REFERENCES

[1] Kurbatkina VV, Levashov EA. Mechanoactivation of SHS. In: Mukasyan AS, Martirosyan KS, editors. Combustion of heterogeneous systems: fundamentals and applications for materials synthesis. Kerala: Transworld Research Network; 2007. p. 131–41.
[2] Eremina EN, Kurbatkina VV, Levashov EA, Rogachev AS, Kochetov NA. Chem Sustain Dev 2005;13:197–204.
[3] Korchagin MA, Grigorieva TF, Barinova AP, Lyachov NZ. Int J SHS 2000;3(9):307–20.
[4] Bernard F, Souha H, Gaffet E, et al. Int J SHS 2000;1(9):1–14.
[5] Maglia F, Anselmi-Tamburini U, Deida C, et al. J Mater Sci 2004;39:5227–30.
[6] Levashov EA, Kurbatkina VV, Kolesnichenko KV. Izv Vyssh Uchebn Zaved, Tsvetn Metall 2000;(6):61–7.
[7] Rogachev AS, Kochetov NA, Kurbatkina VV, Levashov EA, Grinchuk PS, Bernard F, et al. Fiz Goreniya Vzryva 2006;4(42):61–70.

Metal Powders

Suren L. Kharatyan

Preliminary thermodynamic calculations revealed the existence of new routes for reduction of copper, nickel, and cobalt from metal oxygenous compounds (MOC) by organic reducers in combustion mode. The following types of reducers are used: (i) combined organic reducers (COR) representing CH(O)-polymers (e.g., polystyrene: $(C_8H_8)_n$, polyethylene: $(C_2H_4)_n$, polyvinyl alcohol: $(C_2H_4O)_n$, etc.), and CHN(O)-type solid compounds (e.g., melamine: $C_3H_6N_6$, urotropin: $C_6H_{12}N_4$, urea: $(NH_2)_2CO$, etc.), and (ii) complex reducing systems (CRS) representing a mixture of COR and inorganic oxidizer (e.g., ammonium nitrate: NH_4NO_3). In the latter case, a part of the organic reducer plays the role of a fuel. Based on these results, new reaction schemes were developed for obtaining metal powders by combustion reaction [1–5]:

1. MeO + organic reducer → $Me_{(s)}$ + gaseous products + Q
2. MeO + organic reducer + inorganic oxidizer → $Me_{(s)}$ + gaseous products + Q
3. Me oxygenous salts + organic reducer + inorganic oxidizer → $Me_{(s)}$ + gaseous products + Q

In all these cases, the adiabatic combustion temperature was found to be lower than the melting point of metals under consideration, and within certain ranges of parameters metal powder was the only condensed product of reduction process.

It is worth noting that unlike conventional reducers (C, H_2, CO, Me), COR being solid organic compounds easily decomposes before starting the reduction process at relatively low temperatures, forming active intermediate gaseous products "well mixed" with oxide powder. Experiments confirmed the possibility of rapid reaction between metal oxide and decomposition products of organic reducers (gas-solid interaction) at relatively low temperatures ($T \leq 1000°C$) providing reaction proceeding in combustion mode for a number of MOC-COR and MOC-CRS systems with low propagating rates (0.01–0.1 cm/s):

Route I example: for the reduction of copper from CuO:

$$CuO + (C_8H_8)_n \rightarrow Cu + CO_2 + H_2O$$

It has been shown that polystyrene is the preferable reducing agent to obtain copper powder.

Concise Encyclopedia of Self-Propagating High-Temperature Synthesis
http://dx.doi.org/10.1016/B978-0-12-804173-4.00091-0

Route II example: for the reduction of metals from CuO, Cu_2O, NiO, Co_3O_4, and CoO:

$$Co_3O_4 + (C_2H_4)_n + NH_4NO_3 \rightarrow Co + N_2 + CO_2 + H_2O$$

Route III example: for reduction of metals from oxygenous salts of Cu, Ni, and Co (carbonates, basic carbonates, nitrates, etc.):

$$(NiOH)_2CO_3 \cdot 2H_2O + C_3H_3N_6 + NH_4NO_3 \rightarrow Ni + N_2 + CO_2 + H_2O$$

It is important to emphasize that by organizing the reduction reactions in such a way, the target metal was not polluted with byproducts, as all other products were exhaust gases and left the system. The intensive gas evolution taking place during the abovementioned combustion reactions inhibited particle growth and favored synthesis of fine metal powders.

It was shown that through managing the conditions of the reduction process, it is possible to obtain both the metal powder and the foam-like metal with high porosity (up to 90%).

The process was also implemented in pilot-scale in the universal reactor SHS-30 L, and optimal conditions for obtaining up to 15–20 kg of copper, nickel, and cobalt metallic powders were revealed.

After the combustion process, the product represented a slightly sintered briquette, caused by the gas evolution during the combustion process loosening the sample and making further crushing easier.

The purity of the as-derived powders was found to be in the range 99–99.5 wt.% for Cu (main impurity, oxygen) and 99.6–99.8 wt.% for Co and Ni with an oxygen content <0.2 wt.%.

It is safe to suggest that changing the type and amount of the applied reducer, the inert gas pressure, the density of the green mixture, etc., allows obtaining powdered copper, nickel, and cobalt with controllable morphology. Thus the approach described can serve as an effective and simple method for manufacturing copper, nickel, and cobalt fine powders for different applications.

REFERENCES

[1] Yamukyan MH, Manukyan KhV, Kharatyan SL. Copper oxide reduction by combined reducers under the combustion mode. Chem Eng J 2008;137(3):636–42.

[2] Yamukyan MH, Manukyan KhV, Kharatyan SL. Preparation of copper, nickel and cobalt powders by SHS. In: IX International symposium on self-propagating high-temperature synthesis, Dijon, France, 1–5 July; 2007. Book of abstracts, T6-P04.

[3] Yamukyan MH, Manukyan KhV, Kharatyan SL. Obtaining of nickel powder under the combustion mode by reduction of basic nickel carbonate. Chem J Armenia 2008;61(2):159–66.

[4] Kharatyan SL. Metal powders by SHS. Application of organic compounds as effective reducers. In: X International symposium on self-propagating high-temperature synthesis, 6–11 July, Tsakhkadzor, Armenia; 2009. p. 62–3. Book of Abstracts.

[5] Kharatyan SL, Merzhanov AG. Coupled SHS reactions as a useful tool for synthesis of materials: an overview. Int J SHS 2012;21(1):59–73.

Metallurgy

Vladimir I. Yukhvid

According to the definition given by A.G. Merzhanov, SHS metallurgy is a technology based on the high-calorie aluminothermic mixtures that yield molten products at temperatures from 2500 to 3500 K [1]. The metallothermic reactions were discovered by the Russian chemist N.N. Beketov and were adapted for industrial applications by the German engineer H. Goldschmidt around the end of 19th century and beginning of the 20th century. In terms of research in the field of SHS, the SHS metallurgy systems were developed in a separate area [2]. An important contribution in the field of thermite reactions was the development of the use of centrifugal force to control the separation of immiscible melts (phase separation) formed in the processes of SHS metallurgy [3]. Based on this approach, methods were developed for centrifugal SHS-casting [4,5] for manufacturing steel pipes with an inner ceramic coating [6–9].

In recent years, the thermite subject received a new impetus because of developments in the area of nanothermites. The development of various nanotechnologies, including metals and their oxides, allowed the preparation of the reaction of heterogeneous structures with nanostructured reagents. The range of available nanopowders is now quite large (aluminum, boron, carbon, silicon, nickel, and many oxides). Thermite compositions prepared using nanopowders are called nanothermites or superthermites [10]. The metal reducing agent in the superthermites is primarily nanoaluminium, and the latest results in the production of nanoaluminium are described in recent monographs [11,12].

REFERENCES

[1] Merzhanov AG. About self-propagating high-temperature synthesis of refractory materials: theory and applications, In: Experimental heat transfer, fluid mechanics and thermodynamics. Proceedings of the 4th world conference on experimental heat transfer, fluid mechanics and thermodynamics, Brussel, June 2–6PIZA: Edizioni ETS; 1997.

[2] Merzhanov AG, Yukhvid VI, Borovinskaya IP. Self-propagating high-temperature synthesis of cast refractory inorganic compounds. Dokl Akad Nauk SSSR 1980;255(1):120–4 [in Russia].

[3] Merzhanov AG, Yukhvid VI. The self-propagating high-temperature synthesis in the field of centrifugal forces, In: Proc I US-Japanese workshop on combustion synthesis, Tokyo; 1990. p. 1–22.

[4] Yukhvid VI. Modifications of SHS processes. Pure Appl Chem 1992;64:977–88.

[5] Yukhvid VI, Kachin AR, Zakharov GV. Centrifugal SHS surfacing of refractory inorganic materials. Int J SHS 1994;3(4):321–32.

[6] Odawara O. Method for providing ceramic lining to a hollow body by thermite reaction, US Patent 4 363 832; 1982.

Concise Encyclopedia of Self-Propagating High-Temperature Synthesis
http://dx.doi.org/10.1016/B978-0-12-804173-4.00092-2

[7] Odawara O, Ikeuchi J. Ceramic composite pipes produced by a centrifugal-exothermic process. J Am Ceram Soc 1986;69(4):C80–1.
[8] Odawara O. Long ceramic-lined pipes produced by a centrifugal thermite process. J Am Ceram Soc 1990;73(3):629–33.
[9] Kachin AR, Yukhvid VI. SHS of cast composite materials and pipes in the field of centrifugal forces. Int J SHS 1992;1(1):168–71.
[10] Puszynski JA. Processing and characterization of aluminum-based nanothermites. J Therm Anal Calorim 2009;96(3):677–85.
[11] Gromov AA, et al. Combustion of metal nanopowders. Tomsk: Del'taplan; 2008 [in Russia].
[12] Fahrner W, editor. Nanotechnology and nanoelectronics. Berlin, NY: Springer-Verlag; 2005.

Microgravity SHS

Osamu Odawara

SHS mainly consists of two processes: reaction propagation and structure formation of final products. Thermo-gravitational transportations of the gas or liquid phase are considered to affect the reaction propagation rate in the former process, which subsequently affects the latter process. Since such transportation and fluctuation caused by mass disturbance and heat convection are restrained under microgravity environments (MGEs), the propagation rate will be lowered and the products synthesized under MGEs can keep the position where they react even in gas or liquid phase resulting in the uniform composition distribution of product.

Shteinberg et al. [1] performed highly pore material production under a parabolic flight MGE of about 30 s, and they successfully manufactured more than 95% porous TiC with the SHS process. Odawara et al. also carried out the SHS experiments with TiB_2-Al composite formation under free fall, parabolic flight, and TR-IA sounding rocket MGEs [2]. It was confirmed that the lack of mass migration and the improvement of wetting between TiB_2 and Al under MGEs affected the formation of a fine and dense cermet-like structure existing in the products. These exploratory works were mainly performed for solid phase products and successfully showed the difference between products obtained under MGE from those obtained in the terrestrial condition.

The research activities were dedicated to the evaluation of both the physicochemical and structural characteristics of the products obtained by means of the SHS process under MGEs, and also aimed to verify whether the suppression of sedimentation and buoyancy phenomena under microgravity conditions allow the development of new processes whose final products are characterized by microstructural, mechanical, and thermochemical properties unique or superior with respect to terrestrial analogs—in brief, to understand how the processes and related products are dissimilar under microgravity conditions. A major goal up to now has been the definition of fundamental mechanistic differences between how this SHS process behaves under normal and low-gravity conditions, with the ultimate aim of defining approaches to tailoring SHS processes for the production of materials with optimized properties for various end uses [3].

From a technological innovation point of view in applications, ISFR (in-situ fabrication and repair) and ISRU (in-situ resource utilization) should be taken into account.

Concise Encyclopedia of Self-Propagating High-Temperature Synthesis
http://dx.doi.org/10.1016/B978-0-12-804173-4.00093-4

Regarding ISRU, the possibility of exploiting in-situ resources, such as oxygen and nitrogen, will guarantee more flexibility when scheduling and managing future space missions. [4]

Space technology has been developed for frontier exploration not only around the Earth's orbit but also the Moon's, Mars', for example. For performing such explorations, it is not enough to develop the technology with the resources on Earth. They should be investigated to utilize the resources at other places whose nature is different from that of the Earth's. Resources such as metal oxides that exist on planets or satellites could be refined and utilized as a supply of heat energy, and the SHS process is a feasible technology for such requirements. To carry out experiments and remain longer on extraterrestrial planets, permanent habitats are needed that provide a shield from heat, cold, cosmic rays, and meteorite impacts. Because of the negligible lunar atmosphere and scarce Martian atmosphere, spaceship landings and launchings will create frozen ice crystals from the fuel and gravels that travel far away at fast speeds. High-speed flying objects including micrometeorites will threaten nearby devices and inhabitants. With solid landing pads and protective walls, landing and launching will be much safer. Most proposals for construction on the Moon and Mars are based on transporting structural elements from Earth and assembling them at the destination. Such an approach is expensive and infeasible for large-scale implementation. Intense research in ISRU has been conducted on approaches based on the use of in-situ material and digital fabrication techniques to construct the infrastructure elements, and specifically lunar regolith characteristics have been extensively studied with the SHS technology [5].

An important ISRU goal is to reduce the cost of human missions to the Moon and Mars and to enable the establishment of long-duration manned space bases and to return energy or valuable resources to the Earth. When a sustainable settlement is established beyond low Earth orbit (LEO), energy can be supplied mainly by solar cells on its surface, and some fabrication scenario with thin film deposition in moon-based, ultra-high vacuum can be designed [6]. These concepts and technologies will be applicable to both the Moon and Mars, and these capabilities might achieve the maximum benefits of ISRU with the aid of SHS technology applications, which may reduce the mass, cost, and risk of human Mars' missions as soon as possible.

REFERENCES

[1] Shteinberg AS, Shcherbakov VA, Martynov VV, Mikhoyan MZ, Merzhanov AG. Sov Phys Dokl 1991;36:385–7.
[2] Odawara O. Ceram Int 1997;23:273–8.
[3] Mukasyan AS, Lau C, Varma A. AIAA J 2005;43(2):225–46.
[4] Miyazaki E, Odawara O. J Space Technol Sci 2002;18:17–25.
[5] Faierson EJ, Logan KV, Stewart BK, Hunt MP. Acta Astronaut 2010;67:38–45.
[6] Pittman B. Ad Astra 2011;23(3):36–9.

Microheterogeneous (Discrete) Models

Alexander S. Rogachev

Initial reactive mixtures for SHS are commonly composed of powder particles and therefore have a microheterogeneous structure. It can be represented as a network of elementary reactive cells, where each cell can be defined as a minimal volume of the reactive medium that contains reactants in the same concentrations as the whole mixture. In the simplest case, the elemental cell consists of one metal particle with surrounding finer particles of nonmetal reactants (e.g., Ti particle ~10 μm surrounded by carbon black particles ~0.1 μm in size). The elementary reaction cell burns out during the time interval t_R, while the time of transmitting combustion from the burned cell to the neighbor fresh cell is t_T. When $t_R \gg t_T$, the combustion process in the microheterogeneous medium can be described by quasihomogeneous models. However, if $t_R \leq t_T$, microheterogeneous, or discrete, models must be applied. Several mathematical models accounting microheterogeneity have been developed [1–7]. They are analyzed in reviews [6,7].

According to microheterogeneous model (MHM), the combustion process is a sequence of fast-burning reactions inside the reactive cells and stagnation periods when heat from the burned cell warms up the neighbor cell until its ignition. A typical structure of the combustion wave, calculated by MHM, is shown in Fig. 1. The temperature of the reacting cell becomes higher than the adiabatic temperature for a short time during the reaction; this over-adiabatic temperature peak is one of the main distinguishing features of the microheterogeneous combustion mode. Another feature is that thickness of the microheterogeneous combustion wave is equal to one or a few diameters of the reaction cell, unlike the quasihomogeneous mode, when the combustion front includes many reaction cells.

A burning time of the reaction cell is ~0.1–1.0 ms, while the size of the cell is ~10–100 μm. Therefore the steady-state microheterogeneous combustion wave may look similar to homogeneous combustion at the common recording rate of 24 fps. High-speed macro- and micro-video recording, and other methods that allow revealing of the microheterogeneous mode are considered in the review [8]. The MHM allows calculation of effective activation energy of the combustion wave propagating; the calculated values are in good agreement with experimental measurements. A boundary between steady state and oscillating regimes of combustion, calculated by

Concise Encyclopedia of Self-Propagating High-Temperature Synthesis
http://dx.doi.org/10.1016/B978-0-12-804173-4.00094-6

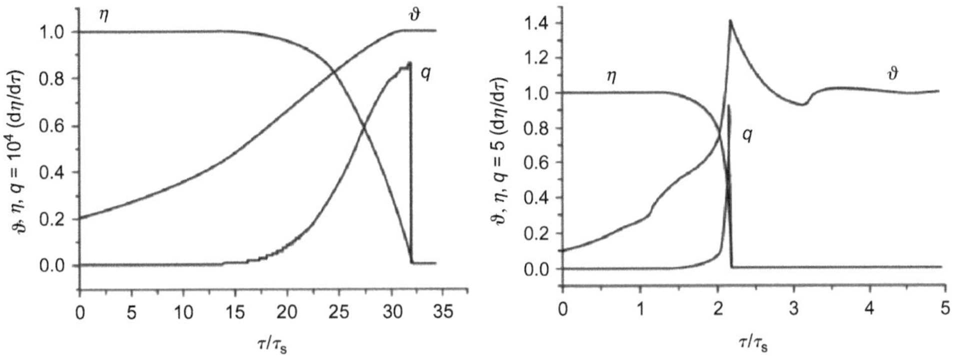

Fig. 1 Profiles of dimensionless temperature (ϑ), the heat release rate (q), and the dimensionless concentration of fuel (η), calculated by MHM for quasihomogeneous (left) and microheterogeneous (right) modes of combustion. The dimensionless normalization time τ_s corresponds to a shift of the combustion wave by one cell, τ is a dimensionless time.

MHM better fits experimental data for gasless systems than a quasihomogeneous approach. Thus one may conclude that MHMs provide an adequate description for a wide range of SHS systems, especially for gasless mixtures.

REFERENCES

[1] Rogachev AS, Merzhanov AG. On the theory relay-race combustion wave propagation in the heterogeneous media. Dokl Akad Nauk 1999;365(6):788–91.
[2] Merzhanov AG, Rogachev AS. Discrete heat waves in active heterogeneous media: basic principles and introduction to the theory. Russ J Phys Chem 2000;74(Suppl.1):S20–7.
[3] Smolyakov VK. "Roughness" of the gasless combustion front. Combust Explos Shock Waves 2001;37 (3):274–84.
[4] Merzhanov AG, Krishenik PM, Shkadinsky GK. A model of transverse flame propagation in alternating layers of combustible and inert substances. Dokl Phys 2001;46(9):619–23.
[5] Rogachev AS. Microheterogeneous mechanism of gasless combustion. Combust Explos Shock Waves 2003;39(2):150–8.
[6] Mukasyan AS, Rogachev AS. Discrete reaction waves: gasless combustion of solid powder mixtures. Prog Energy Combust Sci 2008;34(3):377–416.
[7] Rogachev AS, Baras F. Macrokinetics of combustion in heterogeneous media: comparative analysis of homogeneous, layered and cellular models. In: Mukasyan AS, Martirosyan KS, editors. Combustion of heterogeneous systems: fundamentals and applications for materials synthesis. India: Transworld Research Network; 2007. p. 41–65 [chapter 2].
[8] Rogachev AS, Mukasyan AS. Experimental verification of discrete models for combustion of microheterogeneous compositions forming condensed combustion products (Review). Combust Explos Shock Waves 2015;51(1):53–62.

Molybdenum Borides

Victoria V. Kurbatkina

Interest in molybdenum borides results from their high melting temperatures, chemical resistance, scale resistance, and hardness. A large group of borides of transition materials was obtained through SHS and is widely used in the form of powder, three-dimensional materials, and coatings. Despite the fact that SHS has gained a reputation as a resource- and energy-efficient method, the problem of obtaining refractory compounds with low formation and combustion temperatures restricts the range of synthesized compounds. It is problematic to realize SHS in the weakly exothermic Mo-10%B mixture with a low thermal effect of the reaction (<1 kJ/g) and adiabatic combustion temperature of 2310 K, which is lower than the melting temperatures of molybdenum (2898 K) and boron (2348 K). As a rule, molybdenum borides are obtained through SHS powder or SHS metallurgy technologies from oxide raw materials with a reducing stage. The relatively high chemical activities of Mo and B hamper obtaining chemically pure molybdenum borides because of side effects and admixtures. Preliminary mechanical activation (MA) makes it possible to realize the SHS process in the low-exothermal reaction mixtures of Mo and B with the preset composition along with a high chemical and structural homogeneity of the products being ensured. MA increases the heat released and the heat release rate of the SHS reaction. The mass is transferred more rapidly from one component to the other because of the reduced sizes of the reacting particles when the solid-phase reactions occur. During MA, the morphology of the powder changes, the reaction surface widens, the quantity of structure defects, and the value of the absorbed energy grows, which leads to an increased reaction capacity.

The possibility of obtaining one-phase molybdenum monoboride MoB [1] through SHS MA method was shown. This initial mixture consisted of molybdenum spherical particles sized at 3–5 μm and amorphous boron powder sized at <1 μm. After MA, agglomerates sized at 3–5 μm are formed that consist of Mo particles about 1 μm in size covered with boron formed after the MA. The molybdenum lattice microdeformation value increases from 0.05% in the initial mixture to 0.53%, which corroborates growth of the energy stored as a result of the MA. It was shown that the combustion rate for the MA mixtures goes up by a factor of 3, which is caused both by the specific surface increase by a factor of 1.7 as a result of crushing the molybdenum particles and growth in concentrations of the bound and adsorbed oxygen on the

Concise Encyclopedia of Self-Propagating High-Temperature Synthesis
http://dx.doi.org/10.1016/B978-0-12-804173-4.00095-8

molybdenum particles. Gas phase mass transfer contributes greatly to increasing the combustion rate. The work [2] shows that excess oxygen causes the formation of volatile suboxide MoO_3 and, as a result, to strengthening the role of the exothermal reaction $3B(solid) + MoO_3(gas) \rightarrow MoB(solid) + B_2O_3(solid)$, which takes place within the combustion wave simultaneously with the reaction $Mo + B \rightarrow MoB$. If T_0 exceeds 600 K, the MoO_3 vapor's partial pressure grows, which leads to purification of the molybdenum particles of oxygen and reducing the influence of the gaseous reagent on the combustion rate. After that, the interaction takes place mainly as solid-phase reaction diffusion.

The contribution of MA to the SHS kinetics and the activation energy were calculated using the equation proposed by Ya.B. Zeldovich for heat release rates in approaching narrow zones of combustion [3]. The combustion process activation energy (E_a) when using the mechanically activated Mo-10%B mixture is reduced by 24 ± 5 kJ/mol. The activation energy was also assessed with the help of traditional methods using the Arrhenius equation. In order to do this, the dependences of the combustion rate (U_c) on the starting temperature (T_0) for the initial mixture and the mixture after MA were built. The U_c (T_0) dependence for the after-MA mixture consists of two areas, which is indicative of the combustion mechanism change. In the first area (at $T_0 < 600$ K), the activation energy equaled $E_a = 191 \pm 6$ kJ/mol and $E_a = 780 \pm 90$ kJ/mol in the second one. It should be noted that when $T_0 < 600$ K, the combustion process activation energy value for the MA mixture was found to be lower than for the initial mixture ($E_a = 229 \pm 14$ kJ/mol). The difference is 30 ± 5 kJ/mol, which is perfectly consistent with the value calculated according to the methodology [3].

Through forced SHS consolidation, a dense material with the 100% MoB content, porosity lower than 5%, microhardness of (HV^{200}) 20 ± 3 GPa, grain size of 3–4 μm was obtained.

REFERENCES

[1] Eremina EN, Kurbatkina VV, Levashov EA, Rogachev AS, Kochetov NA. Chem Sustain Dev 2005;13:197–204.
[2] Egishyan AV, Manukyan KhV, Harutyunyan AB, Kharatyan SL. Int J SHS 2006;15(1):33–9.
[3] Kurbatkina VV, Levashov EA. Mechanoactivation of SHS. In: Mukasyan AS, Martirosyan KS, editors. Combustion of heterogeneous systems: fundamentals and applications for materials synthesis. Kerala: Transworld Research Network; 2007. p. 131–41.

Nanoenergetic Gas Generators

Karen S. Martirosyan

Nanoenergetic gas generators (NGG) are thermite composites that rapidly release a large amount of gaseous products having high PV (pressure × volume) values and that generate a fast-moving thermal wave during a reaction [1]. Nanostructured materials are subjects of increased interest for many energy-related applications because they have improved properties compared with their microscopic and macroscopic counterparts. The long-term research in this area has focused on the finding and characterization of new NGG formulations. This research involved a systematic study of various nanoenergetic reactants, their synthesis and fabrication and determination of the peak gas pressure evolution and rate of gas released [2,3]. Nanoscale particles have a significantly higher surface area to volume ratio than microparticles, providing closer contact between solid particles in energetic mixtures. It was shown that among common thermite reactions, the Al/Bi_2O_3 and Al/I_2O_5 mixture generated the highest pressure pulse [4–7]. A possible explanation for the high pressure rise during the combustion of Al/Bi_2O_3 and Al/I_2O_5 nano systems is that the reaction product (bismuth or iodine) boils at a temperature of 1560°C and 184°C, respectively, which is lower than the maximum reaction temperature ~2000°C. This causes bismuth or iodine evaporation and increases the released gas pressure. A comparison of energy densities shows that the volumetric energy of thermite-based materials can easily exceed the best existing molecular energetic materials by an approximate factor of 2 and is 25.7 kJ/cm^3 for iodine-based thermite.

The values of the peak pressure obtained during the combustion of various reactant mixtures are shown in Fig. 1A. It can be seen that the values of pressure generated by bismuth trioxide nanoparticles synthesized by combustion synthesis more than three times larger than that by commercial Bi_2O_3 nanoparticles. The Bi_2O_3-Al and I_2O_5-Al mixtures generated the highest pressure peak of about 11 MPa. IR images of the surface temperature during the reaction of the thermite reactant mixture showed that the average rate of temperature rise was up to ~40000°C/s. The experiments showed that the pressure peak depended on the bismuth trioxide concentration in the mixture. The largest pressure peak was obtained for 80% weight percentage of highly crystalline Bi_2O_3. The dynamic features of the pressure discharge during the detonation of mixtures of Al (100 nm) and as-synthesized Bi_2O_3 are shown in Fig. 1B. The data showed that the pressure release for the crystalline bismuth oxides powder with particle size ~40 nm

Concise Encyclopedia of Self-Propagating High-Temperature Synthesis
http://dx.doi.org/10.1016/B978-0-12-804173-4.00096-X

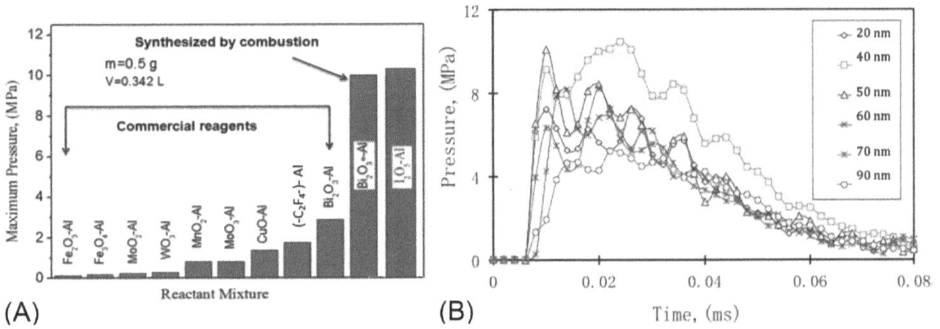

Fig. 1 (A) The pressure obtained during the explosion of different nanoenergetic thermites. (B) Pressure profiles during the reaction of Al with Bi_2O_3 of various particle sizes.

were higher than for either the amorphous-like powder with 20 nm or the 100 nm highly crystalline bismuth trioxide particles. One possible explanation for the high pressure rise during the detonation of these crystalline bismuth trioxide nanoparticles is that the electron/ionic conductivity of those particles are higher than that of the amorphous-like particles. The higher electron/ionic mobility may accelerate the interchange between species during the reduction of bismuth trioxide that can increase the reaction rate and generate higher pressure [5].

The pressure release of the Bi_2O_3/Al. reaction using commercial Bi_2O_3 nanoparticles (100 nm) was approximately 3 MPa. This is about one-third of the pressure peak using the nanopowders we synthesized. To estimate the total energy release (E) the adiabatic ideal gas relation $E=PV/(\gamma-1)$ can be used, where γ is the ratio of specific heat in the system, P is the pressure, and V is the reactor volume. For complex polyatomic gases, $\gamma=1.2$, and for diatomic gases, $\gamma=1.4$. Based on the measured peak pressure and assuming $\gamma=1.3$, the energy released by the Bi_2O_3/Al reaction was calculated to be ~20 MJ/kg. The particle size of reactants, aluminum, and metal oxides has a very strong effect on detonation front velocity. The highest front velocity ~2500 m/s for NGG composites based on bismuth trioxide were achieved. The estimated Mach number $M=V_f/V_s \sim 7.35$ (where V_f thermal front velocity and $V_s \sim 340$ m/s speed of sound in the medium) revealed that the high temperature and pressure promote the supersonic detonation mode.

REFERENCES

[1] Martirosyan KS. J Mater Chem 2011;21:9400–5.
[2] Martirosyan KS, Wang L, Vicent A, Luss D. Propell Explos Pyrotech 2009;34:532–8.
[3] Martirosyan KS, Luss D. Chem Eng Technol 2009;32(9):1376–83.
[4] Martirosyan KS, Wang L, Luss D. Chem Phys Lett 2009;483:107–10.
[5] Martirosyan KS, Wang L, Vicent A, Luss D. Nanotechnology 2009;20:405609.
[6] Wang L, Luss D, Martirosyan KS. J Appl Phys 2011;110:074311.
[7] Martirosyan KS, Zyskin M, Jenkins CM, Horie Y. J Appl Phys 2012;112:094319.

Nanoenergetic Materials

Karen S. Martirosyan

Major progress has been made during the past two decades in developing energetic materials that can rapidly release temperature and pressure waves and that have extensive potential applications [1–3]. Nanoscale particles have a significantly higher surface area to volume ratio than micro-particles, providing closer contact between solid particles in a mixture. The small size of nano-particles increases the homogeneity of the reactant mixture and improves the uniformity of a propagating reaction front. Nanostructured highly exothermic reactive mixtures, referred to as Nanoenergetic Materials (NMs) or metastable intermolecular composites (MIC), can release energy much faster than conventional energetic materials [4,5]. The size reduction of reactant powders from micro- to nano-size increases the reaction front propagation velocity in some systems by two to three orders of magnitude [6]. This is presumably because of the reduction of the diffusion distance between the nanostructured grains of the intermixed reactants. NMs have various potential military applications and are likely to become the next-generation explosive materials. They enable adjustment of the energy density and power release by a suitable choice of fuel and oxidizer composition and control of the particle size distribution. The combination of NMs and MEMS (microelectromechanical systems) technology may enable a reduction in the size and mass of energetic systems.

Well-established nanoenergetic thermite reactions include those mixtures of nanosize aluminum and nanopowders of metal oxides:

$$Fe_2O_3 + 2Al = Al_2O_3 + 2Fe \quad (1)$$

$$3Fe_3O_4 + 8Al = 4Al_2O_3 + 9Fe \quad (2)$$

$$3MoO_2 + 4Al = 2Al_2O_3 + 3Mo \quad (3)$$

$$WO_3 + 2Al = Al_2O_3 + W \quad (4)$$

$$3MnO_2 + 4Al = 2Al_2O_3 + 3Mn \quad (5)$$

$$MoO_3 + 2Al = Al_2O_3 + Mo \quad (6)$$

$$3CuO + 2Al = Al_2O_3 + 3Cu \quad (7)$$

$$3(-C_2F_4-) + 4Al = 4AlF_3 + 6C \quad (8)$$

$$Bi_2O_3 + 2Al = Al_2O_3 + 2Bi \quad (9)$$

Concise Encyclopedia of Self-Propagating High-Temperature Synthesis
http://dx.doi.org/10.1016/B978-0-12-804173-4.00097-1

$$3I_2O_5 + 10Al = 5Al_2O_3 + 6I \tag{10}$$

The main distinguishing features of these reactions are their significant enthalpy release and tunable rate of energy discharge, which gives rise to a wide range of combustion rates, energy release, and ignition sensitivity [7]. Reduction of the reactant particles' size from microsize to nanosize generates a more intimate contact between the particles. This increases the reaction front propagation velocity and rate of release energy because of the reduction of the mass transport limitations. MIC may have higher energy densities than conventional explosives and may generate shock waves with velocities of up to 2500 m/s [8]. Among numerous possible MIC formulations, the most widely investigated are Al/Fe_2O_3, Al/MoO_3, Al/WO_3, and Al/CuO nano systems. Most previous research in this area focused on the development of suitable fuel/metal oxidizer nanocomposites and study of the impact of the particle size on the corresponding amplitude and velocity of the propagating temperature front and ignition features.

Several research teams have successfully investigated chemical and technological ways to tailor NMs to the targeted applications either to produce gas, heat, or chemical compounds. Nanofabrication as well as advanced nanoscale patterning technology provide novel opportunities to get a new class of tailored NMs. The elaboration of energetic multilayer nano foils by physical vapor deposition or chemical vapor deposition or sputtering techniques is a promising approach for the integration of nanoenergetic systems within micro-electro-mechanical systems because of compatibility with mass production processes, the high quality of the fabricated layer, and the possibility to control precisely each metal and oxide thickness layer and interdiffusion region [9]. More recently, a reactive self-heating model [10] of aluminum spherical nanoparticles was developed that provides a better understanding of the multiscale, multiphysics oxidation processes in higher energy density nanostructured materials. Such modeling includes moving boundaries, describing the dynamics of oxidation of particles, discontinuities, allowing the modeling of shock wave propagation, and heat transfer in nanostructured mixtures undergoing highly exothermic chemical reactions [11].

REFERENCES

[1] Dlott DD. Mater Sci Technol 2006;22(4):463.
[2] Puszynski JA, Bulian CJ, Swiatkiewicz JJ. J Propuls Power 2007;23:698.
[3] Perry WL, Smith BL, Bulian CJ, Busse JR, Macomber CS, Dye RC, et al. Propellants Explos Pyrotech 2004;29(2):99.
[4] Umbrajkar SM, Schoenitz M, Dreizin EL. Propellants Explos Pyrotech 2006;31:382.
[5] Moore K, Pantoya ML. Propellants Explos Pyrotech 2006;31(3):182.
[6] Prakash A, McCormick AV, Zachariah MR. Adv Mater 2005;17(7):900.
[7] Martirosyan KS. J Mater Chem 2011;21:9400–5.
[8] Martirosyan KS, Wang L, Vicent A, Luss D. Nanotechnology 2009;20:405609.
[9] Rossi C, Zhang K, Estève D, Alphonse P, Tailhades P, Vahlas C. J Microelectromech Syst 2007;16(4):919–31. 36.
[10] Martirosyan KS, Zyskin M. Appl Phys Lett 2013;102:053112.
[11] Martirosyan KS, Zyskin M, Jenkins CM, Horie Y. J Appl Phys 2014;115:104903.

New Directions

Alexander S. Rogachev

Continuous development of SHS generates new scientific and technological directions in this field. We can outline several fast-developing subtypes of the method that appeared over the past 15–20 years that can be characterized as new directions for SHS. A common feature of all these approaches is that initial reactants are intermixed at a submicrometer, a nanometer, or even an atomic level. In some cases, they allow the production of nanocrystalline products and of obtaining unusual characteristics of the combustion process, such as very fast propagation of the reaction wave. One new direction is solution combustion synthesis [1]. According to this method, oxidizers (metal nitrates) and organic fuels (glycine, urea, and others) must be dissolved in water to produce a homogeneous solution. A relatively low combustion temperature and a large amount of gases evolving during the process result in the formation of nanosize products [2,3]. Another direction is based on the layer-by layer magnetron deposition of altering layers of solid reactants (such as Ni and Al) to form reactive multilayer nanofilms [4]. This direction is commonly named "thin film SHS." A gasless combustion reaction can propagate in such films with a velocity up to 90 m/s [5]. A third new direction is mechanical activation of SHS [6,7]. Intensive mechanical treatment, such as high-energy ball milling in the planetary mills leads to the formation of lamellar composite particles with a submicrometer and nanometer thickness of the reactant layers. The temperature of the reaction initiation decreases significantly, and the SHS process can be carried out easier in the activated systems.

Since the SHS is a living technology, we can expect the appearance of more novel directions in this field in the near future.

REFERENCES

[1] Patil KC, Hedge MS, Rattan T, Aruna ST. Chemistry of nano-crystalline oxide materials: combustion synthesis, properties and applications. Hackensack, NJ: World Scientific; 2008.

[2] Mukasyan AS, Epstein P, Dinka P. Solution combustion synthesis of nanomaterials. Proc Combust Inst 2007;31(2):1789–95.

[3] Aruna ST, Mukasyan AS. Combustion synthesis and nanomaterials. Curr Opinion Solid State Mater Sci 2008;12:44–50.

Concise Encyclopedia of Self-Propagating High-Temperature Synthesis
http://dx.doi.org/10.1016/B978-0-12-804173-4.00098-3

[4] Barbee TW, Weihs T. Ignitable heterogeneous stratified structure for the propagating of an internal exothermic chemical reaction along an expanding wavefront and method of making same, US Patent No 5,538,795. Jul. 23, 1996.
[5] Rogachev AS. Exothermal chemical waves in multilayer nano-foils. Russ Chem Rev 2008;77(1):22–38.
[6] Grigorieva T, Korchagin M, Lyakhov N. Combination of SHS and mechanochemical synthesis for nanopowder technologies. KONA Powder Part J 2002;20:144–58.
[7] Bernard F, Gaffet E. Mechanical alloying in SHS research. Int J SHS 2001;10(2):109–32.

Nickel Aluminides

Victoria V. Kurbatkina

NiAl alloys possess valuable physicochemical and mechanical properties, have better processibility than ceramic materials, are resistant to corrosion, and are thus considerably superior to ordinary metals. The most widely used are Ni_3Al and NiAl alloys. The interest in these alloys is determined by their high-temperature strength of up to 0.7–0.8 T_M, their heat resistance up to 1100°C, and their low density compared with nickel. Aluminides hold a record among intermetallics in specific strength characteristics. The main disadvantage is their brittleness at room and high temperatures for multiple reasons, including phase nonuniformity of the derived materials. The specific features of the reaction of intermetallic compound formation from elements are a relatively low-heat reaction and low-adiabatic reaction temperatures in comparison with other SHS systems. However, the NiAl compound is an exclusion, as SHS reactions from elements in this case are possible at room temperature; therefore the Ni-Al system is one of the most explored SHS systems. The adiabatic combustion temperature is 1611°C, which is higher than the melting temperature of basic components and approximately equal to the NiAl melting temperature. The phase composition and the structure of the chemicals in Ni-Al system depend on the porosity of the charge briquette, initial temperature, heating rate, and dimensions of the samples. A big difference in the Ni and Al melting temperatures causes partial evaporation of the mixture's components. Phase nonuniformity of the combustion products is also caused by multiple intermediate phases formed in this system.

According to combustion mechanism studies [1–3], the interaction of the Ni-Al system occurs in the mechanism shown in Fig. 1. At the first stage, contact melting is observed at 640°C in the Ni and Al eutectic reaction with formation of the Al_3Ni phase. At heating beyond 660°C (aluminum melting point), the quantity of the melt increases, fine particles merge, and reaction-induced spread of Ni particles on the surface causes formation of an Al_3Ni_2 film. Then the interaction is in the form of reactive diffusion through the layer of the formed product. At a temperature higher than 854°C, mass transfer processes are observed in the melt, which leads to an increase in the Ni concentration and formation of the Al_3Ni_2 phase. As a result of the dissolution, Ni particles shrink, and the NiAl phase is formed (temperature close to 1133°C). At a temperature higher than 1133°C, NiAl crystallites grow, and at 1638°C, a single-phase product forms.

Concise Encyclopedia of Self-Propagating High-Temperature Synthesis
http://dx.doi.org/10.1016/B978-0-12-804173-4.00099-5

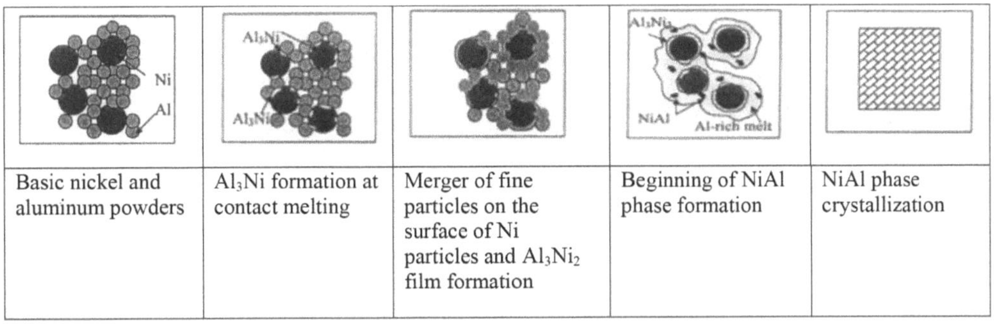

Fig. 1 Interaction mechanism in combustion wave for the Ni-Al system.

An effective way to enhance the structural and phase uniformity is to apply preliminary mechanical activation (MA) to the reaction mixture [2,4]. Following MA, deformation flow of components is observed with the formation of multilayer agglomerates. The more plastic material serves as a cementing layer, which wraps and unites more solid particles. The reaction surface of such a mixture multiplies, as the structure of agglomerated particles consists of multiple layers with a width of several microns, and the combustion rate is higher. The combustion product of MA mixtures is a single-phase NiAl material.

To produce NiAl, the following conditions are recommended: to achieve NiAl porous fillet in the process of layer by layer combustion, T_0 should be 0.2–0.4 of the end-product melting temperature (T_M); to create a NiAl ingot, T_0 should be 0.5–0.7 of the T_M; to produce NiAl in the process of thermal explosion, T_0 should be close to the melting temperature of the lowest melting eutectics; pressure of inert gas should be 0.1–0.2 MPa; the minimum diameter of the reaction mixture briquette should be 3 cm; and relative porosity of the briquette should be 30%–50%. Fineness, morphology of basic powders, and concentration of alloying elements vary for the production of NiAl with a preset composition and properties.

REFERENCES

[1] Rogachev AS, Varma A, Merzhanov AG. Int J SHS 1996;2(1):25.
[2] Rogachev AS, Kochetov NA, Kurbatkina VV, Levashov EA, Grinchuk PS, Bernard F, et al. Fiz Goreniya Vzryva 2006;4(42):61–70.
[3] Zhu P, Li JCM, Liu CT. Mater Sci Eng A 2002;329-331:57–68.
[4] Kurbatkina VV, Levashov EA. Mechanoactivation of SHS. In: Mukasyan AS, Martirosyan KS, editors. Combustion of heterogeneous systems: fundamentals and applications for materials synthesis. Kerala: Transworld Research Network; 2007. p. 131–41.

Nitrides and Nitride Ceramics

Inna P. Borovinskaya

G. Hegg is known as the first to determine the regularities of nitride structure. He defined the notion of simple structures in which nonmetal atoms are embedded into the metal crystal lattice, which he named interstitial phases. The following types of crystal lattice are the most widespread ones for interstitial structures: face-centered cubic, volume-centered cubic, hexagonal compact, and hexagonal simple. According to his concept, transition metal nitrides (metal-like nitrides) are typical interstitial phases. These interstitial phases are often considered to be heterodesmic compounds with a complicated imposition of covalent, metal, and ionic bonds. Many authors believe that it is the bonds' composition that defines the unusual behavior and unique properties of nitrides. Nonmetal nitrides, for example, BN, Si_3N_4, and AlN, are mainly characterized by covalent bonds and appear to be true chemical compounds.

The self-propagating high-temperature synthesis (SHS) method allowed synthesizing homogeneous stoichiometric, nonstoichiometric, and super-stoichiometric nitrides of transition metals of Groups IV–VI of the periodic system as well as all known single-phase nitride solid solutions and nonmetal boron, silicon, and aluminum nitrides. The following SHS nitrides are particularly interesting: super-stoichiometric $TaN_{1.2}$, diamond-like cubic boron nitride BN_{cub}, cubic modification of complex nitride BPN_x, carbonitrides, hydronitrides based on transition metals such as titanium; zirconium, tantalum, nitride-based alloys and cermets, and composite materials obtained from nitrides with other refractory compounds.

The SHS nitrides are obtained by filtration combustion of bulk powder initial reagents or compact samples of low porosity in gaseous nitrogen. The influence of combustion terms on the chemical and phase composition of the nitrided products was studied for elements of different groups of the periodical system, and the results allowed defining at least three types of the systems:

- Systems with metals melting at combustion and with the products of a wide range of homogeneity. They include systems based on metals of Group IV-a (Ti, Zr) and those based on Al and Si.
- Systems with metals that do not melt at combustion and with multiphase products of nitriding (Group IV-a, mainly Nb, Ta). Mo and W nitrides were unstable and formed at high nitrogen pressures.

Concise Encyclopedia of Self-Propagating High-Temperature Synthesis
http://dx.doi.org/10.1016/B978-0-12-804173-4.00100-9

– Systems resulted in the formation of several phases, e.g., α-Si_3N_4, β-Si_3N_4; BN_{hex} and BN_{cub}; AlN_{wurt}, and AlN_{cub}.

Metal-like nitrides (based on metals of Groups IV–VI, transition metals) are chemically stable in aggressive media and metal vapors and are used in tool materials that require wear and heat resistance at rapid cutting and machining of cast iron and steel. Titanium and zirconium nitrides were used to develop "golden alloys" imitating gold and applied as coatings, e.g., for the domes of churches, watches, bracelets, crowns of teeth, and jewelry [1]. The most remarkable ones are metal and nonmetal nitrides synthesized from elemental oxides at their reduction with carbon in nitrogen medium, and also nitrides obtained at combustion of elemental mixtures with metal azides in nitrogen. In both cases, very fine (including nanosize) nitride powders with narrow particle size distribution are obtained. In the case of nitride synthesis with oxide or other compound reduction, nanoparticle formation is connected with either the atomic state of initial metals and nonmetals released at oxide reduction or NH_3 formation. When azides are used, not only nitrogen but also ammonium participates in the nitriding reaction, gaseous products are formed and direct solid-phase nitriding of coarse particles is practically absent. Nitrides are mainly formed as a result of the gas-phase reaction. The synthesis of transition metal nitrides from metal alloys was well studied within research and practical development in powder metallurgy with the aim of production of cermets based on nitrides and nitrided hard alloys. Such alloys were also obtained by the SHS method. This research and applied work is being actively developed in Russia [2]. The SHS method of nitride production by reagent combustion in gaseous or liquid nitrogen is one of the most complicated processes since it is connected with gas hydrodynamic supply to the reaction zone due to the pressure difference at the front and in the sample absorbing nitrogen. So SHS nitriding has peculiarities such as the existence of two stages in the combustion process (burning-afterburning), the possible dissociation of combustion products, and the various modes of nitrogen supply to the combustion zone (kinetic-diffusion). For example, the diffusion supply can be carried out at gaseous nitrogen dilution with inert gas. There can be filtration difficulties with the gas supply because of the reagent or intermediate product melting, and so on. All these peculiarities are described in paper [3].

Synthesis of nitride-based ceramics at the SHS mode is well developed for nonmetal ceramics (based on silicon, boron, aluminum, and SiAlONs) [3]. It can be carried out either by the methods of powder metallurgy using SHS nitride powders with their following compaction or by direct one-stage synthesis at the combustion mode (SHS gasstatting). The properties of SHS nitride ceramics obtained by sintering or hot pressing usually differ from those of their analogs [4,5].

Excellent results were achieved at investigation of catalytic properties of sialon ceramics of various compositions at CO_2 afterburning. But the striking example of catalytic properties of SHS sialon powders was observed during the tests of honeycomb afterburners for automobiles by Sango Co. Ltd., with Honda (Fig. 1).

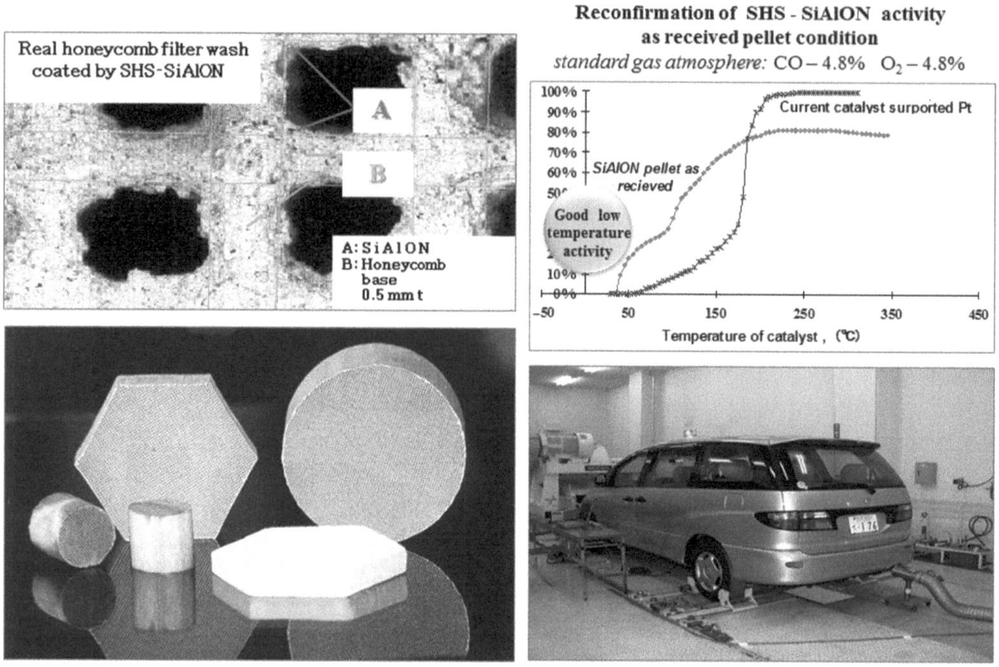

Fig. 1 Testing of SHS-SiAlON (Sango Co. Ltd., Honda).

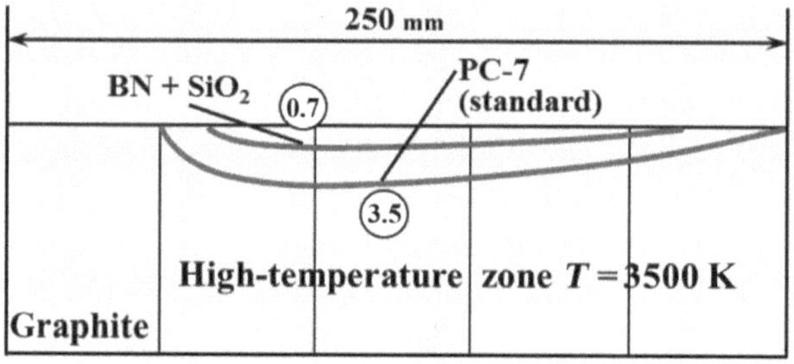

Substance carry-over in MHD-generator canal

Operation terms:

Flow temperature . . . . . 3500 K
Heat flow . . . . 10^3 W/cm^2
Density of Al_2O_3
Liquid drops flow . . . . . 10 kg/m^2 s

Fig. 2 BN-based electroinsulating material for MHD generator lining.

Material	Items and their application
BN, BN-TiB_2, BN-SiO_2, BN_xC_y, BN-Al_2O_3, BN-ZrO_2, AlN-TiB_2-BN, SiAlON-SiC-BN, Si_3N_4-SiC-TiN, SiAlON-TiN, etc.	Molds and crucibles for alloy casting (Fe, Co, Ni, Cr, Ti); locking washers, parts of casting machine "Sirius" (50 cycles); parts of equipment for single crystal growing; protection in laser devices (CO_2, 3 kW); protection tubes for measuring temperature of aluminum and other metal melts; parts of MHD generators, aircraft and rocket engines

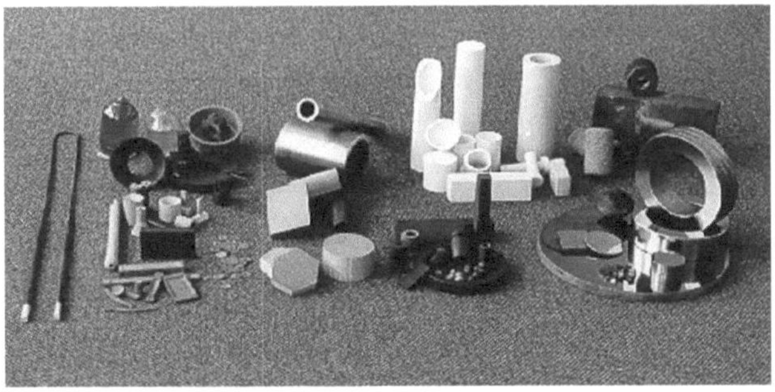

Fig. 3 Ceramic items of various purposes.

In other experiments, SHS ceramics with honeycomb structures were synthesized in one stage by the SHS gas-statting method; the shape of afterburners and cell size being kept constant. Fig. 2 demonstrates the high-temperature strength of BN-based ceramics in the aggressive medium. This example proves that the combination of SHS with different physical and mechanical effects is the optimum way from a technological and ecological viewpoint and can become one of the main methods for manufacturing materials and items with high operation characteristics [6]. This statement is confirmed by the ceramic items based on Si_3N_4, BN, SiAlONs, and so on that are already used in industry.

Ceramic items obtained by sintering or direct SHS gas-statting are shown in Fig. 3.

REFERENCES

[1] Kislyi PS, Bondarchuk NI, Borovikova MS. In: Kislyi PS, editor. Cermets. Kiev: Naukova dumka; 1985.
[2] Maksimov YuM, Chukhlomina LN, Braverman BSh, Smirnov LA. Self-propagating high-temperature synthesis. Novosibirsk: Nauka; 2014.

[3] Borovinskaya IP, Loryan VE, Zakorzhevsky VV. Combustion synthesis of nitrides for development of ceramic materials of new generation. In: Gromov AA, Chukhlomina LN, editors. Nitride ceramics: combustion synthesis, properties, and applications. Weinheim: Wiley-VCH Verlag GmbH & Co. KGaA; 2014 [chapter 1].
[4] Pampuch R, Lis J, Stobierski L. Self-propagating high-temperature synthesis of ceramic powder. Sci Ceram 1988;14:15–26.
[5] Pampuch R. Processing of particulate SHS products to sinterable powders. Int J SHS 1997;6(2):187–202.
[6] Borovinskaya IP. SHS-ceramics: synthesis, technology, application. Nauka-proizvodstvu 2001;10 (48):11–8.

Nitrides and Oxynitrides Synthesis in N_2/O_2 Mixtures

Alexander A. Gromov

Metal particles combustion in air with N_2 competing with O_2 as the oxidizer has been known about since P.F. Pokhil and I. Glassman et al. [1,2], although related experimental data cannot be explained by thermodynamics; that is, metals should react with O_2 under the burning temperature $T > 1000°C$ as a significantly more active oxidizer than N_2. If the formation of nitrogen-containing intermediates is probable, the intermediates must immediately be oxidized by O_2 (on the example of TiN, Eq. 1), especially at high temperatures caused by the kinetics of this reaction [3].

$$TiN(s) + O_2(g) \rightarrow TiO_2(s) + \tfrac{1}{2}N_2(g) + 621\,kJ \qquad (1)$$

Thermodynamic calculation of AlN formation and its subsequent afterburning was first performed for the system "Al/air" [4]. Experimentally, AlN formation (>50 wt.%) at SHS of nanoAl in air was first confirmed in Ref. [5]. Nitrides formation from SHS of metal/air or metal/(N_2/O_2) significantly reduces the enthalpy of metal combustion (in a factor 2.6 for AlN in comparison with Al_2O_3), and the AlN content in condensed combustion products increases linearly with N_2 concentration in O_2. A maximal AlN content is ~55 wt.% at 94 vol.% N_2 (Fig. 1). The AlN/Al_2O_3 ratio in SHS products of Al/air (1 atm) is ~4–8 and increases with the sample mass. The combustion products form a 3D needle-shaped structure (Fig. 2).

SHS in air for B, Al, Ti, and Zr is similar [6]. The combustion spots are propagated through the surface and overlap (spin or fingering combustion). The bright stage of combustion was followed by cooling. The maximum burning temperature is lower than the boiling point of B, Ti, and Zr, but not of Al. SHS for those powders was accompanied by a flame torch of the volatile products, most probably BO, B_2O, Al_2O, AlO_2, and AlO [7], TiO and hydrogen, accumulated by powders storage. According to XRD, nitrides are the major phase of combustion products for SHS (B, Al, Ti, Zr)/air (Fig. 2).

The SHS of Ti/TiO_2 (TiO_2—"inert" diluent) and Ti/Al (Al—"active" diluent) in air results in up to 85 wt.% TiN. The yield of TiN obtained by SHS of Ti/air (1 atm) was higher than the SHS in N_2. The mechanism of nitride formation by SHS in air or N_2/O_2 is probably determined by the formation of intermediate volatile suboxides, which have a high reducibility, along with the direct metal nitridation. The SHS products

Concise Encyclopedia of Self-Propagating High-Temperature Synthesis
http://dx.doi.org/10.1016/B978-0-12-804173-4.00129-0

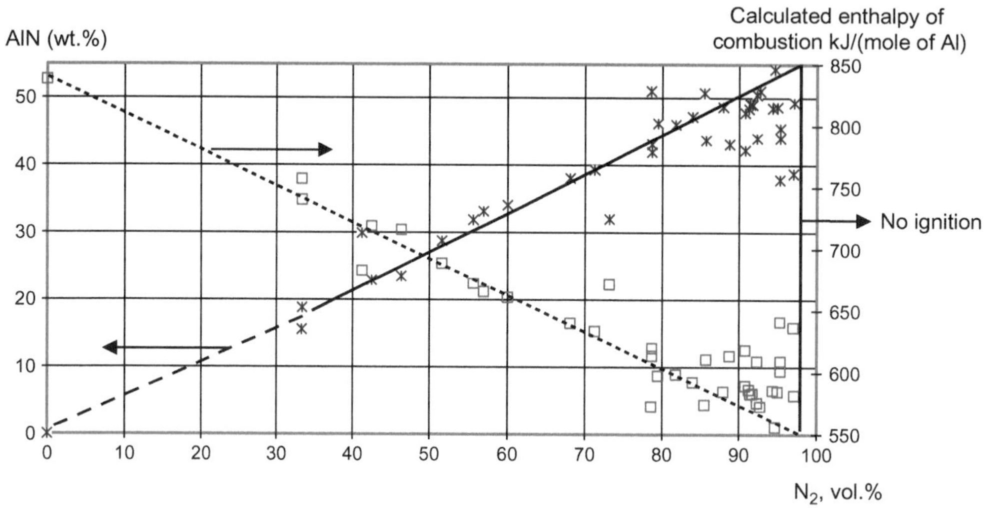

Fig. 1 AlN content for SHS of Al burned in O_2/N_2 at 1 atm and the corresponding calculated enthalpy of combustion.

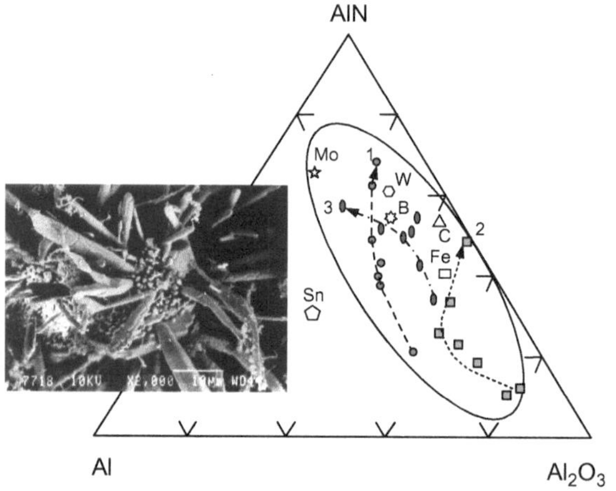

Fig. 2 Diagram of phases composition of SHS Al/air: (1) combustion on open air; (2) AlH_3 combustion; (3) combustion in closed vessel at initial pressure 1 atm; influence of additives Sn, Mo, W, Fe, C, B. Phase composition change at sample mass increasing is shown with dotted arrows. Inset: SEM image of SHS products nanoAl/air (2000 ×).

of microAl/nano-Al_2O_3 in air contain a significant amount of Al_3O_3N (Eq. 2). Large Al particles acquire a coating layer of fine Al_2O_3 particles during preliminary mixing. Insulated aluminum particles are easily ignited. Propagation of the combustion wave ignites the majority of aluminum particles.

$$Al(l) + Al_2O_3(s) + \tfrac{1}{2}N_2(g) \rightarrow Al_3O_3N(s) \tag{2}$$

The mechanism of AlN (Al_3O_3N) SHS in air of the (Al/Al_2O_3) mixture's combustion is similar to the combustion of Al: it has a similar combustion scenario and composition of condensed combustion products.

Recently the synthesis of HfN, GaN, Mg_3N_2, Ca_3N_2, and the other nitrides by SHS in air was reported. Thus the formation of a new class of SHS reactions for nonequilibrium products (nitrides and oxynitrides) in the II–IV group for metal burning in air was experimentally confirmed.

REFERENCES

[1] Pokhil PF, Belyaev AF, Frolov YuV, Logachev VS, Korotkov AI. Combustion of powdered metals in active media. Moscow: Science; 1972.
[2] Glassman I, Yetter RA. Combustion. 4th ed. Cambridge, MA: Academic Press; 2008. p. 495–550.
[3] Swihart MT, Catoire L. Combust Flame 2000;130:210–22.
[4] Boborykin VM, Gremyachkin VM, Istratov AG, Kolesnikov-Svinarev VI, Kuznetsov GP, Leipunskii OI, et al. Combust Explos Shock Waves 1983;19(3):270–6.
[5] Il'in AP, Proskurovskaya LT. Rus Powder Metal 1990;9:32–5.
[6] Gromov A, Pautova Yu, Lider A, Korotkikh A, Teipel U, Chaplina E, et al. Powder Technol 2011;214 (2):229–36.
[7] Leo B, Searcy AW. J Am Chem Soc 1951;73:5308–14.

Nonoxide Ceramics

Alexander S. Mukasyan

Nonoxide ceramics include carbides, nitrides, *borides*, silicides, and others. Their applications include superhard abrasives (B_4C, SiC), rocket nozzles (TiB_2), electrodes for metal melts (ZrB_2), and heating elements ($MoSi_2$). The most important structural nonoxide ceramics are *silicon carbide* SiC, *silicon nitride* Si_3N_4, and the *sialons*, nitride-based ceramics with varying oxide content. Almost all these compounds can be produced by the combustion synthesis method [1–5] (see also articles on carbides, silicide, nitrides, and borides).

The benefits of CS are fully exploited in direct production of the nonoxide ceramic materials in the combustion wave. Three main features need to be outlined. The first feature is purely technological and is related to the method for achieving high temperatures. The conventional method for production of ceramics includes several steps: (i) synthesis of boron nitride powders, (ii) milling these powders prior to preparing the initial compacts, and (iii) using high-temperature furnaces for long-term sintering of powders. In CS, high temperature is achieved at the reaction front because of self-heating of the medium, and it is logical to use these temperature not only for synthesis of the BN and Si_3N_4 phase but also for sintering the obtained nitride. Thus, instead of the three-stage technology (powder production + milling + traditional sintering), we have a one-stage technology (SHS sintering of materials/products). The second and third features are fundamental. The strong and predominantly covalent (i.e., directional) atomic bonds in nonoxide ceramics inhibit atom mobility so that solid-state sintering below the decomposition temperature is slow. Thus typically liquid-phase sintering with additives or reaction-bonding techniques is necessary for densification. High temperatures are required to intensify the mass transfer and achieve the desired consolidation. Currently, no furnaces allow sintering at temperatures above 2400°C. However, temperatures of ~3000°C can be achieved in the combustion wave. In addition, because of the characteristics of the crystal structure, it appears that under equilibrium conditions, many ceramic powders cannot be sintered at high temperatures because the corresponding phases (BN, Si_3N_4) dissociate under such conditions. It turns out that the nonequilibrium conditions of the combustion wave may result in temperatures that are equal to the dissociation temperature of the products at which synthesis can be carried out. Indeed, the combustion temperatures in the boron–nitrogen system can be equal to the dissociation temperatures of BN at the corresponding gas pressure. Thus the CS approach may

Concise Encyclopedia of Self-Propagating High-Temperature Synthesis
http://dx.doi.org/10.1016/B978-0-12-804173-4.00101-0

provide extremely high temperatures for sintering, which are considerably higher than the temperatures that can be reached in advanced conventional furnaces.

REFERENCES

[1] Merzhanov AG. Self-propagating high-temperature synthesis and powder metallurgy: unity of goals and competition of principles. In: Particulate materials and processes: advances in powder metallurgy and particulate materials. Princeton, NJ: Metal Powder and Ind. Federation; 1992. p. 341.

[2] Merzhanov AG. Advanced SHS ceramics: today and tomorrow morning, In: Invited paper presented at international symposium 100th anniversary of ceramic society of Japan, Yokohma; 1991.

[3] Varma A, Rogachev AS, Mukasyan AS, Hwang S. Combustion synthesis of advanced materials: principles and applications. In: Wei J, editor. Advances in chemical engineering. vol. 24. New York, NY: Academic Press; 1998. p. 79–226.

[4] Gromov AA, Chukhlomina L, editors. Nitride ceramics: combustion synthesis and applications. Weinheim: Wiley, VCH; 2014. ISBN: 978-3-527-33755-2. 360 p.

[5] Mukasyan AS. Gerhardt R, editor. Properties and applications of silicon carbide. Rijeka: InTech, Janeza Trdine; 2013. p. 389–409.

Nonstationary Combustion Wave

Oleg V. Lapshin

If all points move in the stationary combustion wave in the same direction and at the same speed, then in certain circumstances, there is a loss of stability. In this case, the reaction zone and the heating zone start moving at different speeds, and the flame front propagates in the nonstationary mode. At present, the most interesting phenomena (self-oscillating combustion, spin, and chaotic waves) occurring during unstable propagation of the exothermic reaction are found and described in the SHS processes [1].

In the self-oscillating mode of combustion, the instantaneous velocity of the flame front makes periodic pulsations, that is, there is an alternation of flashes and depressions of an average value of the velocity (Fig. 1). K.G. Shkadinsky, B.I. Khaikin, and A.G. Merzhanov obtained the interpolation formula that can be used to separate the stationary and self-oscillating modes [2]:

$$\alpha_{st} = 9.1\gamma - 2.5\beta = 1 \tag{1}$$

When $\alpha_{st} > 1$, the combustion process develops in the stationary mode; when $\alpha_{st} < 1$, the combustion process develops in the self-oscillating mode. In Eq. (1) $\gamma = cRT_g^2/(EQ)$, $\beta = RT_g/E$, c is the heat capacity, R is the universal gas constant (Boltzmann constant), Q is the heat of reaction, E is the activation energy, and T_g is the combustion temperature. The loss of stability is also shown to occur when the excessive enthalpy S meets the inequality

$$S/S_* > 7.14\alpha_{st}^{-1} \tag{2}$$

where S_* is the enthalpy for initial substances or products of combustion.

In the spin mode, the combustion front is at the center of a chemical reaction that helically moves along an initial substance and can occupy both the surface layer of the sample and the inner regions (Fig. 2). Here v is the direction of the reaction front movement along the sample, and the reaction front (shaded area) is directed at an angle φ to the combustion propagation. The rotation of this area forms a helical track typical for spin on the sample surface. The spin combustion mode was observed for various SHS system types (hybrid, thermite, and gas-free).

The paper [3] provides a common theoretical approach describing both (self-oscillating and spin) phenomena and based on the concept of the thermal instability of the longitudinal and lateral disturbances, as well as the first mathematical modeling of

Concise Encyclopedia of Self-Propagating High-Temperature Synthesis
http://dx.doi.org/10.1016/B978-0-12-804173-4.00102-2

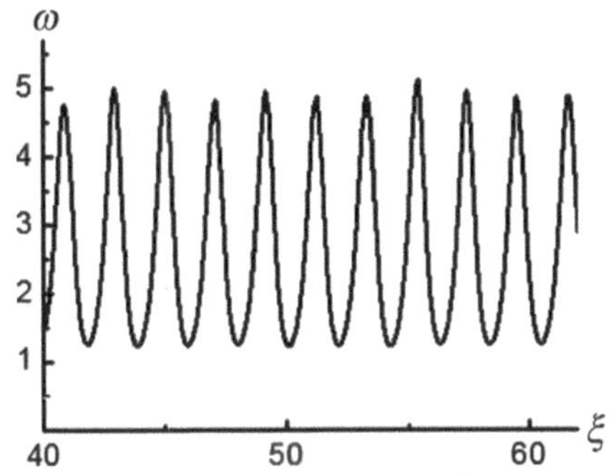

Fig. 1 Distribution of the combustion rate along the sample length.

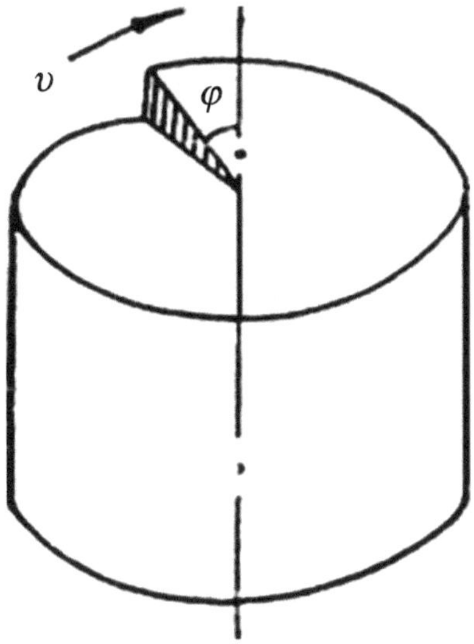

Fig. 2 Schematic illustration of the reaction front movement during spin combustion.

spin combustion. In the chaotic modes, many chemical reaction centers are formed with elevated temperatures and move in different directions at different rates, so there is total decay of the combustion wave (Fig. 3).

The combustion modes can also be changed during the wave propagation of the reaction front; for example, in certain conditions, the stationary mode may pass into the self-oscillating one and then into the multicentric and spin mode. The main difficulty in studying the processes of unstable combustion is the use of complex experimental and computer equipment.

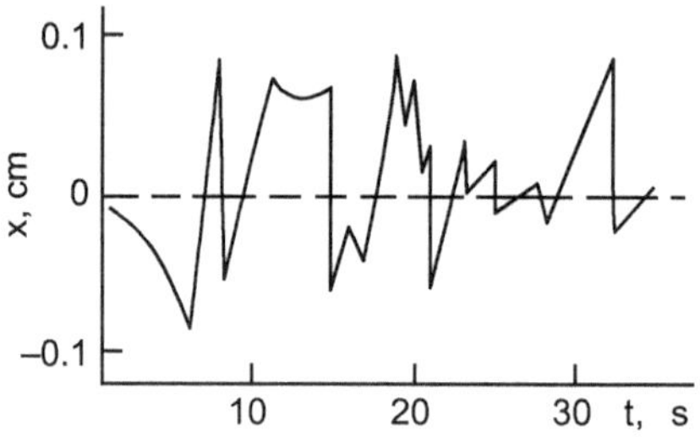

Fig. 3 Dynamics of the reaction front for the chaotic mode of combustion.

REFERENCES

[1] Merzhanov AG, Mukasyan AS. Solid-phase combustion. Moscow: Torus Press; 2007. 336 p.
[2] Shkadinsky KG, Chaikin BI, Merzhanov AG. Combust Explos Shock Waves 1971;(1):19–28.
[3] Ivleva TP, Merzhanov AG, Shkadinsky KG. Proc USSR Acad Sci 1978;239(5):1086–8.

Organic SHS

Evgeny G. Klimchuk

Organic SHS (organic self-propagating high-temperature synthesis, OSHS) is an autowave mode of combustion-like exothermic organic synthesis in solid-phase dispersion (powdered) intermixed with compositions containing organic compounds. As is typical of SHS processes, the combustion wave occurs after local initiation with the subsequent formation of a final product (Fig. 1).

The OSHS mode was confirmed to exist for ~50 systems, about 10 of which were studied in detail. Thermochemical studies demonstrate that OSHS is quite prevalent, and not only in oxidation or decomposition reactions, which are more characteristic for pyrotechnical compositions.

OSHS was first suggested in 1986 as an application of the known SHS method for obtaining useful condensed organic substances, materials, and items. The term was officially presented in 1988 and has been used as an independent term since 1990. A short review [1] of the concept of the scientific direction and its future development has been published [2].

OSHS differs fundamentally from SHS processes in inorganic and pyrotechnical systems because of low maximum temperatures in the synthesis wave (70–250°C compared to 1200–4100°C for inorganic systems), a wider range of dispersity of reacting powders, a lower temperature of initiation and propagation of the process, smaller heat evolution, and activation energies that are characteristic of organic reactions. All of which may be attributed to relatively small bond energies in organic molecular crystals.

Based on these features, both experimental and technological equipment and OSHS techniques were simplified and physical and mathematical modeling became more evident, with research conclusions being applicable for inorganic systems as well. Thus ESR, NMR, and optical spectroscopy have been employed for in situ study of OSHS processes for the first time. Free radicals and temperature fluctuations in a synthesis wave were found with their assistance. Visualization of OSHS processes is more informative because they are not accompanied by visible light emission and can be observed in reflected light. Embedding of electrodes in an organic dielectric mixture enables registering electrical parameters of the wave (conductivity and electromotive force), providing information on the OSHS mechanism. Reacting medium motion in OSHS (formation

Concise Encyclopedia of Self-Propagating High-Temperature Synthesis
http://dx.doi.org/10.1016/B978-0-12-804173-4.00103-4

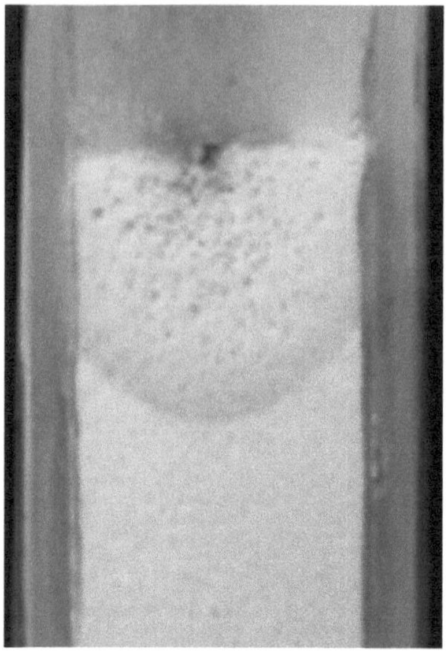

Fig. 1 Typical view of the reacting sample (initiation point is visible above).

of cracks, pores, bubbles, etc.) is accompanied by acoustic emission that has become a basis for an acoustic method for studying the OSHS mechanism and products structure.

The features of the OSHS macrokinetic mechanism are most studied in piperazine/malonic acid and 8-hydroxyquinoline/chloramine-B systems [3]. This study includes the following stages: (1) insignificant (<10%) reaction on particle surface of contact at mixing, (2) melting of a fusible component at ignition, acting as a solvent, (3) exothermic interaction accompanied by phase changes of reagents, (4) capillary spreading of a melt in pores, and (5) formation of products structure.

Mass transfer in sample areas with a temperature below reagents' melting points is caused by solid- and gas-phase diffusion and at higher temperatures is due to liquid-phase diffusion. Molecular mechanisms for OSHS are rather various, as is characteristic of organic reactions. These are, for example, proton transfer in acid-base reactions, radical substitution and addition in oxyhalogenation, electrophilic substitution in Friedel-Crafts acylation, and so on.

Products of OSHS are low-molecular-weight compounds of different chemical classes, applicable as intermediate reagents for syntheses.

The useful products, which have currently been synthesized, are salts of piperazine (malonate, citrate, adipinate, etc.) that are parts of cheap antiparasitic and antigout drugs. In OSHS, *o*-carboxybenzoylferrocene and its salts are synthesized in the ferrocene/phthalic anhydride/$AlCl_3$ system, which is a basic component in the drug ferrocerone used in the treatment of iron-deficit anemia. Mono- and dibrommalonic and -acetic acids (production of vitamins) are obtained in the malonic acid/*N*-bromosuccinimide system. By oxidation of hydroquinone with $KBrO_3$ or chloramine-B, quinhydrone (analytical chemistry and photography) and quinone (dyes production) were synthesized.

In oxidative halogenation of triphenylphosphine and 8-hydroxyquinoline with chloramine-B, triphenyl-*N*-(phenylsulfoil)-phosphinimin and oxy- and Cl-substituted 8-hydroxyquinoline, accordingly, intermediates in the production of physiologically active compounds are produced. A combination of OSHS with mechanochemical activation enables obtaining transition-metal acetylacetonates.

OSHS products differ from the substances obtained with conventional methods of organic chemistry by their higher purity because solvents, yielding impurities or by-products, are fully or partly excluded from a synthesis technique (dry synthesis). The impurities absence are also promoted by the known effect of self-purification caused by thermodesorption of volatile impurities in SHS.

Specific conditions of heat- and mass-transfer in OSHS result in products with a unique microstructure that is especially useful for synthesis of solid-phase drugs, where structure-property relation is important.

The limited use of solvents has resulted in a preference for OSHS technologies that promote explosion, fire, and ecological safety that are especially useful for synthesis under extreme conditions, where solvents are inaccessible or undesirable (space and polar stations, etc.). The combination of simplicity and rapidity, inherent in the absence of bulky furnaces, temperature controllers, and mixers (which are typical in "click-synthesis"), results in decreased costs for products, increased reliability, as well as resource and power-saving properties.

OSHS systems can be used as a working substance for independent sources of heat (chemical furnace) in temperatures ranging from 70°C to 300°C.

OSHS in solid-phase monomers (acrylamide derivatives, caprolactam, etc.) mixed with polymerization initiators (peroxides, amines) formulate plastics. The term historically assigned to such processes is "frontal polymerization," initially applied as a designation for a liquid-phase process (1972) [4].

Contact of the OSHS wave with polymers can lead to a surface modification that is used for coloring and texturing.

In some OSHS systems, the mode of natural radial shrinkage of the products enables hollow product formation from organic materials (tubes, beakers, etc.) to occur in one technological stage, without the use of special shaping devices.

REFERENCES

[1] Klimchuk EG. Autowave exothermic organic synthesis in the mixes of organic solids. Macromol Symp 2000;160:107–14.

[2] Klimchuk EG. Organic SHS. In: Merzhanov AG, editor. Conception of SHS development as a field of scientific and technical progress. Chernogolovka: Territoriya; 2003. p. 112–8.
[3] Klimchuk EG, Avetisyan GM, Minasyan VT, Khodak AA, Gazaryan KG, Mukasyan AS, et al. Russ Chem Bull 1999;48(12):2245–58.
[4] Chechilo NM, Khvilivitskii RJ, Enikolopyan NS. Dokl Akad Nauk SSSR 1972;204:1180–1.

Orthopedic and Dental Implants

Yury R. Kolobov, Boris V. Trifonov, Evgeny V. Golosov, Mikhail B. Ivanov, Vladimir N. Sanin

In recent years, with the use of SHS for the preparation of cast materials and alloys, a technology was developed that combines in one process SHS and metallothermy, called SHS metallurgy. Technological features of this method allow obtaining polymetallic alloys with uniform distributed components in volume; however, realization of the high temperatures in the combustion process (higher than the melting point of the components) does not allow full control of the formation of the microstructure during crystallization and cooling.

This leads to a significant lack of uniformity in the structure and composition of cast workpieces, which essentially limits the use of SHS alloys in practice. An impact on the resulting materials by severe plastic deformation (SPD) in combination with traditional methods of mechanical-thermal treatment can be used as a promising method to eliminate this disadvantage.

This action makes it possible to form a bulk submicrocrystalline (grain size less than 1 μm) and nanostructured (grain size less than 100 nm) metals, alloys, and composites for technical and medical purposes. Interest in such metallic materials is due to their unique physicochemical and mechanical properties, which are substantially different from those of conventional coarse-grained materials [1]. Low-cost SPD method being actively developed is a combination of cross-helical and conventional varietal rollings.

Nonporous alloys of the specified system with a homogeneous ultrafine-grained structure were obtained by SHS-metallurgy and different methods of rolling on the example of the system Co-Cr-Mo-based alloys. Such alloys are widely used in orthopedics and stomatology. It was found out that the synthesized SHS-cast alloy has a heterophase structure consisting of a solid solution on a cobalt base with uniformly distributed precipitations of the particles containing high concentration of molybdenum in the matrix. Subsequent exposure to SPD allowed the formation of a homogeneous submicrocrystalline structure (grain size less than 700 nm), providing increased value in strength and plasticity to the material's bulk samples [2].

This approach is promising for further development in new technologies related to creating materials with high physical-mechanical characteristics for medical purposes. As an example of the use of SHS of materials for medical purposes, the possibility of producing materials based on Ti-Ta-Al exists. The results of this work are of interest for

Concise Encyclopedia of Self-Propagating High-Temperature Synthesis
http://dx.doi.org/10.1016/B978-0-12-804173-4.00104-6

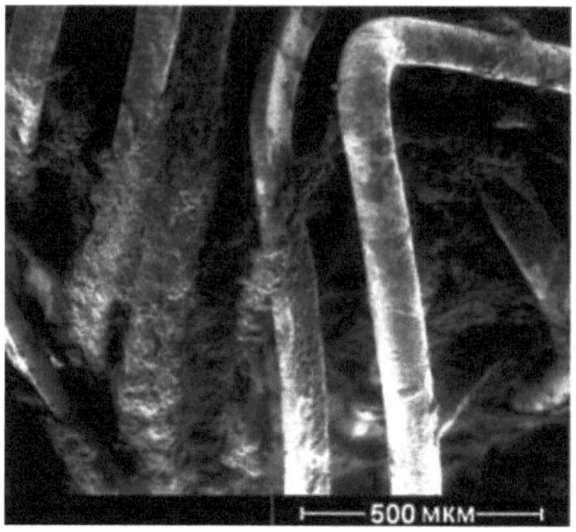

Fig. 1 The titanium fiber with an emerged bone tissue.

further research to obtain materials with predetermined structures and properties on the basis of titanium-tantalum alloys [3].

Another example of the use of such materials is the production of the final products and constructions based on biocompatible metals and alloys. Metal materials, which were obtained by SHS, can be used for the subsequent production of metallic fiber by utilizing the hanging drop method for extraction of the melt and for manufacturing composite materials. The authors used a titanium fiber, forming a volume framework and impregnated with bioactive osteoplastic composition, was manufactured composite material with high osteoconductive properties due to the presence of nanosized hydroxyapatite with the addition of a resorbable polymer binder, antiinflammatory and antiseptic preparations. Bone tissue can be easily integrated into the composite material to form a single monolithic matrix reinforced with titanium fibers (Fig. 1) [4].

REFERENCES

[1] Kolobov YuR. Nanotechnol Russ 2009;758–75. 11–12.
[2] Kolobov YuR, Bozhko SA, Sanin VN. Fundam Res 2014;24–8. 12–1.
[3] Kamynina OK, Bozhko SA, Boyarchenko OD, et al. Powder Metall Multifunctional Coat 2015;2:21–9.
[4] Trifonov BV, Nadezhdin SV, Kolobov, et al. Comp Nanostruct 2013;2:59–64.

Oxides

Karen S. Martirosyan

The rapidly growing market demand for nanoparticles calls for cost-effective and environmentally friendly technologies for their large-scale production. Recently a novel, simple, and energy-efficient synthesis of nanoscale and submicron particles of complex oxides referred as carbon combustion synthesis of oxides (CCSO) was invented [1]. In CCSO the exothermic oxidation of carbon ($C + O_2 = CO_2$, $\Delta H^{298}_{CO_2} = -393.5\,\text{kJ/mol}$) generates a thermal reaction wave with relatively low temperatures from 600°C to 1200°C that propagates at a velocity of 0.1–4 mm/s through the solid reactant mixture converting it to the desired oxide product [2–4] by the reaction (1)

$$\sum_{i=1}^{n} \mu_i X_i^{(s)} + \alpha C^{(s)} + \beta O_2^{(g)} = \sum_{j=1}^{m} P_j^{(s)} + \delta CO_2^{(g)} + (-\Delta H) \tag{1}$$

where $X_i^{(s)}$ is a solid compound (such as an oxide, a superoxide, a nitride, a carbonate, a chloride, or an oxalate) containing the metal needed to form the oxide, $P_j^{(s)}$ is a solid complex oxide product, μ_i, β,and δ are the stoichiometric coefficients, and $(-\Delta H)$ is the heat of the reaction

$$\alpha = \frac{x/12}{(100-x)/\sum \mu_i M_i^{(s)}} \tag{2}$$

where α is the exothermic coefficient, x is the carbon weight percent in the mixture, and $M_i^{(s)}$ is the molecular weight of the ith reactant.

Unlike SHS of metal carbides (TiC, ZrC, and others) or glass-carbide composites, the carbon used in the CCSO is not incorporated into the product and is emitted from the sample as carbon dioxide [5,6]. The extensive emission of CO_2 in CCSO affects the product's structure, changes the local thermal physical properties (such as heat capacity and conductivity), and causes a loss of heat to the surroundings [7]. The thermal front motion and gas flow dynamics as well as the rate of particle grow during the CCSO were predicted using numerical simulation [8]. The high rate of CO_2 released helps to form a highly porous (up to 70% porosity) and friable product having a particle size ranging from 50 to 800 nm with a specific surface area up to 10 m^2/g. Fabrication by CCSO homogeneous nanostructured complex oxides such as $BaTiO_3$, $Mn_{0.25}Zn_{0.75}Fe_2O_4$, $LaGaO_3$,

Concise Encyclopedia of Self-Propagating High-Temperature Synthesis
http://dx.doi.org/10.1016/B978-0-12-804173-4.00105-8

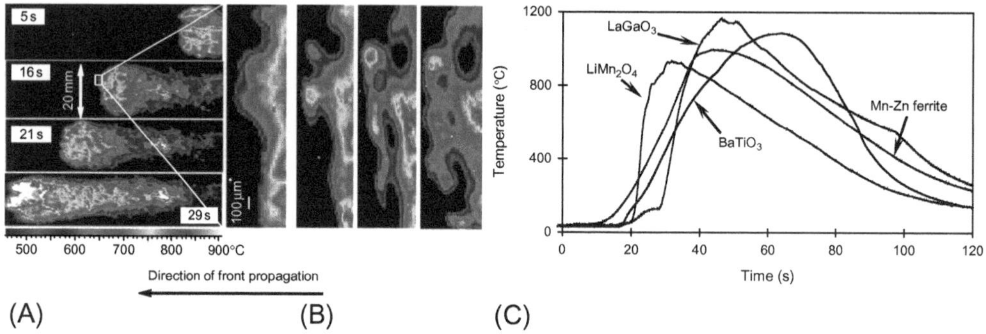

Fig. 1 IR thermal images obtained during the carbon combustion synthesis of Mn-Zn ferrite: (A) macro level and (B) micro level; the images were recorded every 16.6 ms. (C) Temperature at the center of the sample generated by carbon combustion synthesis of several complex oxides.

$CoFe_2O_4$, $LiMn_2O_4$, $LaGaO_3$, $LiMn_2O_4$, $YBa_2Cu_3O_{6.5}$, and several magnetic rare-earth ferrites [5–7] was reported.

IR thermal images in Fig. 1A and B showed that a sharp boundary did not exist between the reaction and postreaction zones [3]. The temperature evolution during carbon combustion synthesis of several oxides is shown in Fig. 1C. The maximum temperatures ranged from 900°C to 1200°C, well below the melting or sublimation temperature of both the reactants and the final products [7]. Increasing the graphite particles' size from 45 nm to 100 μm decreased the maximum reaction temperature. A complete conversion to a complex oxide structure was obtained only for the carbon concentration exceeding 9wt.%. Solid-state interactions between the components and the nucleation of the crystalline particles started in the early period of combustion and continued in the postcombustion zone.

The effluent gas composition during the carbon combustion synthesis of barium titanate, lithium manganese, and lanthanum gallium oxide was determined by a three-channel gas micro chromatograph analyzer. The data revealed that the effluent gas contained only oxygen and carbon dioxide. No carbon monoxide CO was detected in the effluents during the carbon combustion synthesis of complex oxides.

The CCSO process has the potential to enable a more economic and energy-efficient production of nanostructured perovskite materials, which have wide applications in materials science. This method does not require the use of expensive precursors and equipment. The process can be easily scaled up for a continuous production of complex oxides.

REFERENCES

[1] Martirosyan KS, Luss D. US Patent 7,897,135; 2011.
[2] Martirosyan KS, Luss D. AIChE J 2005;51(10):2801–10.
[3] Martirosyan KS, Luss D. Ind Eng Chem Res 2007;46:1492–9.

[4] Martirosyan KS, Iliev M, Luss D. Int J SHS 2007;16:36–45.
[5] Martirosyan KS, Chang L, Rantschler J, Khizroev S, Luss D, Litvinov D. IEEE Trans Magn 2007;43(6):3118–20.
[6] Martirosyan KS, Galstyan E, Xue YY, Luss D. Super Sci Tech 2008;21:065008.
[7] Martirosyan KS, Luss D. Chem Eng Tech 2009;32(9):1376–83.
[8] Markov A, Filimonov I, Martirosyan KS. J Comp Phys 2012;231:6714–24.

Perovskites

Tatiana V. Barinova, Inna P. Borovinskaya

Perovskite $CaTiO_3$ is a constituent component of the multiphase ceramics used for immobilization of high-level radioactive wastes. Perovskite is capable of isomorphic substitution for strontium, trivalent actinoides, rare earths, and sodium [1].

To date, the method of SHS compaction has been applied [2–4] to synthesize the following perovskite-containing ceramics: (a) $CaTiO_3$-ulvöspinel Fe_2TiO_4-rutile TiO_2-Fe, (b) $CaTiO_3$-zirconolite $CaZrTi_2O_7$-Fe, and (c) $CaTiO_3$-zirconolite $CaZrTi_2O_7$-pollucite $CsAlSi_2O_6$-Fe. The ceramics were tested in experiments with solid imitators of HLWs in the form of calcinates and oxides added in amounts of up to 10–25 wt.%.

Doped perovskite is a modified natural mineral $CaTiO_3$ in which the Ca sites are partially replaced by strontium, trivalent actinoides, or rare earths. Doped perovskite may also contain Tc^{4+}at the Ti^{4+}sites.

The chemical and physical properties of perovskite oxides vary with composition from dielectric to superconductivity, and they find application as sensors, ferroelectrics, and catalysts. To date, various perovskites have been fabricated by the SHS method [5] and also by solution-combustion synthesis [6], for example:

$$\begin{aligned} &Ca(NO_3)_2(aq) + TiO(NO_3)_2(aq) + C_4H_{16}N_6O_2(aq) \\ &\rightarrow CaTiO_3(s) + CO_2(g) + H_2O(g) + N_2(g) \end{aligned}$$

SCS-produced powders had the following characteristics: powder density 3.46 g cm^{-3}, specific surface 21 m^2 g^{-1}, and particle size 80 nm.

Pure nanostructured and submicron particles of $BaTiO_3$, $SrTiO_3$, and $LiNbO_3$ with a perovskite structure were prepared [7] by using the technique of carbon combustion synthesis of oxides in which carbon was used as a fuel.

REFERENCES

[1] Ringwood AE. Disposal of high level nuclear wastes: a geological perspective. Miner Mag 1985;49:159–76.

[2] Glagovskii EM, Kuprin AV, Pelevin AP, et al. Development of immobilization process of long-life radionuclides to mineral-like matrix systems by self-propagating high-temperature synthesis. Vopr Atom Nauki Tekh Ser Tekh Prikl Fiz 1999;(2):54–7.

Concise Encyclopedia of Self-Propagating High-Temperature Synthesis
http://dx.doi.org/10.1016/B978-0-12-804173-4.00106-X

[3] Barinova TV, Borovinskaya IP, Ratnikov VI, Ignat'eva TI. SHS of mineral-like ceramics for immobilization of radioactive nuclear wastes. Int J SHS 2007;16(2):92–5.
[4] Barinova TV, Borovinskaya IP, Ratnikov VI, Ignat'eva TI. SHS immobilization of cesium in mineral-like matrices. Int J SHS 2007;16(3):133–6.
[5] Boldyrev VV, Aleksandrov VV, Novikov NP, Smirnov VI. On combustion of oxide mixtures. Dokl Acad Nauk SSSR 1977;233(2):395–7.
[6] Patil KC, Hegde MS, Rattan T, Aruna ST. Chemistry of nanocrystalline oxide materials. Combustion synthesis: properties and applications. Singapore: World Scientific Publishing Co. Pte. Ltd.; 2008
[7] Martirosyan KS, Iliev M, Luss D. Carbon combustion synthesis of nanostructured perovskites. Int J SHS 2007;16(1):36–45.

Phase Separation in SHS-Casting

Vladimir I. Yukhvid

A distinctive feature of thermite (or metallothermic) systems is an extremely high combustion temperature. As a result, the main products appear in the combustion wave in the form of melts, often immiscible. The final solid product structure forms by crystallization from melt. An important role is played here by the processes of macroscopic phase separation under the action of buoyancy forces, that is, because of the buoyancy of lighter droplets of another melt and solid particles and the descent of heavier phases. For example, the metal melt descends down, while the melt oxide rises upward in the field of gravity. Three main stages can be distinguished in the processes, as shown in Fig. 1.

At the first stage, the combustion wave propagates across the mixture. Chemical conversion of the initial mixture into the two-phase product (1 and 2) melt occurs in the combustion front. Behind the combustion front is a medium of the oxide product with metal phase drops distributed in it. During the second stage, the metal and oxide phases are separated because of gravitation. Layers of immiscible melts appear. The third stage involves cooling of the product. As a result, the final composition and structure of the metal and oxide layers are formed.

The process of the metal phase accumulation can be described by these equations [1]:

Equation of the metal phase accumulation in the ingot $M = mNSL\dfrac{\rho_1}{\rho_1 - mN}$; distance that can be covered by a drop in the melt $L = \displaystyle\int_0^{t_k} v\,dt$; equation of a metal drop motion in the oxide melt $m\dfrac{dv}{dt} = P_1 - P_2 - F$; equation of two-phase melt cooling $-\rho c\dfrac{dT}{dt} = \dfrac{S}{V}\alpha(T - T_0)$; starting conditions $t=0$: $T = T_c$, $v=0$.

M is the ingot mass at the moment when phase separation is over ($t = t_k$); t is the current time; m and v are drop mass and velocity; N is drop concentration in the volume unit; mN is metal phase suspension density; $S = 2\pi r^2 + 2\pi r L_0$ and $V = \pi r^2 L_0$ are the total surface area and volume of the melt; P_1 is a drop weight; P_2 is buoyant force; F is drop motion resistance force (Stokes force) equal to $3\Pi d_r \mu v$, where $\mu = \mu_0 \exp(E_B/RT)$ is the oxide melt viscosity; E_B is energy of activation of viscous flow; ρ and c are density and thermal capacity of the two-phase melt; ρ_1 is density of the metal

Concise Encyclopedia of Self-Propagating High-Temperature Synthesis
http://dx.doi.org/10.1016/B978-0-12-804173-4.00107-1

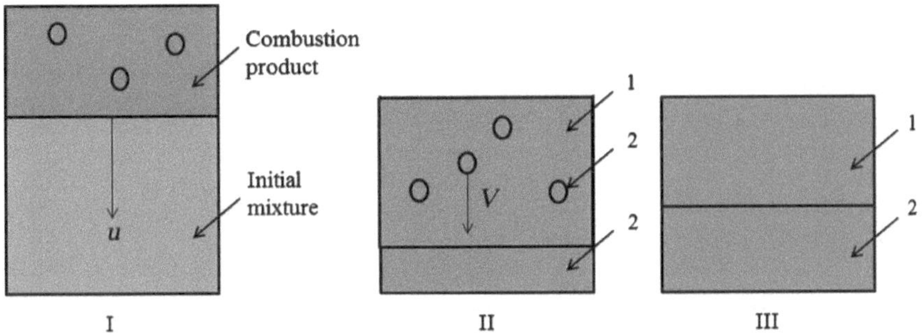

Fig. 1 Combustion of the thermite mixture and phases separation. (1) light molten product (e.g., oxide); (2) heavy molten product (metal, carbide, etc.).

phase (the target product); ρ_2 is density of the oxide phase melt; T_c is the melt initial temperature (combustion temperature); T, T_0 are the melt current temperature and the vessel wall temperature, correspondingly; and α is coefficient of heat transfer from the melt to the vessel walls.

Therefore the completeness degree of phase separation η_{ph} as a function of the process parameters can be represented as follows:

$$\eta_{ph} = A[gr/\alpha\mu_0(r + L_0)]\exp(-E/RT_c)\,[1 - \exp - E(T_c - /RT_c^2)]$$

The combustion temperature T_c appears to be the strongest parameter that allows regulating the completeness degree of phase separation. The other parameters influencing the completeness degree are centrifugal acceleration, coefficient of heat transfer, the system dimensions (r and L_0), and preexponential coefficient of viscosity (it can be decreased by flus introduction). It can be concluded that essential overheating of the melt (300–400°C higher than its melting point, T_*) is required for complete phase separation under the terms of intensive cooling.

Simplified theoretical models of the phase separation were developed in the works [2–4]. A more complete mathematical modeling of combustion in aluminothermic systems with an account of gravity-assisted impregnation was reported [5,6]. Taking into account the role of the gravity in phases separation, it might be assumed that in microgravity conditions, when the gravity-driven separation does not occur, a synthesized product is to be a composite material with oxide particles uniformly distributed all over a metal matrix. However, experiments in microgravity [7,8] have shown that phase separation occurs even in the absence of gravity. The explanation is an alternative mechanism of phase separation under the influence of capillary forces.

Artificial increases of gravity forces by means of centrifugal machines, promotes the process of separation of phases. This approach is widely used in centrifugal SHS-casting [9].

REFERENCES

[1] Lyakishev N.P., Pliner Yu.L., Ignatenko G.F., Lappo S.I. Aluminotermiya, Metallurgiya. Moscow (in Russian); 1978.
[2] Yukhvid VI. Conditions for phase separation in metallothermic processes. Russ Metall 1980;(6):57–61.
[3] Yukhvid VI. Dynamic interaction of high-temperature multi-phase melt with the metal matrix. Russ Metall 1988;(6):122–7.
[4] Bulaev AM. Self-acceleration of the process of phase separation in burning metallothermic mixtures. Combust Explos Shock Waves 1992;28(4):395–9.
[5] Shkadinsky KG, Shkadinskaya GV, Matkowsky BJ. Gravity linduced separation in "liquid flames" Combust Sci Technol 1996;115:229–58.
[6] Raymond CS, Shkadinsky KG, Volpert VA. Gravitational effect on liquid flame thermite systems. Combust Sci Technol 1998;131:107–29.
[7] Lau C, Mukasyan AS, Varma A. Reaction and phase separation mechanisms during synthesis of alloys by thermite type combustion reactions. J Mater Res 2003;18(1):121–9.
[8] Mukasyan AS, Lau C, Varma A. Influence of gravity on combustion synthesis of advanced materials. AIAA J 2005;43(2):225–46.
[9] Yukhvid VI. Tsvetnaya Metallurgiya 2006;5:62–78 (Russia).

Phase Transitions During Combustion of Alloys in Nitrogen

Yury M. Maksimov, Boris Sh. Braverman

It is known that phase transitions can accelerate heterogeneous reactions in solids [1,2]. Acceleration of the oxidation reaction was found for the Ti-Zr and Ti-Hf alloys because of the polymorphous β-α-phase transition under isothermal conditions at high temperatures [3].

The influence of phase transition on the heterogeneous reactions proceeding under substantially nonisothermal conditions was first found by studying the combustion of the Fe-V alloy in nitrogen [4]. A peculiarity of the Fe-V system was the presence of the σ-phase with a homogeneity area of 45–65 wt.% Fe (Fig. 1) formed because of the ordering of the α-solid solution.

Ferrovanadium alloys were investigated for cases when the iron content was 30 wt.% (α-FeV) and 50 wt.% (σ-FeV). The alloys were prepared in a vacuum oven, milled to the powder dispersity <50 μm and compressed into tablets. The tablets were burned in a nitrogen atmosphere.

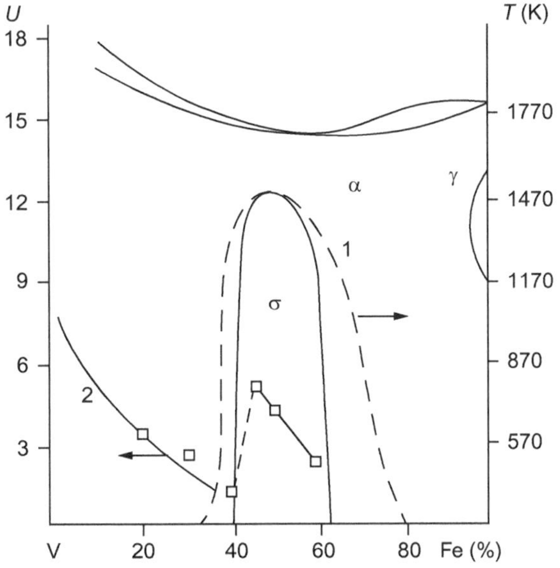

Fig. 1 Phase diagram of Fe-V (1) and the combustion rate (2) versus the alloy composition.

Concise Encyclopedia of Self-Propagating High-Temperature Synthesis
http://dx.doi.org/10.1016/B978-0-12-804173-4.00077-6

The combustion rate of the σ-FeV alloy was found to be 1.5–2 times higher than the combustion rate of α-Fe at all pressures under study. A significant aftercombustion of α-Fe took place after the propagation of the reaction front leading to the formation of a multiphase product consisting of α-Fe and the lower vanadium nitrides $\beta(V_3N)$, $\beta(V_9N)$, and σVN. There was no aftercombustion for the σ-FeV alloy, only vanadium mononitride and free iron could be found in the combustion products.

Analysis of the combustion products and qualitative experimentation conducted under isothermal conditions showed that an increase in the combustion rate of the σ-FeV alloy was caused by the phase transformation $\sigma \rightarrow \alpha$. Fig. 1 demonstrates the combustion rate of the alloy versus the iron on the background of the phase diagram for the Fe-V system (at a nitrogen pressure of 60 atm). It is seen that an increase in the iron content up to 40% leads to a monotonous decrease in the rate. However, a further increase in the concentration of iron in the alloy leads to the sharp increase in the combustion rate (increase in the content of Fe by 5% (from 40 to 45 wt.%) results in the increase in the combustion rate by 3 times).

It should be noted that the phase transition significantly influences the formation of the final product structure. After completion of the transition $\sigma \rightarrow \alpha$, iron was melted, and the melt dispersed the nitride base of the sample during cooling of the dense composite material. In contrast to the σ-alloy, the nitriding of α-FeV was not explosive, but was gradual. There was a wide region of aftercombustion. After combustion, the α-FeV samples represented a loose and poorly sintered material. It should be noted that the effect of activating the combustion process in the presence of the phase transition was also found during nitriding of vanadium-nickel alloys.

In this system, an alloy with an equiatomic composition was also α-phase and formed the two-phase well-sintered products during the interaction with nitrogen in the combustion mode. This effect can be observed during the combustion of a Ti-Cr alloy in nitrogen (Fig. 2) for a Cr content of 66% in the alloy, which corresponds to the area of homogeneity for the $TiCr_2$ compound [5]. In this case, the combustion product consisted of the TiN and Cr phases. The combustion products of the powder mixture with the same content of the reagents were multiphase and consisted of titanium nitride, chromium, chromium seminitride, and chromium nitride during combustion. Analysis of the phase diagram for the Ti-Cr system [6,7] and the combustion rate of the alloy in nitrogen indicated that the area of the sharp change in the combustion rate of the alloy containing 66% Cr corresponded to the area of homogeneity for the λ-phase.

The maximum combustion temperature of the alloy was higher than the transition temperature $\lambda \rightarrow \beta$, so the probable reason for the increase in the combustion rate was the phase transformation of the intermetallic λ-phase into the β-solid solution at $T = 1350°C$.

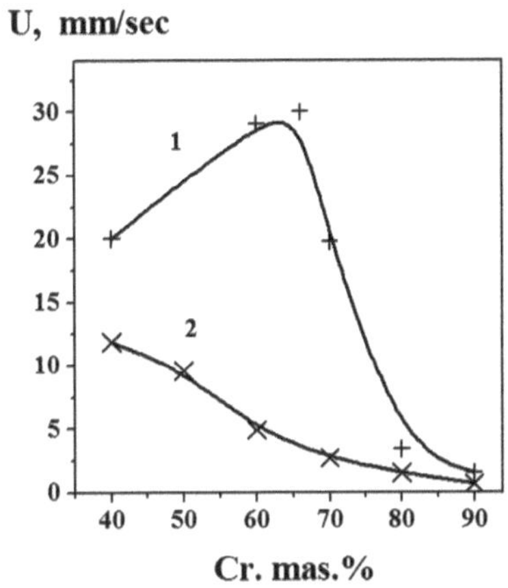

Fig. 2 Influence of the composition on the combustion rate in nitrogen for the alloys (1) and mixtures (2) of Ti and Cr.

REFERENCES

[1] Hedvall JA, Rzen UZ. Z Anorg Allg Chem 1936;229:416.
[2] Simon A, Klugel E. Z Anorg Allg Chem 1952;220:76.
[3] Voytovich RF, Golovko EP, D'yakonova LV. Prot Metals 1976;XII(5):590.
[4] Maksimov YuM, Merzhanov AG, Raskolenko LG, et al. Proc USSR Acad Sci 1982;264:629–32.
[5] Maksimov YuM, Braverman BSh, Raskolenko LG. Combust Explos Shock Waves 1984;20(4):28–31.
[6] Svechnikov VM, Teslyuk TYu, Kocherzhevsky YuO, et al. Proc USSR Acad Sci 1970;9(A):837.
[7] Minaeva SA, Budberg PB. Drits ME, editor. Phase equilibria in metal alloys. Moscow: Nauka; 1981.

Pigments

Galina G. Xanthopoulou

SHS offers significant potential for application to the preparation of many pigments with new compositions, structures, and properties. The primary draw of SHS for production of pigments is the complete or partial elimination of an external energy supply by utilization of the internal heat released in chemical reactions. The rapid autowave combustion-like, self-sustaining reactions that occur during SHS yield products with desired compositions and structure. Varying the rates of heat release and transfer provides control over the process rate, temperature, degree of conversion, composition, and structure of products. For pigment production this means that during SHS it is easy to dope ions of metals in spinel structure, to ions' replacement with ions of other diameters. The size of the metal ion substituted into the crystal lattice, especially spinel-type crystals, is related to the wavelength of reflected light (i.e., the color). Such correlations allow for prediction of color in new SHS compounds. For example, in the Co-Me-O spinel, as the substituted metal ion radius increases from 0.55 to 0.85 A, the color wavelength decreases from 750 nm (violet) to 450 nm (red).

As a result of the specific characteristics of the SHS method, there are many obvious advantages for the manufacture of pigments, as compared to traditional "furnace" methods: synthesis takes place within only a few minutes instead of hours, and the preheating temperature, T_0, never surpasses 1070 K, although final combustion temperatures can be much higher. SHS synthesis of pigments and industrial production was first developed in 1985 at the Institute of Combustion Problems in Almaty, Kazakhstan. Since then a broad range of pigments in almost all colors and hues have been produced. Their low production cost (mainly because of the very low thermal treatment costs) and high thermal stability makes these pigments very competitive in industrial processes such as ceramics manufacturing, and they are widely used in many former Soviet Union countries.

SHS pigments usually are prepared from initial batch mixtures consisting of various salts, metals, and oxides compacted under a pressure of 5–10 MPa in the form of cylinders or other shapes. The samples are preheated in an electric furnace at temperatures of only 770–1070 K for a few minutes prior to initiation of SHS. After initiation, temperatures can reach up to 3000 K (depending on the composition of the system) for up to a few minutes. Final pigment properties depend on the properties of the

Concise Encyclopedia of Self-Propagating High-Temperature Synthesis
http://dx.doi.org/10.1016/B978-0-12-804173-4.00108-3

starting raw materials, concentration, and conditions during SHS (i.e., grain size, combustion temperature, and cooling conditions). Following combustion, the compacts are crushed, ground, and milled similar to furnace-manufactured pigments.

Practically all examined parameters of SHS pigments were found to meet industrial specifications. The melting temperature of SHS pigments is typically in excess of 1670 K, which allows use as glazes for ceramic tiles and porcelain. SHS pigments are also used in paints (water and oil-based), epoxide resins, plastics, ceramic tiles, and decoratively for ceramic artifacts from alumina, porcelain, faience, etc. SHS pigments pass full animal studies for application as cosmetics. The high-temperature interaction between metastable SHS pigments (those that have been cooled at very high rates and did not have time to fully stabilize) and glazes on ceramic tiles can result in exceptionally beautiful designs. A number of SHS pigments produced in glazes are capable of imparting color to marble and pearls and enhancing the reflectivity of glazes and enamel. In addition, they can be used for the production of gemstones by the addition of a single pigment without the need for more complicated processing. On the other hand, fully stable and well-formed crystalline structures result from slow cooling or prolonged heating at temperatures that allow full interaction between the components. This is essential for their use in the production of porcelain enamel, varnishes, emulsions, etc.

Another relevant combustion synthesis method, solution combustion synthesis (SCS), was developed in India. SCS pigments were synthesized by the combustion of aqueous solutions containing stoichiometric amounts of the corresponding metal nitrate and reducer, such as glycine and urea. This method appears to be well suited for the direct preparation of nano-scale spinel pigments. During SCS, pigments are produced in time spans of a few minutes at temperatures ranging from 570 to 870 K resulting in nano-scale particles. This eliminates the need to have a powder-crushing stage, a great advantage in comparison with traditional methods and SHS. However, one disadvantage is the relative difficulty of producing intensely colored spinel pigments. This has been attributed to the lower combustion temperature and the lack of sufficient energy release to produce stoichiometric spinels.

Powders Production by SHS

Valentina K. Prokudina

Powders production was one of the pioneering technological branches of SHS. Its development first started in 1970s. Initially, this technology was developed for production of Ti-based refractory compounds. At first, only elemental powders were used as raw materials for SHS. As the investigations broadened, other raw materials were used, including oxide ones. Depending on the type of raw materials used and techniques that were used to organize the combustion, different types of SHS powder technology developed. These include synthesis from elements (SHS-E), magnesium-thermal synthesis (SHS-M), and metallurgical or casting synthesis (SHS-C). SHS powder technology is based on combustion of the reaction mixture in an inert (in the case of gasless combustion) or active (filtration combustion with gas-solid interaction) gas environment. The powder production process is similar to all mentioned types of SHS and includes basic operations: preparation for synthesis, followed by synthesis and consequential treatment of the synthesized products. Functional additives are designed to regulate the temperature and speed of combustion, gas emission, and so on and are used in all types of SHS powder technologies, as well as reactive and inert gases. In the preparation for synthesis, bathing and mixing of raw materials is necessary for all technologies, whereas mixture drying is unnecessary in the case of oxide raw materials. Postsynthesis treatment depends on the type of synthesis.

In the case of SHS-E, the product only goes through mechanical treatment: grinding, crushing, milling, and separation. Coproducts in SHS-E do not form. Waste after the grinding can be used as a diluting agent. In the case of SHS-M, the synthesized product goes through milling and acidic separation. Separation is necessary to remove magnesia from the product. After separation, the product is dried. In the case of SHS-C, the synthesized product is a two-layer casting, with a top layer of alumina and a bottom layer of the desired product. The layers can be easily separated. Castings undergo crushing and milling until the desired particle size is obtained.

During supporting operations (preparation for synthesis and postsynthesis treatment), standard powder metallurgy equipment is used: crushers, ball mill and vibration mill, sieves, screens, and so on.

The primary operation in SHS powder technology is synthesis. Reactors for synthesis must satisfy the extreme parameters of SHS and ensure safety from exploitation. This is

Concise Encyclopedia of Self-Propagating High-Temperature Synthesis
http://dx.doi.org/10.1016/B978-0-12-804173-4.00110-1

especially important for closed-type reactors where it is necessary to prevent the possibility of emergency depressurization with consequent ejection of the powder mixture into the air and its explosion because of oxidation of mixture. These requirements are met by universal (multipurpose) SHS reactors, which work in periodic regime with a productive capacity of dozens of kilograms per hour. Up to six reactors are united in a single installation with a common remote control [1]. Reactors are made from austenite steel, which contains 12% chromium, 18% nickel, and 10% titanium. These steels are characterized by a yield strength of at least 2 MPa and a tensile strength of at least 5 MPa. These steels are not prone to carburization, nitriding, and hydrogenation, even when exposed to high temperatures of synthesis (up to 4000°C) in an environment that contains active agents like carbon, nitrogen, hydrogen, and so on. Because of their high ductility, these steels sustain cyclic pressure loadings and do not fragmentize in case of rupture.

SHS reactors easily allow variations according to the type of product being manufactured. You just change the raw materials and adjust the technological parameters of the process. That is why they are called "multipurpose"/universal.

A line of reactor types were designed in ISMAN. They differ by constructive elements depending on the purpose and parameters of synthesis and have a swept volume up to 30 L. The designed reactors differ also by their type of catch. Catches are selected depending on operational pressure and the convenience of loading and unloading the powder mixture. Installed reactors and different types of sealing catches are shown in Fig. 1.

Special reactors [2] were designed for azide technology (SHS-Az). These reactors were developed for combustion of mixtures that consist of element powder with sodium azide and halide salt. Combustion of such multicomponent mixtures is followed by the emission of vast amounts of gaseous coproducts. These reactors differ from multipurpose SHS reactors in that they have a special device (filter installation) that was designed to fix the powder mixture and also because they use pipelines and valves that are four times larger in diameter. SHS-Az technology is developed for production of powders such as nitrogen-containing compounds like Si_3N_4, BN, $TiC_{0.5}N_{0.5}$, AlN, and Si_3N_4-SiC.

Fig. 1 Multipurpose SHS reactor with various catches: (A) mechanized bayonet catch, (B) leaf catch, and (C) fungal or screwed catch.

More productive reactors were designed in a number of organizations for commercial production of powders. Reactors with 100 L of swept volume were designed for titanium carbide production [3]. Reactors with 150 L of swept volume were designed for production of nitrided ferroalloys.

A wide variety of powders of different chemical compounds were produced by SHS technology. These include carbides, borides, nitrides, silicides, chalcogenidess, phosphides, hydrides, intermetallics, and various solid solutions and heterogeneous material. The possibility of production of elemental powders, such as titanium, boron, molybdenum, and silicon was also demonstrated.

REFERENCES

[1] Ratnikov VI, Borovinskaya IP, Prokudina VK. Izvestia Vyzov. Poroshkovaya metallyrgia I funkcionalnye pokrytia. Russ J Non-Ferr Met 2013; 1:34–41.

[2] Amosov AP, Borovinskaya IP, Merzhanov AG. Powder technology of SHS-materials. Moscow: Mashinostroenie; 2007. 567 p.

[3] Merzhanov AG, Borovinskaya IP, Prokudina VK, Pesotskaya NS, Nasonova MA. Nauka-Proizvodstvy 1998;8(10):5–12.

Precursors of Ceramic Materials

Tamara A. Khabas

The term "precursor" refers to SHS [1] and has two appreciably different meanings. First, it means a precursor for SHS; second, it means a precursor as an SHS product. The first type of precursor is produced to achieve the minimal distance between the particles of reacting substances. The production of such precursors includes the use of the sol-gel method or salts precipitations or mechanochemistry. In such cases, the particles of one phase are surrounded by another phase, which increases the depth of the combustion wave and provides the maximum homogeneity of its composition (as in the case of SiAlON [2] production). For example, preliminary synthesis of acetate and nitrate salts as precursors provides high purity of an SHS-product and the homogeneous distribution of alloying elements in the product (for instance, yttrium oxide [3], manganite $La_{0.7}Ca_{0.3}MnO_3$ [4]).

The second type of precursor is the incomplete reaction products of SHS, since it is not always possible to immediately (in one phase) obtain a product of the required chemical and phase composition. For instance, such precursors are obtained in cases of low exothermicity involving reaction and process conditions, which makes one of the components of the mixture a chemical stove. This component forms its own reaction product with the gas medium and other components and creates the high temperature for synthesis of other components in the mixture. Such a mixture requires further thermal treatment, leading to a change of its chemical composition and stabilization of phases so that it serves as a precursor of a ceramic material. These processes include SHS types such as the combustion of mixtures containing aluminum nanopowder (nAl). During burning, nAl has a prominent characteristic of not only reacting with air oxygen but also chemically fixing nitrogen. The combustion of aluminum oxide mixtures with nAl in air includes two parallel, and to some extent interconnected, processes. The first one is an energy-liberating process, the interaction of aluminum with components of air. This process results in the formation of two main products, aluminum oxide and nitride (during the combustion of pure nAl, almost in equal fractions). The second process conjugated with the first one and can be solely realized due to the heat liberation by the chemical stove. The products of this process are cordierite, mullite, spinel [5–7], and other complex compounds. By other metals burning in air, ceramic powders were produced containing TiN, ZrN, HfN, and Ni-Zn-ferrites. Such polyphasic products after

Concise Encyclopedia of Self-Propagating High-Temperature Synthesis
http://dx.doi.org/10.1016/B978-0-12-804173-4.00130-7

additional stabilizing of the thermal treatment (sintering) can be used as precursors and components for producing ceramic articles [8,9]. The ratio of the synthesized phases changes in a nonlinear way depending on the content of aluminum in combustible mixtures. The stable formation of aluminum nitride is ensured when the content of nAl in the mixture is more than 50 wt.%, since such conditions guarantee high-temperature, two-stage combustion. The lowest point providing the combustion of metal-mineral (quartz, alumina, and others) mixtures is determined by a metal content of about 30 wt.%. During combustion of the mixture, AlN fibrous crystals and large amount of synthesized compound nucleation sites are formed. To produce precursors that contain phases of complex compounds, such as cordierite, mullite, and spinel, mixtures are used that contain natural minerals (kaolinite, talc, serpentinite), oxides (MgO, Al_2O_3), and metal powders (Al, Mg).

During the combustion stage, along with the formation of aluminum nitride and oxide, the following reactions are possible: kaolinite dehydration and synthesis of primary mullite ($3Al_2O_3{\cdot}2SiO_2$), talc dehydration and cordierite synthesis ($2MgO{\cdot}2Al_2O_3{\cdot}5SiO_2$), and silicon reduction $3SiO_2 + 4Al \rightarrow 2Al_2O_3 + 3Si$.

The mixtures of magnesium oxide and aluminum nanopowder during combustion form a three-component mixture (AlN, Al_2O_3, $MgAl_2O_4$). After the combustion products are sintered in nitrogen by hot pressing the obtained high-strength ceramics was reinforced by nitride fibers. Probably, the reinforcement is connected with the particle morphology of the precursor produced by SHS (Fig. 1), which is presented by elongated strings of AlN covered by small isometric particles of spinel.

Precursors produced by metal-oxide mixture combustion can also serve as solid-phase synthesis activators of complex oxide compounds (cordierite, mullite, spinels). Their small amounts in mineral mixtures result in accelerated reactions and a reduced synthesis temperature, which is conditioned by the presence in the precursor of large amount of new phase grains. Moreover, AlN and Al_2O_3 formed during such SHS and nAl

(A) (B)

Fig. 1 SEM images of precursors produced after combustion of mixtures: (A) nAl-MgO and (B) mineral-metal mixture for cordierite synthesis in air.

actively react in air with mixture components at synthesis temperature. The efficacy of SHS precursor additions, which activate the solid-phase synthesis by sintering, is especially high when introduced into inert mixtures based on pure oxides.

REFERENCES

[1] Merzhanov AG. J Mater Chem 2004;14:1779–86.
[2] Pradeilles N, et al. Ceram Int 2008;34:1189–94.
[3] Balabanov SS, Bykov YuV, Egorov SV, et al. Opt Mater 2013;35:727–30.
[4] Mraković A, et al. Ceram Int 2013;39:3771–7.
[5] Khabas TA, Mel'nikov AG, Il'in AP. Ogneupory i Tekhnicheskaya Keramika 2003;11:14–9.
[6] Khabas TA. Ogneupory i Tekhnicheskaya Keramika 2004;12:5–14.
[7] Esharghawi A, Penot C, Nardou F. Ceram Int 2010;36:231–9.
[8] Gromov A, Chukhlomina L, editors. Nitride ceramics – combustion synthesis, properties, and applications. Wiley VCH; 2014. p. 360.
[9] Fu CM, Syue MR, Wei FJ, et al. J Appl Phys 2010;107:1–3. 09A519.

Principles of Nonunique Structures

Inna P. Borovinskaya

The phenomenon of "nonuniqueness of structure and properties" of composite SHS materials of the same chemical and phase composition was discovered during the synthesis of composite ceramics based on boron nitride with the addition of refractory compounds, for example, titanium diboride [1,2]. Later this effect was observed for ceramic composites based on aluminum nitride, titanium nitride, and so on. The effect of nonunique structure implies that depending on the different terms of combustion synthesis (initial raw materials, SHS parameters, gas pressure), various microstructures are formed including pseudo-homogeneous structures with nitride matrixes that contain particles of refractory compounds of various morphology (uniaxial, needle-like, etc.) or "bridges" of refractory compounds, with intermetallic matrixes and chains of nitride nanoparticles or intermetallic matrixes and nitride spherical particles. At the same chemical and phase composition but depending on microstructure, particle morphology, shape, and distribution in the matrix, the material can have different properties, sometimes of an opposite character. For example, a BN-TiB_2 compound obtained at combustion of the components' mixture in gaseous nitrogen changes its electrophysical properties from dielectric to electroconducting depending on the composition of the raw material (Fig. 1A and B).

Dielectric ceramics BN-TiB_2 with BN matrix and electrical resistance $\Omega > 8 \times 10^{14}\ \Omega$ cm is formed by the reaction $B_{(s)} + TiB_{2(s)} + N_{2(g)}$, but when elemental titanium and BN additives are used as raw materials, the obtained ceramics have the same composition with BN matrix but have electroconducting properties ($\rho = 4 \times 10^{-4}\ \Omega$ cm, which is close to $\rho_{TiB_2} = 10^{-5}\Omega$cm). A change in particle morphology of aluminum nitride (from equiaxial to needle-like ones) at synthesis of AlN-TiB_2 ceramics led to the change of the electrophysical properties of the composite from dielectric to electroconducting ones at the same AlN matrix. This is the most interesting effect of a nonunique structure, considering the fact that in 65% of equiaxial particles, the ceramics have dielectric properties, and when only 15% of needle-like particles are formed, the ceramics becomes electroconducting. This result agrees with the theoretical investigation by Dovzhenko and Zhirkov on "the effect of particle size distribution on the formation of percolation clusters" [3]. Not only electrophysical but also chemical and corrosion characteristics of a material can change in connection with the effect of nonuniqueness. The composite material based on titanium nitride and intermetallic Ni_3Ti can demonstrate

Concise Encyclopedia of Self-Propagating High-Temperature Synthesis
http://dx.doi.org/10.1016/B978-0-12-804173-4.00111-3

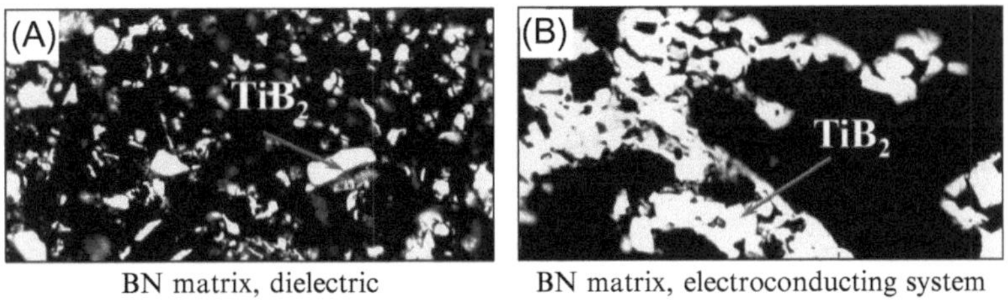

Fig. 1 Ceramics microstructure: (A) pseudo-homogeneous structure and (B) framework structure.

high heat resistance at high-temperature oxidation of the material with intermetallic matrix and spherical particles of titanium nitride in the air. The material obtained by the reaction $Ti + Ni + N_2 \rightarrow Ni_3Ti + TiN$ has high corrosion resistance. In the case when titanium nitride is formed as nanoparticle chains disposed in the intermetallic matrix at synthesis by the reaction $TiNi + N_2 \rightarrow Ni_3Ti + TiN$, corrosion characteristics get worse.

Investigation of the phenomenon of the nonunique structure and properties of SHS materials provides great possibility for synthesizing materials and substances with regulated and preset characteristics.

Previously, the effect of nonuniqueness in the behavior of hydration reaction of transition metals (IV) with nonmetals, for example, carbon [4], was found to be connected with the temperature of the inflammation composition:

$$Zr + {}_xC + H_2 \nearrow ZrC_x + H_2 \rightarrow ZrC_xH_{1-x} + \frac{1-x}{2}H_2 \text{ (high-temperature inflammation)}$$

$$Zr + {}_xC + H_2 \searrow ZrH_2 + {}_xC \text{ (low-temperature inflammation)}$$

This effect is explained by the nonunique character of stationary modes of flame propagation. The theoretical description of the phenomenon of nonunique behavior of SHS reactions depending on the inflammation temperature was given by Khaikin [5].

REFERENCES

[1] Borovinskaya IP. Structure formation in multicomponent SHS-ceramics. In: Proc. 3rd International symposium on SHS, Wuchan (China); 1995. p. 5–6.
[2] Borovinskaya IP, Bunin VA, Vishnyakova GA, Karpov AV. Some specific features of synthesis and characteristics of SHS (TiB_2-AlN-BN)-based ceramic. Int J SHS 1999;8(4):451–7.
[3] Merzhanov AG, Borovinskaya IP. Historical retrospective of SHS: an autoreview. Int J SHS 2008;17 (4):242–65.
[4] Martirosyan NA, Dolukhanyan SK, Merzhanov AG. Non-uniqueness of stationary modes at combustion of zirconium-black carbon mixtures in hydrogen. Fiz Goren Vzryva 1983;5:39–42.
[5] Khaikin BI, Khudyaev SI. Combustion zone propagation in the systems forming condensed reaction products. In: Stesik LN, editor. Gorenie i vzryv. Moscow: Nauka; 1977. p. 121–37.

Products of SHS

Inna P. Borovinskaya

Because of the specific production conditions, SHS products differ from their furnace and plasma-chemical analogs both in structure and purity. Three types of SHS powders are widely used now [1]. They are agglomerated, composite, and single-crystal ones. Agglomerated powders are particles consisting of tightly attached crystals; they can have pores. The particle size in agglomerated SHS powders usually ranges from 10 to 200 μm. SHS products are characterized by high purity, which can be explained by the initial reagent purity as well as by the absence of contamination by the container materials and self-purification process from impurities because of high combustion temperatures. Practically all known refractory compounds (metal and nonmetal carbides, borides, nitrides, oxides, chalcogenides) and intermetallics can be obtained as agglomerates that are further milled and used as abrasive materials of high quality. Milled SHS powders can be applied in various fields of engineering for sintering, coating application, and so on. As a rule, their operational characteristics are very high (high strength, low production expenses, good ecological safety, low cost, etc.). Composite SHS powders are usually mixed-phase and include various substances. Mineral-ceramic (a refractory compound + alumina or magnesia), cermet (a refractory compound + metal), and heterogeneous mixtures of two or more refractory compounds are well known all over the world. A specific feature of SHS composite powders is the high degree at which their components intermix.

Single-crystal SHS powders consist entirely of separate perfect single crystals and are formed as a result of complete milling of a polycrystalline combustion product. The particle size of such powders usually varies from 0.5 to 3.0 μm. When recrystallization processes are suppressed in the combustion products, it is possible to obtain single-crystal powders with a particle size less than 0.1 μm. The main methods for producing SHS nanosize powders are combustion of gases and air-dispersed mixtures resulting in condensed particles [2].

$$R_1(\text{gas}) + R_2(\text{gas}) \rightarrow P_1(\text{solid phase}) + P_2(\text{gas})$$

$$R_1(\text{solid phase}) + R_2(\text{gas}) \rightarrow P_1(\text{solid phase}) + P_2(\text{gas})$$

Cast SHS powders are obtained by mechanical grinding and dispersion of ingots of refractory compounds and corundum or heterogeneous systems consisting of two or more

Concise Encyclopedia of Self-Propagating High-Temperature Synthesis
http://dx.doi.org/10.1016/B978-0-12-804173-4.00112-5

refractory compounds. Examples of the best-known powders are Al_2O_3 doped by Cr_2O_3 (pink corundum), CrB_2, CrB_2-TiB_2, TiC-Cr_3C_2, complicated silicides, titanium and chromium carbides, and $MoSi_2$ [3]. These powders are mainly used for plasma application of coatings, electro-arc surfacing, and abrasive production.

A remarkable group of SHS products contains ceramic and hard-alloyed materials obtained by direct synthesis, that is, mechanical or physical effects on the SHS process (combustion with high pressure of reactive or inert gas, forced SHS densification, processes in a centrifuge, extrusion). High-temperature nitride SHS ceramics based on boron and aluminum and silicon nitrides and nontungsten hard alloys based on titanium carbides and borides with metal binders are well known all over the world. These SHS materials are characterized by low porosity (they can be nonporous) and are produced as items with a preset geometry. They are used in various fields of engineering, including aviation, space, automobile, chemical engineering, electronics, metallurgy, and medicine, and demonstrate high-operational characteristics [4,5].

REFERENCES

[1] Merzhanov AG. Self-propagating high-temperature synthesis and powder metallurgy: unity of goals and competition of principles. In: Capus JM, German RM, editors. Particulate materials and processes: advances in powder metallurgy. Proceedings of the 1992 powder metall. World Congress, vol. 9, San Francisco, CA, USA, 21–26 June 1992. Princeton, NJ: Metal Powder Ind. Fed. Publ.; 1992. p. 341–68.
[2] Dovis KA, Brezinsky K, Glassman I. Chemical equilibrium constants in the high-temperature formation of metallic nitrides. Combust Sci Technol 1991;77:171–8.
[3] Merzhanov AG, Yukhvid VI, Borovinskaya IP. Self-propagating high-temperature synthesis of cast refractory inorganic compounds. Dokl Akad Nauk SSSR 1980;255(1):120–4.
[4] Borovinskaya IP, Loryan VE, Zakorzhevsky VV. Combustion synthesis of nitrides for development of ceramic materials of new generation. In: Gromov AA, Chukhlomina LN, editors. Nitride ceramics: combustion synthesis, properties, and applications. Weinheim: Wiley-VCH Verlag GmbH & Co. KGaA; 2014 [chapter 1].
[5] Merzhanov AG, Borovinskaya IP, Yukhvid VI, Ratnikov VI. New combustion methods of high-temperature material production. In: Scientific backgrounds of materials science. Moscow: Nauka; 1981. p. 193–206.

Quasihomogeneous Models

Peter M. Krishenik

The classical concept of combustion waves in continuous reactive systems gave impetus to the application of the "homogeneous" approach to the analysis of wave structure in heterogeneous systems. In the case of homogeneous combustion, the medium remains homogeneous during the entire process, while mass transfer and intermixing are a necessary condition for occurrence of the reaction and heat release. The existing theoretical models describing combustion of heterogeneous media are based on the assumption that the characteristic time of chemical reaction is much longer than the typical time for heat transfer. A consequence of this assumption is that the width of the preheating and reaction zones is much greater than the particle size of the heterogeneous mixture. This allows you to consider a micro-heterogeneous mixture to act as a quasihomogeneous mixture with averaged thermophysical parameters. Such an approach formed the basis for elaborating on the theories of gasless and infiltration-mediated combustion. Quasihomogeneous models found wide application not only in the analysis of high-temperature synthesis but also in other combustion processes used in chemical engineering.

Concise Encyclopedia of Self-Propagating High-Temperature Synthesis
http://dx.doi.org/10.1016/B978-0-12-804173-4.00113-7

Quenching of the Combustion Wave

Alexander S. Mukasyan, Alexander S. Rogachev

Quenching implies a rapid cooling, or arresting, of the combustion front and is used to study mechanisms of structure formation during combustion synthesis. Following quenching, the phase composition and microstructure of the different zones in the combustion wave can be identified using layer-by-layer analysis of the quenched regions by scanning and transmission electron microscopy and X-ray diffraction.

Several methods have been developed to quench the combustion wave during synthesis (Fig. 1). One approach is quenching in liquid (A). The first attempt to terminate the combustion wave and obtain quenched products was by immersing the sample in liquid argon [1]. It was shown that in this case the cooling rates are not high (up to 250 K/s) because of a gaseous layer that formed on the surface of the sample, which acted as a thermal insulator [2]. The highest cooling rate (10^4–10^5 K/s) was reported for a relatively thin sample quenched with a high-speed water jet [3]. However, the high water velocity (up to 150 m/s) necessary to achieve high rates of quenching often destroyed the sample.

The second approach is based on the use of massive copper blocks to rapidly remove heat from the burning media [4]. Two modifications of this method have been developed. In one case, a reactant mixture is pressed into a wedge-shaped cut in a copper block (B). The sample is ignited, and the reacting mixture is progressively quenched because of a decrease of the reactant mixture volume, which increases conductive heat loss to the walls of the copper block as the combustion wave travels to the apex of the wedge. A series of photodiodes and/or thermocouples can be used to measure the temperature at several locations of the reacting pellet. The continuous acquisition of temperature during the experiment where quenching rates and times can be determined directly from temperature measurements is critical to provide an accurate interpretation of the analysis of quenched samples.

As shown by theoretical analysis and computer modeling, the combustion of a wedge-shape sample is accompanied by the concentration of excess enthalpy brought by the preheating zone [5,6]. By changing the angle of the wedge and sample density, you can ensure that the combustion may change velocity or propagate at a constant speed. The last mode is optimum for quenching experiments, since it introduces minimum distortion to the thermal structure of the combustion wave; consequently, the quenched products adequately reflect the dynamics of structure formation.

Concise Encyclopedia of Self-Propagating High-Temperature Synthesis
http://dx.doi.org/10.1016/B978-0-12-804173-4.00114-9

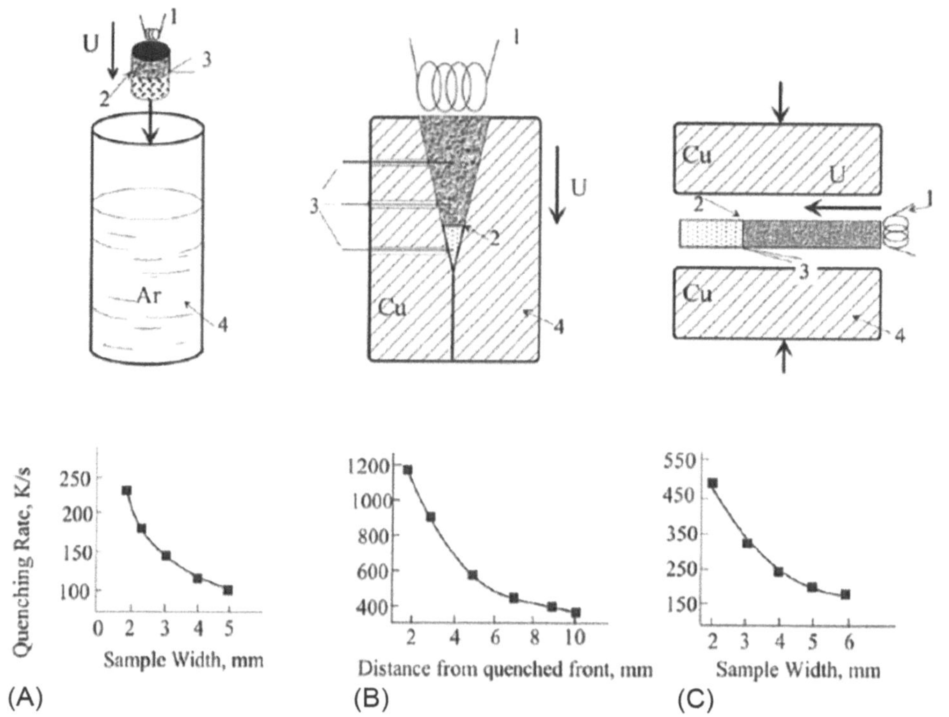

Fig. 1 Schematic drawings of quenching methods and cooling rates in (A) liquid argon, (B) a wedge-shaped cut in a copper block, and (C) between two copper plates. 1, ignitor; 2, quenched front; 3, thermocouples; 4, quenching medium. *(Adapted from Mukasyan AS, Borovinskaya IP. Int J SHS 1992;1:55).*

Another quenching method utilized in gas-solid reacting systems involves a thin sample or reactive foil and two copper plates [2]. At an appropriate instant during the progress of the combustion wave, the reacting sample is "squeezed" between the copper plates. Thus the copper plates serve not only as heat sinks but also isolate the reacting solid compact from the gaseous reactant (C). Shock waves have also been used to quench the combustion synthesis reaction in gas-solid systems [7]. In this method, the metal powder compact was placed in a steel tube, which permitted access of reactive gas (air, nitrogen) to the combustion zone. The steel tube was surrounded by an explosive charge. The sample was ignited by a tungsten coil at one end. When the combustion wave reached a specific location in the sample, which is detected by an embedded thermocouple, the explosive charge was detonated at the other end. This explosive shock wave rapidly altered the heat release conditions in the reacting sample by forcing the sample to contact the steel liner, which was at a lower temperature. In addition, the detonation force increased the sample density, which in turn increased the thermal conductivity and quenched the reacting mixture. A quenching rate of 10^4 K/s for this method was estimated indirectly from microstructural features [7].

REFERENCES

[1] Merzhanov AG, Borovinskaya IP, Volodin YE. Dokl Phys Chem 1972;206(4):905–8.
[2] Khusid BM, Khina BB, Kuang VZ, Bashtovaya EA. Combust Explos Shock Waves 1992;28:389–95.
[3] Mukasyan AS, Borovinskaya IP. Int J SHS 1992;1:55–63.
[4] Rogachev AS, Mukasyan AS, Merzhanov AG. Dokl Phys Chem 1987;297:1240–3.
[5] Stepanov BV. Dokl Akad Nauk SSSR 1992;324(3):571–5.
[6] Stepanov BV, Rogachev AS. Int J SHS 1992;1(3):409–16.
[7] Molokov IV, Mukasyan AS. Int J SHS 1992;1:155–8.

Reaction Chamber

Alexander S. Rogachev

A schematic drawing of the reaction chamber commonly used for laboratory-scale study of SHS processes is shown in Fig. 1 [1]. Several requirements must be taken into account in order to create proper conditions for investigation of the process. First, controlled gas atmosphere must be provided inside the chamber (vacuum, inert gas, or reactive gas). Some solid-gas (hybrid) SHS systems need high pressure of N_2 or other gas, up to hundreds or thousands bars (e.g., B-N, Si-N, Al-N); therefore the chamber must be durable, able to withstand the high pressure in these cases. Vacuum pumping and gas supply systems must be connected with the chamber. Second, electrical inputs and outputs are used for preheating of the sample, initiating reaction (ignition) and data acquisition from thermocouples, electric probes, and other sensors placed inside the chamber. Third, in many studies, transparent window(s) are required for video, thermal vision, and pyrometer monitoring of the process. In cases when infrared radiation is to be detected, special IR-transparent windows must be installed. Free-standing cylindrical or rectangular samples, which are commonly used in order to determine basic regularities of the SHS wave propagation, can be replaced with specialized devices for thermal explosion of quenching.

Design of the reactive chamber depends on the scientific task of the research. Some examples of specified reactive chambers are presented in Fig. 2.

The left photo in Fig. 2 shows a laboratory chamber adapted for macro-photo and video observations at low pressure (from vacuum to 1 bar). Many windows and a binocular microscope system with two-video cameras allow proper illumination and recording of the sample. The right photo shows a portable SHS chamber installed near the Synchrotron beam light channel. This chamber has a U-shaped X-ray transparent window for input and output of the Synchrotron radiation. The previously mentioned three basic requirements are valid for different chambers, including the two examples in Fig. 2.

Concise Encyclopedia of Self-Propagating High-Temperature Synthesis
http://dx.doi.org/10.1016/B978-0-12-804173-4.00115-0

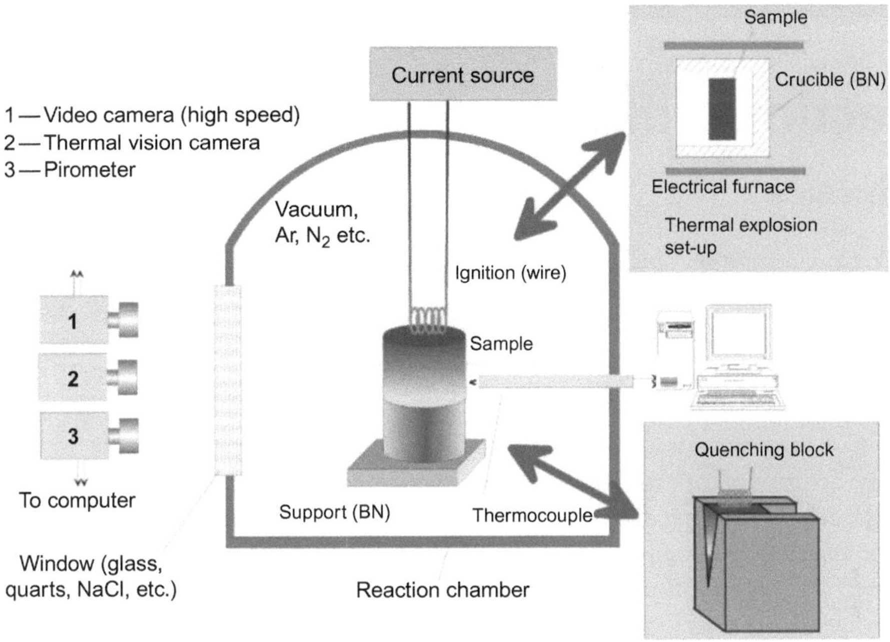

Fig. 1 Schematics of the reaction chamber [1].

Fig. 2 Laboratory-scale reactive chambers.

REFERENCES

[1] Rogachev AS, Mukasyan AS. Combustion for material synthesis. Boca Raton, FL: CRC Press Taylor & Francis Group; 2015. 397 p.

Reaction-Diffusion Model

Florence Baras

In ideal dilute systems, the rate of a given chemical reaction is proportional to the product of appropriate powers of the concentrations of the reactants involved in that reaction. For instance, the reaction $A + xB \rightarrow \alpha$ is described by the kinetics $\frac{dC_\alpha}{dt} = kC_A C_B^x$ where C_i is the concentration of the compound i and k is the rate constant. In the macroscopic description of reactive systems, it is assumed that reaction and mass transfer are decoupled. For nonuniform systems, the preceding evolution equation is replaced by an augmented balance equation in which the mass flux accounts for convection and diffusion. This assumption proves to be inappropriate in condensed phases. When a solid reagent, A, reacts with a solid (or liquid) reagent, B, and a layer product α is formed at the interface, the reaction may be sustained only through the mass transfer toward the interface. Diffusion and reaction are thus strongly coupled. Often, diffusion is the limiting step in the progress of the reaction. In SHS, the reactants are in condensed phase, either in the form of a mixture of solid powders or stacked foils. Furthermore, the SHS wave imposes unique conditions associated with high rates of temperature (10^3–10^6 K/s) and high combustion temperatures ($T_c > 2000$ K). In such conditions, at least one of the reagents melts in gasless combustion of the majority of SHS systems. Among the different models of condensed phase reactions in nonisothermal conditions that have been developed [1,2], a simple and predictive "mean-field" model is briefly discussed later [3].

A "typical" single solid particle surrounded by a liquid is presented in Fig. 1. The whole system is subjected to a propagating temperature front. The time evolution of the temperature (T) is assumed to be given (Fig. 2). The temperature is assumed to be uniform on the spatial scale of the particle. We consider binary systems with components A and B that can react to form compounds α, β, ... and so on. The reaction proceeds through the following stages (see Fig. 1):

- Component B melts and surrounds particles of A, which are assumed to be spherical for simplicity, but other shapes could also be considered. (Minimization of interfacial energy at high temperatures may drive the particles to be spherical.)
- Particle A begins to dissolve into liquid B and, at the same time, B diffuses into A and forms a layer of intermetallic compound, say α.

Concise Encyclopedia of Self-Propagating High-Temperature Synthesis
http://dx.doi.org/10.1016/B978-0-12-804173-4.00116-2

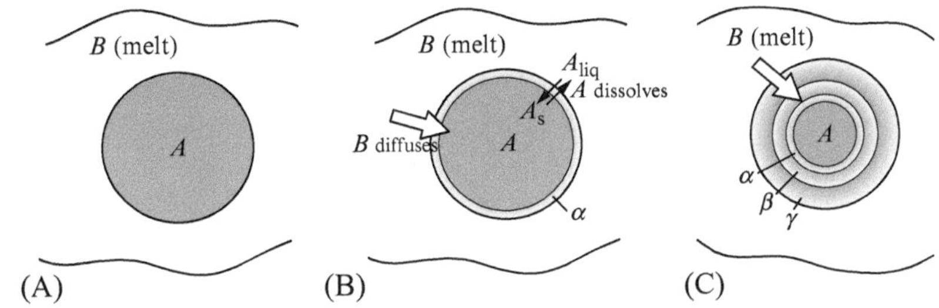

Fig. 1 Different stages of the reaction mechanism between two metals *A* and *B*. The highest temperature the system reaches is higher than the melting point of *B* but lower than the melting point of *A*. (A) At the start, *B* melts and surrounds particles of *A*. (B) During the next stage, *A* begins to dissolve in *B* and at the same time *B* diffuses into particles of *A*. (C) As *B* diffuses into *A*, successive layers of intermetallic compounds, α, β, γ, form while the outermost layer, γ, dissolves into the melt.

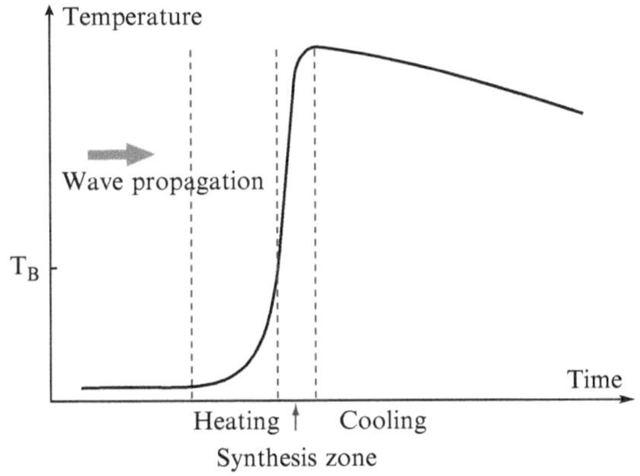

Fig. 2 A typical temperature profile in SHS synthesis. It is the mean *T* variation each particle in the reacting sample is subjected to.

- Since the A–α interface is not in equilibrium, driven by the diffusion of B (more accurately, interdiffusion of A and B) through the α layer, more of A converts to α. Thus the thickness of the α layer grows.
- At the outer surface of the α layer, B may react with the α phase to produce another intermetallic β, resulting in the growth of a β layer. Similarly, more layers, γ, δ, and so on could form, successively, depending on the number of compounds that the components A and B can form.
- The outermost layer dissolves into the melt, thus changing the composition of the melt.

Such a model allows us to follow the time evolution of the layered structure of the particle for a given temperature front. Fig. 3 shows how the combustion temperature value may influence the thickness of the layers, and thereby the microstructure of the product. See also Ref. [4].

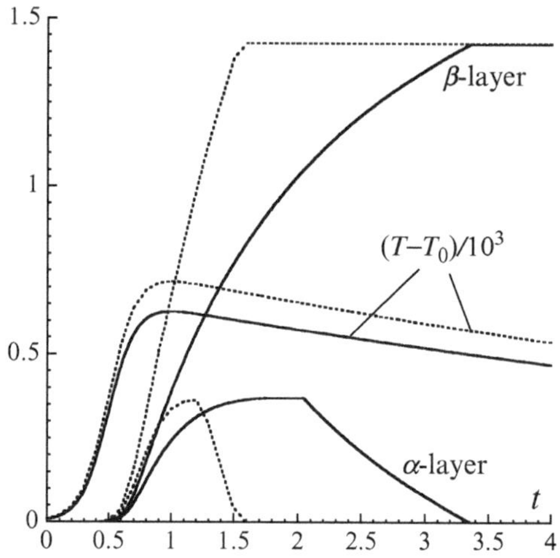

Fig. 3 Comparison of the evolution of the thickness of α-layer and β-layer when the system is subjected to temperature profiles that differ in the maximum value reached. The radii are measured in micrometer, time in seconds and the temperature increase in degree K.

REFERENCES

[1] For a review of reaction-diffusion models, see Rogachev AS, Mukasyan AS. Combustion for material synthesis. Boca Raton, FL: CRC Press Taylor & Francis; 2015. p. 179–88.
[2] Khina BB, Formanek B. Int J SHS 2007;16:61.
[3] Baras F, Kondepudi D. J Phys Chem B 2007;111:6457.
[4] Baras F, Kondepudi D, Bernard F. J Alloy Compd 2010;505:43–53.

Reactive Forging

Irena Gotman, Elasar Y. Gutmanas

Over the past two decades, a gradual shift from basic to applied research has led to the development of a number of unconventional SHS processes that allow simultaneous synthesis and consolidation of inherently porous combustion products. One such method is reactive forging (RF) where a thermal explosion (TE) mode of SHS is employed in combination with applied pressure in order to produce dense near-net-shape items.

In RF, 70%–80% dense reagent powder compacts are placed between the preheated rams (T_{ram}) of a press or a mechanical testing machine (e.g., Instron), and a very low pressure (<1 MPa) is applied in order to provide good contact between the specimen and the rams [1]. Rapid heating of the compact takes place by unidirectional heat transfer from the heated rams through the flat faces. When the ram temperature is sufficiently high, the TE is ignited, immediately after which a moderate uniaxial pressure (not exceeding 200 MPa) is applied. The specimen is held under pressure for not more than one minute and then unloaded. Plastic flow of the SHS products under pressure at the high combustion temperature, T_{comb} results in effective densification of the material. The utilization of exothermic heat instead of external heating makes RF a highly energy-saving processing technique. The presence of liquid at T_{comb} promotes densification, and it is essential that the pressure is applied when the product is in a mushy state. In order to limit the outward flow of the product under pressure at T_{comb} and to obtain near-net-shape parts, the compacts are confined within hollow steel or superalloy cylinders ("dies") of a larger diameter, hence the name reactive forging. In RF, the heating of the compact is very fast (hundreds of degrees per minute) because of the rapid heat transfer from the heated rams, and the ignition temperature of TE, T_{ign}, is typically reached within 1–2 min. These high heating rates distinguish RF from other methods of uni-axial pressure application during TE where a much slower furnace heating is used. Rapid heating to T_{ign} suppresses pre-combustion processes that impede ignition, and eliminates the need for a protective atmosphere during processing. Because of the technical ease of the RF procedure, rapid heating and short pressure-temperature cycle, it is possible to fabricate up to 10 dense near-net-shape specimens in 30 min.

A variation of RF is a method called short distance infiltration (SDI), and it combines pressure-assisted TE of nondense powder blends with pressure-assisted infiltration. In SDI, a nondense compact is rapidly heated under pressure, which results in the melting

Concise Encyclopedia of Self-Propagating High-Temperature Synthesis
http://dx.doi.org/10.1016/B978-0-12-804173-4.00117-4

of the lower-melting component of the blend (Al, Ti-Ni, etc.) [2]. Under the applied pressure, the liquid "lakes" formed are squeezed into pores leading to densification of the compact. SDI can be performed in any porous multicomponent compact provided that the processing temperature, T_0, is above T_m of at least one component, or above T_m of a eutectic phase that can be formed by any two components, and at least one component remains solid during processing. Compared to traditional melt infiltration, SDI has the advantage of considerably shorter (several orders of magnitude) infiltration distances (microns versus millimeters/centimeters). For highly exothermic reactive systems, SDI is followed by TE, and the low-melting liquid disappears while reacting with other components of the blend. Compact densification during the SDI process reduces the porosity that will be carried over to the combustion product, as well as promotes SHS synthesis by increasing the contact area (because of higher density) of the self-sustained reaction. In fact, SDI (at least partial) takes place in any RF procedure where TE is preceded by massive melting of one of the reactants. A wide range of monolithic, composite and gradient materials and 3D-shaped parts were fabricated by RF, including Al_2O_3-TiAl/Ti_3Al [1] and $MgAl_2O_4$(spinel)-TiAl/Ti_3Al [3] composites with interpenetrating architecture, Ti_2AlC, Ti_3AlC_2 and Ti_3SiC_2 MAX phases [4,5], a MgB_2 superconducting intermetallic [6], and many others. The feasibility of fabricating industrial-scale components employing RF has been demonstrated for 75-mm-diameter light armor tiles (40 vol.% of boron carbide) and for 100-mm-diameter ceramic bond superabrasive (diamond and CBN) grinding wheels. Furthermore, the RF process is amenable to computer simulation [7], which allows you to gain better control over the processing conditions and the product microstructure. Moreover, differentiation of temperature evolution curves recorded during RF provide a simple approach to estimate the activation energy of TE. As shown in Ref. [8], the slope of a straight line obtained by plotting $\ln(dT/dt)$ versus $1/T_{ram}$ for a set of RF experiments conducted at different ram temperatures, T_{ram} (sufficiently close to T_{ign}) is proportional to the activation energy.

REFERENCES

[1] Horvitz D, Gotman I, Gutmanas EY, Claussen N. J Eur Ceram Soc 2002;22:947–54.
[2] Gutmanas EY, Gotman I. J Eur Ceram Soc 1999;19:2381–93.
[3] Horvitz D, Gotman I. Acta Mater 2002;50:1961–71.
[4] Khoptiar Y, Gotman I. J Eur Ceram Soc 2003;23:47–53.
[5] Khoptiar Y, Gotman I, Gutmanas EY. J Am Ceram Soc 2005;88:28–33.
[6] Zlotnikov I, Gotman I, Gutmanas EY. J Eur Ceram Soc 2005;25:3517–22.
[7] Klinger L, Gotman I, Horvitz D. Mater Sci Eng A 2001;302(1):92–9.
[8] Horvitz D, Klinger L, Gotman I. Scr Mater 2004;55(5):631–4.

Role of Contact Melting in Gasless Combustion

Yury M. Maksimov, Alexander G. Merzhanov, Larisa G. Raskolenko, Olga K. Lepakova

It is known that gas-free combustion processes are accompanied by the appearance of a liquid phase in the reaction front. The liquid phase provides contact between reagents and contributes to the development of reactions because of the liquid-phase diffusion. In addition, the liquid phase can be formed due to contact melting of eutectic alloys that are the original source of the liquid phase at temperatures much lower than the melting temperatures of initial reagents.

The role of contact eutectic alloys in the combustion wave was studied during combustion of the Ti, B, and Fe powders [1]. The powders were burned in a constant pressure bomb as described in Ref. [2]. The amount of the initial mixtures was calculated taking into account the TiB_2 formation with an iron content in the final product from 21 to 71 wt.%. The powders were mixed and compressed into tablets 20 mm in diameter and height with a relative density of 0.6. In the experiments, the temperature distribution was measured, and the combustion products were analyzed using X-ray diffraction and metallographic analysis. Analysis of the contact eutectic melts was conducted by comparing the temperature combustion profiles and the phase composition of the quenched samples.

The six eutectic melts may exist in the Ti-B-Fe system [3–5]: Ti-FeTi ($T_m = 1087°C$), Fe-Fe_2B (1200°C), Fe-Fe_2Ti (1280°C), Fe-TiB_2 (1350°C), FeB-FeB_n (1500°C), and βTi-TiB_2 (1650°C). A necessary condition for the formation of these eutectic melts in the combustion wave is constant contact between the initial powder particles, the formation of the phases required for obtaining a melt according to the solid-phase mechanism and the achievement of the eutectic melting temperature in the front. The studies showed that the eutectic liquid in the Ti-B-Fe system was formed in the contacts of Ti-B and Fe-B at the initial stage.

Analysis of the temperature profiles in the combustion wave of the Ti-B-Fe system with different Fe content (Fig. 1) showed that the eutectic melts could be formed both in the Stefan mode and in the chemical conversion mode [6]. In the first case, eutectic melting was achieved mainly due to a heat flow from the higher temperature regions of the combustion wave, which led to the inflection of a curve on the profile $T(x)$. For

Concise Encyclopedia of Self-Propagating High-Temperature Synthesis
http://dx.doi.org/10.1016/B978-0-12-804173-4.00119-8

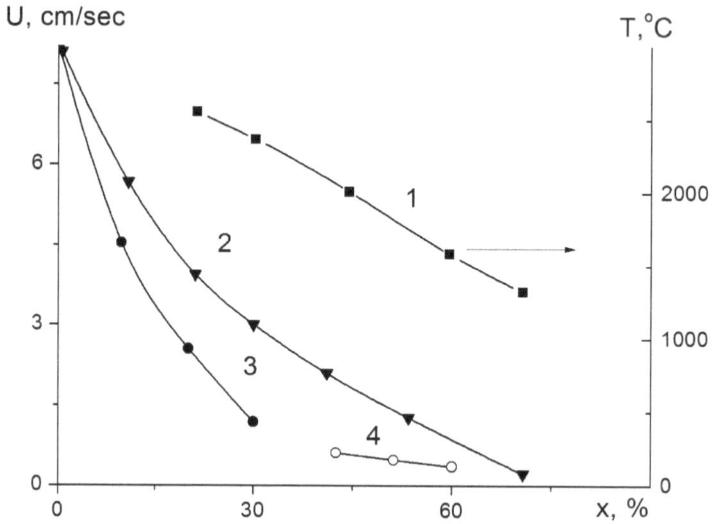

Fig. 1 Temperature distribution in the combustion wave of the compositions: (1) Ti-2B-75% Fe, (2) [Ti-2B-30% TiB_2]–20% Fe, (3) Ti-2B-62% Fe, (4) Ti-2B-45% Fe.

example, in Fig. 1(4), the temperature T_{m2} corresponds, apparently, to the TiB_2-Fe eutectic alloy. The combustion rate, according to Ref. [6], was determined by the maximum combustion temperature. In Fig. 1(2), the temperature $T_{max}=1200°C$ corresponds to the melting of Fe-Fe_2B eutectic alloy.

In Fig. 1, the isothermal plots emerging during the addition of iron corresponded to eutectic melting in the chemical transformations due to the heat release from the reacting components. The plots $T_{m2}=1340°C$ and $T_{m3}=1500°C$, in this case, indicated formation of the eutectic melts TiB_2-Fe and FeB_n-FeB in the combustion wave. In this case, the combustion rate was determined by the phase transition temperature.

The contact eutectic melts detected were the primary sources of the liquid phase in the front. An increase in the temperature led to the coalescence of eutectic melts and the formation of a uniform melt in which the intermediate products were dissolved. Thus the eutectic liquid accelerated diffusion processes at the initial stage of the system reactions, increased the contact area, and activated the combustion process. Contact eutectic melts were of great importance in the combustion mechanism, which was confirmed by the fact that the rate and the maximum temperature of combustion depended on the addition of the final product TiB_2 and Fe to the Ti-2B system (Fig. 2). The rate and the temperature of combustion were decreased during the addition of the final product to the system. For the composition Ti-2B–30% TiB_2, combustion was limited to the maximum combustion temperature $T_{max}=1800°C$. The addition of 10 wt.% Fe to this composition allowed the combustion to be observed at $T_{max}=1700°C$.

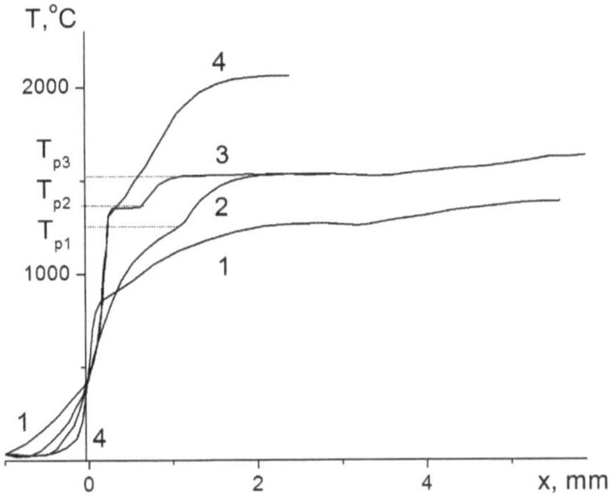

Fig. 2 Rate (2–4) and the maximum combustion temperature (1) versus the degree of doping. The additives α: (1) Fe, (2) Fe, (3) TiB_2, (4) 30%TiB_2-Fe.

The upper concentration limit for iron was 30 wt.%, and the maximum combustion temperature $T_{max} = 1500°C$. In this case, quenching of the samples fixed the FeB_n-FeB eutectic melt. The lower limit of combustion for iron was connected with a small amount of the eutectic melt, and the upper combustion limit was connected with an excess of iron performing the role of inert. The addition of iron to the Ti-2B mixture led to the fact that the combustion limit occurred for the composition Ti-2B–70% Fe, and the maximum combustion temperature was 1350°C, which was significantly below the melting point of the initial reagents and corresponded to the melting point of the TiB_2-Fe eutectic alloy.

Thus formation of the contact eutectic melts in the front significantly intensifies the interaction process, and sometimes, the combustion wave can propagate only as a result of this fact. Our studies showed that conducting the combustion process in the presence of a certain eutectic allowed us to control the structure of the combustion zone and the regularities of propagation.

REFERENCES

[1] Maksimov YuM, Merzhanov AG, Raskolenko LG, Pak AT, Lepakova OK. Proc USSR Acad Sci 1986;286(4):911–4.
[2] Merzhanov AG, Borovinskaya IP. Proc USSR Acad Sci 1972;204(2):366–9.
[3] Vol AE. Structure and properties of double metal systems, vol. 2. Moscow: Fizmatgiz; 1962. 982 p.
[4] Samsonov GV, Vinitsky IM. Refractory compounds. Moscow: Metallurgy; 1976. 558 p.
[5] Shurin AK, Panarin VB. Izv Akad Nauk SSSR Met 1974;15:235–9.
[6] Aldushin AP, Merzhanov AG. Proc USSR Acad Sci 1977;236(5):1133–6.

Shape-Memory Alloys

Volya I. Itin

SHS of alloys with shape memory effects and superelasticity was first performed in Russia (1975–1977) on the basis of the Ti-Ni, Cu-Al-Mn, and Cu-Al-Ni systems. The systems are weakly exothermic, and the calculated adiabatic combustion temperatures of compositions are close to the melting point of the products. At low initial temperatures, layer-by-layer combustion of the Ti-55 wt.% Ni composition (stoichiometry of TiNi) is realized in a pulsing mode that is replaced by a stationary mode with an increase in temperature. The increase in initial porosity leads to the fact that the rate and the maximum temperature of combustion are initially increased and then decreased, passing through a maximum. The ratio between the combustion rate under adiabatic conditions and the critical combustion rate is less than 1.67, which is fully consistent with the broad zones of reactions and the auto-oscillating combustion [1].

The maximum temperature for combustion of the Ti-55 wt.% Ni composition at 773 K does not depend on the dispersity of the titanium particles and is almost equal to melting temperature of titanium nickelide. The combustion rate gradually decreases with the increase in titanium particle size, and the stationary mode is replaced by a pulsing mode.

An increase in the initial temperature of the process or a decrease in the degree of composition dilution by the final product results in a constant increase in the combustion rate, and the maximum combustion temperature increases at first, and then does not depend on these parameters. The maximum combustion temperature is equal to the liquidus point. Combustion of this composition corresponds to the kinetic law for the interaction of reagents $\varphi(\eta)=\eta^{-4}$ [1].

The main controlling SHS parameters are the initial temperature of the process, the dispersity of initial reagents, and the degree of the dilution composition by the final product. The change of these parameters provides two types of products with the specified phase composition: a porous sintered sample and an ingot [1,2].

Further application of hot pressure shaping (pressing, extrusion, rolling) allows us to produce various products and items from the shape memory alloys. At present, the additive technological processes are proposed to conduct selective laser sintering in combination with SHS [3,4]. Milling and grinding of SHS products allow us to obtain the powders suitable, for example, for application of coatings.

Concise Encyclopedia of Self-Propagating High-Temperature Synthesis
http://dx.doi.org/10.1016/B978-0-12-804173-4.00121-6

Nickel-titanium based alloys obtained by the SHS, including the subsequent pressure treatment, are characterized by high chemical inhomogeneity and contain a greater amount of impurities, both in the form of solution and in the form of phases compared to similar alloys produced by conventional melting methods with hot manufacturing [1,2]. In fact, they are dispersion-strengthening materials and have the higher strength properties and stresses developed during shape reconstruction, and the reduced ductility as well. The temperature range of the phase transformations is usually shifted toward lower temperatures and is significantly broadened.

The use of SHS instead of the conventional melting methods leads to a significant reduction in energy consumption, a decrease in cost of complex equipment, and a release of the labor force. However, widespread use of this method requires significant reduction in cost of raw materials (powders of nickel and titanium), which are more expensive compared to that used for production of NiTi, using the known methods.

The SHS method is widely used for the development of porous shape memory materials for medicine [5,6]. A new class of functional materials is developed for implantation into the body and represents porous permeable materials with shape memory and superelasticity of titanium nickelide. The mechanical behavior of these materials in the body is similar to the behavior of living tissues representing poroelastic media, which have elasticity and easily transfer the load.

Body tissue grows well in the porous permeable space of SHS titanium nickelide that has a morphologically developed surface of pores, and the direct connection is formed between the bone and the implant [5,6]. During replacement of the tissue defect, the implant is "packed" completely into the bone matrix to form a lasting connection with the maternal bone. After ingrowth of living tissue in a matrix of porous titanium nickelide, the system retains the properties of elasticity, that is, meets the condition of biomechanical compatibility.

Porous permeable SHS intermetallides based on NiTi are materials that are successfully used to solve complex problems of medicine, in particular, to create the implants for long-term functioning in the body and replacing the defects of hard and soft tissues [5,6].

REFERENCES

[1] Itin VI, Nayborodenko YuS. High-temperature synthesis of intermetallic compounds. Tomsk: Tomsk University; 1989.
[2] Itin VI, Gyunter VE, Shabalovskaja SA, Sachdeva RL. Mater Character 1994;32(3):179–87.
[3] Bansiddhi A, Surgeant TD, Stupp SI, Dunand DC. Acta Biomater 2008;4:773–82.
[4] Elahinia MH, Hashemi M, Tabesh M. Prog Mater Sci 2012;57:911–46.
[5] Gyunter VE, Itin VI, Monasevich LA, Paskal YuI, et al. Shape memory effects and their application in medicine. Novosibirsk: Nauka; 1992.
[6] Gunther VE, Sysolyatin PG, Dambaev GTs, et al. Delay law and new class of materials and implants in medicine. Northampton, MA: STT; 2000.

Shock-Wave Synthesis in SHS Mixtures

Sergey A. Zelepugin, Oksana V. Ivanova

The direction of explosive processing that has not yet reached the level of industrial technology is the shock-wave synthesis in the mixtures capable of chemical transformations. In 1935, Bridgman investigated the effect of high pressures on the rate of solid-phase transformations and showed that solid-phase processes (phase transitions, chemical reactions) developed faster in the solids subjected to high pressures with shear deformation [1]. That fact initiated extensive investigations of the observed effect. In 1956, in the Soviet Union, Riabinin published an article devoted to describing experiments where the powder materials were placed into steel cylindrical containers surrounded by an explosive and were subjected to compression by the shock wave [2]. The studies showed various physical and chemical material transformations occurring under explosive loading of mixtures.

The pioneering work of DeCarli and Jamieson [3] significantly increased interest in studying the practical application of shock-wave synthesis in Japan, Russia, and the United States. In 1965 Russian scientists Batsanov, Deribas together with coworkers published the results on the synthesis in shock waves [4]. Simple experimental equipment was developed by Dremin and Kanel [5] for measuring the particle velocity during explosive compaction of powders. Pyrometric methods were widely used for the temperature measurements during shock-wave compaction and detonation of condensed substances. Experimental studies of shock-wave synthesis were also conducted by Adadurov, Doronin et al., Al'tshuler, and Dolgoborodov et al.

Extensive studies of shock-wave synthesis were conducted by Thadhani and reported in the review article along with the studies of other authors [6]. In the article, the mechanical and kinetic aspects of the synthesis were considered under explosive loading. For a more complete understanding of chemical transformations, Thadhani proposed focusing on the study of solid-phase diffusionless processes causing the atomic rearrangements and, perhaps, the initiation of chemical reactions. Using the SHS mixture (Ti-C) in the experiments, Shteinberg [7] estimated the time required for the completion of the reaction under shock-wave loading, which was about 10^{-7}–10^{-6} s.

To numerically simulate the ultrafast chemical reactions in the mixtures under shock-wave loading, a phenomenological model of chemical reactions was proposed on the basis of the zeroth-order kinetics without reverse transformations [8].

Concise Encyclopedia of Self-Propagating High-Temperature Synthesis
http://dx.doi.org/10.1016/B978-0-12-804173-4.00123-X

A multicomponent medium model was developed for numerical simulation of reacting mixtures under explosive loading, which allowed the properties of each component in a mixture, the variability of process parameters, and the presence of chemical reactions to be taken into account [9].

Interesting studies have been conducted to investigate the possibility of solid-phase chemical reactions to develop in the detonation-like mode [10]. During shock-wave compaction, exothermic reactions proceed in reaction mixtures with considerable heat release that can result in the self-sustaining detonation-like propagation of chemical reactions. The high reaction rate can create the conditions for occurring solid-phase (or low-gas) detonation. Dolgoborodov et al. developed a method for the mechanical activation of mixtures to form mechanically activated energetic composites (MAEC). The MAEC reactivity was significantly higher than conventional mixtures, which led to the substantial increase in the propagation velocity of the detonation-like shock-wave processes and the combustion rate.

It should be noted that almost all these studies of shock-wave synthesis were aimed at demonstrating the presence of chemical reactions in different systems. However, apart from the fact that many organic and inorganic substances can be obtained by the shock-wave synthesis, the understanding of basic reaction mechanisms and kinetics, as well as the influence of various factors on the initiation of reactions is still incomplete. The possibility to combine SHS and shock-wave loading provides a great opportunity for controlled synthesis, which will allow us to develop new methods and technological processes for the production of materials with unusual and unique properties.

REFERENCES

[1] Bridgman P. Effects of high shearing stress combined with high hydrostatic pressure. J Phys Rev 1935;48:825.

[2] Riabinin YuN. Certain experiments on dynamic compression of substances. Tech Phys 1956;26:261.

[3] DeCarli PS, Jamieson JC. Formation of diamond by explosive shock. Science 1961;133(3467):1821–2.

[4] Batsanov SS, Deribas AA, Kutolin SA. Effect of an explosion on matter: thermodynamics of shock compression of powders. Combust Explos Shock Waves 1965;1(2):39–44.

[5] Dremin AN, Kanel' GI, Gluzman VD. Experimental investigation of the pressure profiles associated with the irregular reflection of a conical shock in plexiglas cylinders. Combust Explos Shock Waves 1972;8(1):85–8.

[6] Thadhani NN. Shock-induced chemical reactions and synthesis of materials. Prog Mater Sci 1993;37:117–226.

[7] Gryadunov AN, Shteinberg AS, Dobler EA. High-speed impact initiation of chemical reaction in Ti–C powder mixtures. Dokl Phys Chem 1991;321(5):1009–13.

[8] Zelepugin SA, Nikulichev VB. Numerical modeling of sulfur—aluminum interaction under shock-wave loading. Combust Explos Shock Waves 2000;36(6):845–50.

[9] Ivanova O, Zelepugin S, Yunoshev A, Silvestrov V. A multicomponent medium model for reacting porous mixtures under shock wave loading. J Energ Mater 2010;28(Suppl. 1):303–17.

[10] Dolgoborodov AYu. Mechanically activated oxidizer-fuel energetic composites. Combust Explos Shock Waves 2015;51(1):86–99.

SHS Centers

Inna P. Borovinskaya

There are three stages in the development of SHS: exploration in the research institutes of the USSR Academy of Sciences, mainly in Chernogolovka and Moscow; joining of other scientific institutions of the USSR to the problem; and propagation of SHS all over the world.

The main results obtained in the first stage are connected with the beginning of the experimental and theoretical realization of the discovery of solid flame: development of experimental diagnostics of combustion of powder mixtures of transition metals (Ti, Zr, Hf, Nb, Ta) and nonmetals (B, C) and appearance of gasless combustion theory, discovery of two threshold combustion modes (layer-by-layer and surface) of porous metal samples in gaseous nitrogen, development of filtration combustion conception and mathematical modeling, and discovery of new types of combustion—oscillatory combustion and spin waves. This stage contributed greatly to the practical application of combustion processes. It developed the methods of chemical synthesis and the production of a wide range of refractory inorganic metal and nonmetal carbides, borides, nitrides, silicides, and so on; it investigated links between chemical and phase compositions and combustion parameters; and it established the optimum terms of the material production. The Institute of Structural Macrokinetics at the USSR Academy of Sciences (ISMAN), which was created to develop SHS, became the first SHS center and initiated the organization of various SHS centers in Russia country and abroad. Later it became the lead organization of the All-Union Interbranch Scientific and Technological Unit.

The first Russian SHS center started developing new fields of combustion application, a novel method for the production of practically all known inorganic compounds and materials, and highly efficient technologies of powder, and item and material production. One of the results of the research activity of the SHS center in Chernogolovka was elaboration of the boundary between combustion science and material science, called "structural macrokinetics" by Merzhanov [1–3]. The thorough description of the scientific work and practical application of the technologies and materials of the SHS center in Chernogolovka can be found in these articles: SHS in *Russia*, *ISMAN*, *Industrialization*, and other thematic sections.

The important step of SHS development appeared to be "self-propagation" of the method in the former USSR. In the 1970s the Scientific Center of the USSR

Concise Encyclopedia of Self-Propagating High-Temperature Synthesis
http://dx.doi.org/10.1016/B978-0-12-804173-4.00124-1

Academy of Sciences stopped being the only place where SHS was studied. An SHS center was founded in Armenia on the base of a laboratory of the Institute of Chemical Physics, the Armenian AS, which was headed by a famous scientist in the field of chemical physics, academician A.B. Nalbandyan with involvement of S.K. Dolukhanyan and S.L. Kharatyan. New centers appeared in Tomsk (Tomsk State University), Ukraine (Kiev Institute of Material Science), Uzbekistan (Institute of Nuclear Physics of the Uzbek AS), and so on. The most active centers were included in the Interbranch Scientific and Technological Unit, which was called "Termosintez." All the participants worked actively and contributed greatly to the successful implementation of SHS technologies in industry.

One of the first SHS centers was the SHS Engineering Center organized in 1988 by A.K. Amosov who was a former student at the Department of Engineering Chemical Physics at Kuibyshev Polytechnical Institute (now called Samara Polytechnical University). The main focus of the center was the investigation and practical application of azide-based technologies of SHS ceramic nitrogen-containing powders, filtration technology of nitrides, forced SHS compaction technology, SHS alloying additives for aluminum-based alloys, and the study of SHS sensitivity to mechanical effects.

Soon the Scientific-Educational Center of SHS was organized by State Resolution No.744/119 of Sept. 21, 1989. It was based on the Moscow State Institute of Steel and Alloys (MISIS) and Institute of Structural Macrokinetics. This center was the first to unite the resources of the educational and scientific institutes to carry out fundamental and applied investigations and train specialists in various aspects of the scientific problem. The creative role in the organization of the center belonged to the director of ISMAN, academician A.G. Merzhanov, and the vice-director of MISIS, Professor N.N. Khavsky. The first scientific supervisors of the center were Professors I.P. Borovinskaya and N.N. Khavsky. The director of the center (until now) was the young specialist E.A. Levashov, a talented, creative, and sociable man. He appeared to be not only a good scientist but also an excellent organizer. This center united the leading specialists of MISIS and ISMAN in the fields of chemical physics, combustion and explosion, structural macrokinetics, material science, powder metallurgy, pressure-treatment of metals, and theory of metallurgical processes. The tasks of the center are very broad and include training of specialists in SHS, development of hard-alloyed materials of various aims, development of theoretical models of combustion processes and structure formation at SHS, development of new methods of vacuum rolling of SHS products, elaboration of scientific backgrounds of the method of electrospark deposition of coatings, study of the processes of magnetron spraying and development of new targets, synthesis of biocompatible and bioactive films, and manufacturing of diamond tools and other materials by SHS.

The team of scientists from Tomsk State University (later it became the Department of Structural Macrokinetics of the Scientific Center in Tomsk), headed by

Yu. M. Maksimov, a student and follower of academician A.G. Merzhanov, has developed studies on combustion mechanism and regularities, SHS product structure and phase formation, and synthesis of new materials. The scientists at the center are engaged in computer-assisted modeling of macrokinetics of solid-phase chemical transformations at the point of dynamic compression, elaboration of macroscopic theory of mechanochemical synthesis and background of nonthermal regulation of combustion processes and heat explosion of condensed systems, study of emission in SHS processes, and development of block catalysts for associated gas conversion to syngas.

Kazakh SHS Research Center in Alma-Ata started its activity under the leadership of G.I. Ksandopulo, who was the head of the Department of Kinetics and Combustion at Kazakh University. The main subject under study of the center was R&D of oxide compounds and materials (refractories, pigments, catalysts). Then the center was reorganized as the Institute of Combustion headed by Z. Mansurov. But the interest of SHS was preserved and expanded. The main achievement of the center was industrial assimilation of "furnon" refractories based on complex oxides and refractory compounds. They are characterized by high adhesion bonding strength at high-operation temperatures and are used in shaft, rotatory and cement kilns, coke calciners, metallurgical roofs, and so on. They are synthesized in the place where it is necessary to apply the protective coating. Very important scientific investigations of SHS for development of high-porous oxide materials for catalysts were carried out at the center and now are continued in Greece. The main specialist responsible for the work is G.G. Ksandopulo, who extended the range of oxide SHS materials and developed the methods of SHS pigment application on plastic and glass as well as novel highly efficient catalysts.

Georgian SHS Research Center in Tbilisi was organized in the Institute of Metallurgy of the Georgian Academy of Sciences. The institute was headed by academician F.N. Tavadze, a creative person known all over the world. The first investigations were connected with the production of amorphous alloys synthesized by hardening of the melt injected on the rotating drum. They also studied manganese-containing ferroalloys and used local manganese ores to synthesize them. Young researchers led by G.F. Tavadze (now director of the institute) and G.Sh. Oniashvili expanded the themes under study by the development of SHS analogs of some processes of ferrous and nonferrous metallurgy, SHS processing of local raw materials, boron production using radioactive isotope B^{10}, forced densification of hard alloys with phosphate binders, and so on [4].

In 1980 active SHS investigations started abroad in Japan and the United States. They were induced by a review issued by Joy Crider, who was entrusted to write it by J. McCauley, a pioneer of SHS in the United States [5]. In Japan a team of specialists interested in SHS appeared upon the initiative of Professor M. Koizumi. A wide range of research work was carried out at the Center of Advanced Science and Innovations

(Osaka) and other Japanese institutions and companies (O. Odawara, E. Miyamoto, M. Ohyanagi, Sh. Otani, A. Makino, and others).

Two SHS centers were founded in China. One of them was at the University of Science and Technology in Beijing, the other was in Wuhan. The former was headed by Professor Ho-Yi Lai, the latter by Professor R.Z. Yuan. Later Professor Ge and Dr. Fu joined them. As a result, many scientists from different educational institutions in China became interested in SHS. They also began training specialists in some research institutions and industrial enterprises in different cities and provinces of China.

Today SHS investigations are carried out in more than 50 countries all over the world by separate groups of scientists or single activists.

Detailed discussion of SHS activity around the world is presented in the following entries: *Armenia, Belorussia, China, France, Georgia, India, ISMAN, Italy, Japan, Poland, Republic of Kazakhstan, Russia, South Korea, Spain, Turkey*, and *United States of America.*

REFERENCES

[1] Merzhanov AG, Borovinskaya IP. Historical retrospective of SHS: an autoreview. Int J SHS 2008;17 (4):242–65.
[2] Merzhanov AG. Self-propagating high-temperature synthesis: twenty years of search and findings. In: Munir ZA, Holt JB, editors. Combustion and plasma synthesis of high-temperature materials. New York: VCH Publishers; 1990. p. 1–53.
[3] Merzhanov AG, Rogachev AS. Structural macrokinetics of SHS processes. Pure Appl Chem 1992;64 (7):941–53.
[4] Merzhanov AG. 40 years of SHS: fortune of a scientific discovery. vol. 4. Chernogolovka, Russia: Ann Inst Chemical Physics, RAS; 2007. p. 30–88 [in Russian].
[5] Crider JF. Self-propagating high-temperature synthesis: a Soviet method for producing ceramic materials. Ceram Eng Sci Proc 1982;3(9–10):519–28.

SHS Effects

Yury M. Maksimov

SHS processes have different effects related to capillary spreading, centrifugal impregnation, dispersion, phase separation, degasification and self-cleaning, X-ray radiation, as well as physical fields of electrical, magnetic, and electromagnetic nature.

There is a crucial interest in work that reports on detection of self-produced magnetic fields induced by combustion, for example, during the synthesis of complex ferrites. Weakly oscillating magnetic fields are also induced by combustion of some metals: Ti, Zr, Mn, Nb, Fe, and so on.

The work directed to the study of the gravitational field's impact on SHS are of great importance, since different processes are not equally sensitive to gravitation. SHS processes where melts form are the most sensitive to gravitation. In connection with the development of work in space materials, to optimize the technology of SHS material production, there is a need to conduct studies on microgravity combustion.

Most of these effects mentioned play a useful role in solving the major problems concerning SHS synthesis of substances and materials. The most comprehensive review of studies directed toward understanding the SHS effects is provided in the fundamental works [1,2].

REFERENCES

[1] Merzhanov AG. Combustion and synthesis of materials. In: Telepa VT, Hachoyan AV, editors. Chernogolovka: ISMAN; 1998. 512 p.
[2] Merzhanov AG, Mukasyan AS. Solid phase combustion. Moscow: Torus Press; 2007. 336 p.

Concise Encyclopedia of Self-Propagating High-Temperature Synthesis
http://dx.doi.org/10.1016/B978-0-12-804173-4.00125-3

SHS in Armenia

Suren L. Kharatyan

The first discussion on SHS in Armenia was initiated by A. G. Merzhanov during the second All Soviet Union Conference on Combustion and Explosion, which was held in 1969 in Armenia.

Two years later, A.G. Merzhanov gave an invited talk about SHS processes at the Armenian Academy of Sciences. At that time, it was decided to send young Armenian scientists and graduate students to Chernogolovka, Merzhanov's laboratory at the branch of the Institute of Chemical Physics for specializing in the field of combustion science.

The year of 1971 was historical for the Armenian SHS community, as in December of that year, the Government of Armenia adopted a resolution to establish an SHS sector in the Laboratory of Chemical Physics of Armenian Academy of Sciences with the appointment of new staffs, a research area, and appropriate governmental funding. A.G. Merzhanov was appointed as an informal head of the SHS sector. During the next two years, about 10 junior students from Yerevan State University and Polytechnic Institute were sent to Merzhanov's laboratory in Chernogolovka on a mission of conducting their undergraduate research. In following years, some students were enrolled in PhD programs in Merzhanov's laboratory, and others returned to Yerevan and took positions at the SHS sector. From 1972 to 1975, a number of qualified young scientists and engineers were recruited to conduct pioneering SHS-related research in Yerevan.

At that point, the Laboratory of Chemical Physics became a big and diverse research unit. Therefore in 1975, the Government of Armenia adopted a decision to reorganize the Laboratory of Chemical Physics as the Institute of Chemical Physics. One of the key motivations for creating a new Institute related to the significance of SHS research and the development of innovative technologies for synthesis of materials.

Shortly after the Institute of Chemical Physics was organized, a laboratory of SHS Technologies was founded there, and Professor S. K. Dolukhanyan became the leader of the laboratory. Since that time, Dolukhanyan's laboratory has engaged in intensive research of SHS processes in metal—hydrogen and metal—silicon systems. Many metal hydrides, including intermetallic hydrides, were synthesized and new SHS technologies of hydrides were developed. Among the significant contributions of Dolukhanyan's

Concise Encyclopedia of Self-Propagating High-Temperature Synthesis
http://dx.doi.org/10.1016/B978-0-12-804173-4.00009-0

laboratory in SHS is the development of self-sustaining synthesis technology of molybdenum disilicide, which was implemented in high temperature heaters, making it the first example of SHS-related commercialization effective. Over the next decade, hundreds of tons of $MoSi_2$ powder were produced. For their contribution to the Armenian economy, A.G. Merzhanov and members of S. K. Dolukhanyan's laboratory were awarded the State Prize of Armenia in 1980.

In 1977 a research group of SHS Thermodynamics and Kinetics was organized at the Institute of Chemical Physics, which in 1982 was reorganized as the Laboratory of Kinetics of SHS Processes, and S. L. Kharatyan became the leader of the laboratory. Many young scientists joined this laboratory, which began investigating kinetics of SHS processes initially in metal-gas systems. Great deal of research in this laboratory during 1980s and 1990s focused on modeling SHS processes and developing new reactive diffusion models. The laboratory also greatly advanced stimulation methods of low-caloric reactive systems. Particularly, innovative approaches to activate low-exothermic mixtures were developed through the use of minor additives—often referred to as chemical activation of SHS. In 2008 a group of scientists from this laboratory received the Presidential Prize and Gold Medal of Armenia for its outstanding contribution in combustion and materials sciences.

In 1980 a new engineering division at the Institute of Chemical Physics was organized, which was solely engaged in development of SHS materials for different commercial applications. During 14 years of activity (1980–94), this division produced and supplied different USSR companies with ~40 tons of carbides, silicides, hydrides, and complex compounds of transition metals.

A great contribution in the development of Armenian SHS community was organized by the A. G. Merzhanov regular All Soviet Union workshops on SHS in Armenia, where hundreds of scientists from all over country came to give talks, exchange ideas, and socialize.

Two symposia were organized later in Armenia. The first one, dedicated to the 75th anniversary of A. G. Merzhanov in 2006, had a broad topic covering all aspects of nonisothermal processes. The second one was the 10th International Symposium on SHS (2009).

In the post-Soviet years, the SHS community in Armenia has started to collaborate internationally with European, US, South Korean, and Chinese scientists.

More than 40 PhD dissertations on SHS-related topics were defended during past decades. Six people became Doctors of Sciences, and one person became a corresponding member of the National Academy of Sciences of the Republic of Armenia.

Today the SHS community in Armenia is active in both research and education. SHS-related works are conducted also in the Yerevan Polytechnic Institute. Several SHS-related undergraduate courses are now taught at the Yerevan State University. During the past 15 years, more than 50 students conducted their undergraduate and master research projects in the SHS-related laboratories of the Institute of Chemical Physics of National Academy of Sciences of the Republic of Armenia. Some of these students continued their research and later accepted positions at the Institute of Chemical Physics. Now more than 10 of them are working and continuing their research in the leading universities and companies of the United States, China, South Korea, and other countries.

SHS in Belorussia

Boris B. Khina

Research of SHS in Belorussia began in the mid-1980s under the influence of and in cooperation with the scientific school of Professor A.G. Merzhanov at the Institute of Structural Macrokinetics and Materials Science RAS (ISMAN). The studies have focused on the theoretical aspects of phase and structure formation mechanisms that operate in SHS waves and in filtration combustion waves, and on materials synthesis by SHS in an ultrasound field and mechanically activated (MA) SHS. At present the main institutions involved in R&D activities on SHS are the A.V.Luikov Heat and Mass Transfer Institute of the National Academy of Sciences in Belarus (HMTI NASB, Minsk), the Physico-Technical Institute (PhTI NASB, Minsk), the Institute of Powder Metallurgy (IPM NASB, Minsk), the Institute of Technical Acoustics (ITA NASB, Vitebsk), and the Belorussian National Technical University and Belorussian State Technological University (both in Minsk).

Summary of the main achievements of the Belorussian SHS-Centers:

- Theoretical and experimental investigation of filtration combustion of gases in porous media, which yielded new knowledge on heat and mass transfer accompanied with homogeneous and heterogeneous reactions, and the thermal structure and behavior of combustion waves (HMTI NASB).
- Experimental investigation and computer modeling of filtration-combustion synthesis of graphite fluoride [1], which resulted in the development of a controllable low-temperature regime, ensuring high degree of conversion, up to 90%, in a reactor (HMTI NASB).
- Thermodynamic study of SHS of tantalum carbide [1], which revealed the role of gas-phase mass transfer and oxygen impurities in a system where the adiabatic combustion temperature is below the melting points of the reactants, product, and lowest-temperature eutectics (PhTI and HMTI NASB).
- A model of filtration combustion in the Ti–N_2 system taking into account the flow of gas into a porous specimen and its adsorption on the surface of titanium particles because of formation of the refractory product, TiN_x, which permits determination of an effective depth of nitrogen penetration into the specimen as well as the plotting of a diagram that describes experimentally observed structure-forming mechanisms (solid-state, solid-liquid, and liquid-phase) and SHS wave propagation modes (layer-by-layer and surface combustion) [1] (PhTI and HMTI NASB).

Concise Encyclopedia of Self-Propagating High-Temperature Synthesis
http://dx.doi.org/10.1016/B978-0-12-804173-4.00013-2

- Stochastic model of SHS [1], including a probability of heterogeneous reaction in the high-temperature zone of combustion wave, which revealed the dynamic behavior of SHS waves in the parametric domain of unstable combustion (PhTI and HMTI NASB).
- Percolation model of SHS [2], which described combustion of a randomly packed mixture of reactive and inert particles in nonadiabatic conditions. The model revealed the transition from frontal (layer-by-layer) combustion to a regime where SHS propagates over percolation clusters formed by combustible particles; the latter situation corresponds to strongly diluted and/or MA powders (HMTI NASB).
- Theoretical models of SHS in metal-carbon (Ti-C) and metal-metal (Ni-Al) systems, which revealed the limitations of the diffusion model of SHS, proved the nonequilibrium nature of interactions in the SHS waves and permitted plotting of phase-formation-mechanism maps [3] with the coordinates of "particle size-heating rate" (PhTI NASB).
- Experimental studies of SHS in an ultrasound field, which revealed the influence of ultrasound on both the combustion characteristics (temperature and velocity) and the materials' features (particle size distribution and crystal lattice parameters of refractory products) (ITA and PhTI NASB).
- Theoretical study of MA SHS [3,4], which demonstrated that the observed effects cannot be ascribed to nonequilibrium point defects (vacancies and self-interstitials) generated in metals during MA nor to the stored energy of plastic deformation, and are connected mainly to microstructural features of composite particles formed because of periodic plastic deformation during the MA stage of production (PhTI NASB).
- Experimental investigation of MA-SHS in a number of metal-oxide systems [5], which disclosed specific features of phase and structure formation during synthesis. This permitted tailoring the regime of mechanical activation and powder composition to the structure and properties of a target SHS product. The results were applied to the development of nanostructured powder compositions for spray-coating deposition used in a wide range of applications, including protection of internal walls of heat exchangers at power plants from erosion by flying ash at elevated temperature (IPM NASB).

REFERENCES

[1] Khusid BM, Khina BB, Rabinovich OS. Comp Chem Engineer 1994;18(11–12):1183–95.
[2] Rabinovich OS, Grinchuk PS, Khina BB, Belyaev AV. Int J SHS 2002;11(3):257–70.
[3] Khina BB. Combustion synthesis of advanced materials. New York, NY: Nova Science; 2010.
[4] Mukasyan AS, Khina BB, Reeves RV, Son SF. Chem Eng J 2011;174(2–3):677–86.
[5] Talako TL, Letsko AI, Grigor'eva TPh. Combust Sci Technol 2010;182(11):1598–609.

SHS in China

Zhengyi Fu

SHS research in China began in the late 1980s at Wuhan University of Technology, headed by Professor Run Zhang Yuan, and Beijing University of Science and Technology, headed by Professor Heyi Lai and Professor Changchun Ge.

In 1988 Professor Munir, from the University of California, Davis in the United States, visited China. He gave lectures at Wuhan University of Technology and Beijing University of Science and Technology and introduced in detail the SHS research progress at that time. In 1991 a Chinese delegation took part in the First International Symposium on SHS in Alma-Ata. Based on the information they knew, the Chinese scientists persuaded their government that SHS was a valuable and important technology, after which the Ministry of Science and Technology of China supplied "863 High Technology Project" funds to support SHS research at the two Chinese universities. Before 1991 the SHS research in different units in China were supported mainly by university self-funding.

Professor Merzhanov from Institute of Structural Macrokinetics, Russian Academy of Science together with his colleagues visited Wuhan University of Technology in 1993. Since then, they have visited the university a number of times, which helped the SHS research at the university significantly. The two institutions signed an agreement to establish the Sino-Russia Joint Research Center for SHS research and a memorandum to organize the international SHS symposium.

In 1993 Wuhan University of Technology organized the First Chinese Symposium on SHS. In 1995 Wuhan University of Technology organized the Third International Symposium on SHS. Since then SHS research in China has expanded rapidly.

Wuhan University of Technology: Research is headed by Professor Zhengyi FU. The main research directions are SHS-based fast synthesis and sintering, SHS for nano materials, and SHS for thermoelectric materials.

Beijing University of Science and Technology: Research is headed by Professor Changchun Ge. The main directions are SHS for functional graded materials, centrifugal SHS for tube coatings, and combination of sol-gel and SHS.

Harbin Institute of Technology: Research is headed by Professors Jiecai Han and Xiaodong He. The main directions are SHS for high-temperature materials and SHS for multiphase ceramics.

Concise Encyclopedia of Self-Propagating High-Temperature Synthesis
http://dx.doi.org/10.1016/B978-0-12-804173-4.00029-6

Shijiazhuang Mechanical Engineering College: Research is headed by Professor Zhongmin Zhao. The main direction is high-gravity field SHS for ceramic-based composites.

Technical Institute of Physics and Chemistry, Chinese Academy of Sciences: Research is headed by Professor Jiangtao Li. The main direction is high-gravity field SHS for cemented carbides.

Fu's group developed a fast sintering method that combined the SHS process and a quick pressing (called SHS/QP) to prepare dense nanocrystalline ceramics. In their work, the heat generated by SHS reaction acted as a high-temperature source, which supplied fast heating to the nano-size powders. Significant mechanical pressure was applied to the sample, when the temperature reaches the maximum [1–3]. Densification of nano-grained alumina was achieved within 2 min, while the grains had almost no growth, as shown in Fig. 1. Fully dense magnesia and yttria nanocrystalline ceramics were also obtained, and the concurrent grain growth was restrained.

Some institutes in China fabricate ceramic-based composites by combustion synthesis in a high-gravity field [4–6]. In Yang's work, transparent Y_2O_3-Al_2O_3-SiO_2 ceramics were prepared by high gravity combustion synthesis with heating assistance. The high-gravity field with an acceleration of $1200 \times g$ was induced by the centrifugal effect. The ceramics made by this method shows the maximum transmittance of 54% in visible to near-infrared region and hardness of 8.1–8.5 GPa.

Yan and Geng focused on the synthesis of fine ceramic powders by a sol-gel autoignition synthesis process [7,8]. In Yan's work, highly photoactive nanocrystalline Ce-doped TiO_2 photocatalysts with anatase single-phase were formed. In Geng's work, as-prepared CoCuMnOx powders were then used as solar-absorbing pigment to fabricate a novel kind of spectrally selective solar absorber coatings by spray-coating technique.

Luminescent nanoparticles were synthesized by combustion synthesis [9,10]. In Tang's work, the mixture reactants of AH_2PO_4 (A=Na, K), $Ba(NO_3)_2$, $Eu(NO_3)_3$, and $CO(NH_2)_2$ were stirred and preheated in a furnace at 600°C. The slurry was ignited and a spontaneous combustion reaction started quickly. Photoluminescence results show that the $NaBaPO_4$:Eu and $KaBaPO_4$:Eu phosphors emitted a strong blue light under UV excitation. In Qiu's work, Eu^{2+}, $R3^{+}$ co-doped alkaline earth aluminates MAl_2O_4: Eu^{2+},

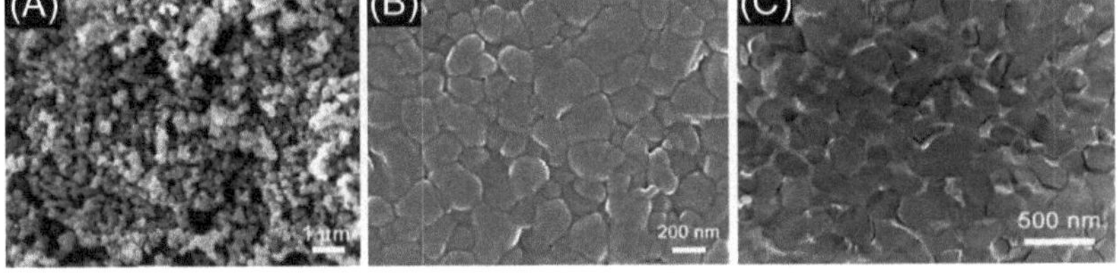

Fig. 1 Morphology of alumina samples: (A) raw materials in 200 nm; (B) and (C) SEM and TEM images of sintered specimens.

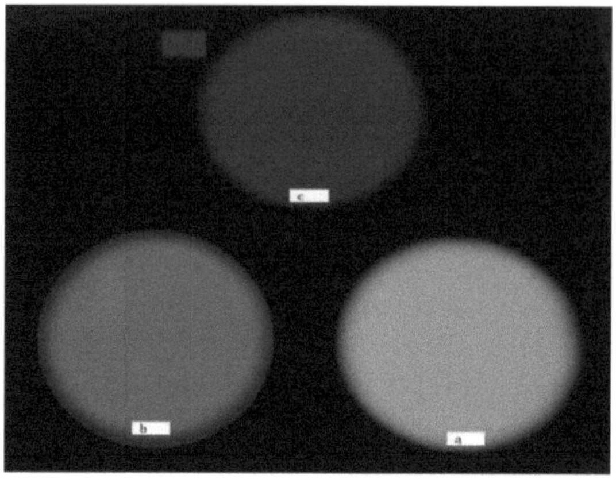

Fig. 2 Photos of as-prepared long-persistent luminescence nanoparticles: (A) $SrAl_2O_4$: Eu^{2+}, Dy^{3+}; (B) $BaAl_2O_4$: Eu^{2+}, Nd^{3+}; and (C) $CaAl_2O_4$: Eu^{2+}, La^{3+}.

R^{3+} (M = Sr, Ba, Ca; R = Dy, Nd, and La) nanoparticles with high brightness have been prepared by SCS without a postannealing process. Analytical results show that every nanoparticle has the perfect crystal phase and the luminescence changes from yellow-green to blue-green and then to blue-purple in the visible range with the changing composition from $SrAl_2O_4$: Eu^{2+}, Dy^{3+} to $BaAl_2O_4$: Eu^{2+}, Nd^{3+}, and then to $CaAl_2O_4$: Eu^{2+}, La^{3+}, as shown in Fig. 2.

Su et al. reported a successful synthesis of single-phase thermoelectric (TE) materials by SHS. The stoichiometric amounts of high purity Cu (4N) and Se (4N) powders were pressed into a pellet; then the pellet was sealed in a silica tube under a pressure of 10^{-3} Pa and brought to the ignition point using a torch flame. The reacted pellet was then sintered by SPS at 973 K. The TE performance of samples prepared by SHS is comparable and on many occasions better than materials prepared by the traditional synthetic routes [11–13].

REFERENCES

[1] Meng F, et al. Rapid densification of nano-grained alumina by high temperature and pressure with a very high heating rate. J Am Ceram Soc 2007;90(4):1262–4.
[2] Liu J, et al. Ultra-high heating rate densification of nanocrystalline magnesia at high pressure and investigation on densification mechanisms. J Eur Ceram Soc 2014;34(12):3095–102.
[3] Liu J, et al. Grain growth stagnation in the dense nanocrystalline yttria prepared by combustion reaction and quick pressing with an ultra-high heating rate. J Eur Ceram Soc 2014;34(10):2475–82.
[4] Yang Z, et al. Preparation of transparent Y_2O_3-Al_2O_3-SiO_2 glasses by high-gravity combustion synthesis with heating assistance. J Am Ceram Soc 2012;95(6):1799–802.
[5] Yin X, et al. Combustion synthesis of Ti_3SiC_2/TiC composites from elemental powders under high-gravity conditions. J Am Ceram Soc 2010;93(8):2182–7.
[6] Zhao Z, et al. Al_2O_3/ZrO_2 (Y_2O_3) self-growing composites prepared by combustion synthesis under high gravity. Scr Mater 2008;58(3):207–10.
[7] Geng Q-F, et al. Sol-gel combustion-derived $CoCuMnO_x$ spinels as pigment for spectrally selective paints. J Am Ceram Soc 2011;94(3):827–32.

[8] Qiu Z, et al. Combustion synthesis of long-persistent luminescent MAl_2O_4: Eu^{2+}, R^{3+} (M = Sr, Ba, Ca, R = Dy, Nd and La) nanoparticles and luminescence mechanism research. Acta Mater 2007; 55(8):2615–20.
[9] Tang W, Chen D. Photoluminescent properties of $ABaPO_4$:Eu (A = Na, K) phosphors prepared by the combustion-assisted synthesis method. J Am Ceram Soc 2009;92(5):1059–61.
[10] Yao S, et al. Preparation and luminescence property of red-emitting hardystonite phosphors by near-ultraviolet irradiation. J Am Ceram Soc 2011;94(5):1477–82.
[11] Xianli S, et al. Self-propagating high-temperature synthesis for compound thermo-electrics and new criterion for combustion processing. Nat Commun 2014;5:4908.
[12] Yang D, et al. Manipulating the Combustion Wave during Self-Propagating Synthesis for High Thermoelectric Performance of Layered Oxychalcogenide $Bi_{1-x}Pb_xCuSeO$. Chem Mater 2016; 28(13):4628–40.
[13] Liang T, et al. Ultra-fast synthesis and thermoelectric properties of Te doped skutterudites. J Mater Chem A 2014;2(42):17914–8.

SHS in France

Frédéric Bernard

About 20 years ago, French researchers decided to federate themselves in order to build a SHS French group. Such collaboration has enriched the French scientific progress. Both in situ TRXRD and IR thermography, on the one hand, and the different levels of modeling, on the other hand, have generated strong progress in the knowledge and control of many SHS processes, including mechanically activated SHS; SHS of simple binary materials, such as intermetallics or nitrides, as well as more complex materials, MAX phases, composites (SiO_2-Al_2O_3, NiAl-ZrO_2, ...); SHS of end products such as powders up to rather more complicated bulk materials such as well fully densified intermetallics.

In order to identify the intimate mechanisms that control SHS reactions—such as the role of liquid formation, the existence and effects of transitory phases, and their parameters that may induce changes in the microstructure or nature of the end products—real-time in-situ investigations of structural changes in the combustion area were developed through the use of synchrotron radiation [1]. Various intermetallic systems were studied, such as FeAl, $MoSi_2$, $NbAl_3$, and Ti_3SiC_2 [2].

High-energy ball milling of powder mixtures was reported to be an efficient technique for the preparation of nanocrystalline materials. Besides direct synthesis of materials, high-energy ball milling is a way of modifying the conditions in which chemical reactions usually take place either by changing the reactivity of as-milled powder mixtures [3], or by introducing an additional energy to the reactant system that effectively lowers the activation barriers for reactions. Various significant effects of the mechanical activation (MA) step have been shown, such as a decrease of ignition temperature, an increase of the combustion front velocity (for example, in the Mo/2Si system [4], MA can multiply the combustion front velocity by a factor of three compared to the classical value obtained under the similar ignition condition). In some cases, MA conditions can overcome the thermodynamic obstacle and allow the ignition of a combustion front in the process of cold pressing intermetallic disilicides.[5]

Great efforts in the field of modeling have led to both a better understanding of the reaction mechanisms and better control of the microstructure of end products. These efforts were carried out at different levels: (i) finite element simulation was undertaken to evaluate the consequences of a mechanical strain field on the geometry of a TiN bar

Concise Encyclopedia of Self-Propagating High-Temperature Synthesis
http://dx.doi.org/10.1016/B978-0-12-804173-4.00058-2

synthesized by SHS [6], (ii) sub-micrometer modeling of high-temperature synthesis of silicon carbide SiC was undertaken to find parameters controlling the dynamics of formation and the microstructure of the obtained ceramics [7], and (iii) mean field modeling was undertaken by using the phase diagram of the binary system, a general theory that describes the reaction taking place in a single solid of one component surrounded by the melt of the second component [8]. As the reaction progressed, various intermetallic compounds of the two metals created a multilayer system. Such layered systems were observed in experiments with metal foils [8].

There are different ways how one can use SHS process:

(i) Heat generation is the most obvious one of them. A "chemical oven," where the SHS reaction is used only to generate heat, is a realistic application because of the low cost of the initial precursors [9]; (ii) SHS of composite such as dense NiAl-ZrO_2 composite can be produced by combining thermal explosion with a post thermal cycle [10], and dense $MoSi_2$_Al_2O_3_based functionally graded materials (FGM) with different alumina contents have been prepared using an SHS reaction between molybdenum trioxide, aluminum, and silicon or silica at room temperature [11]; (iii) repairing and coatings, to simultaneously repair and coat the Ni-based superalloy. The values previously optimized for the "SHS build-up process" [12] were taken as the starting point of this study. The maximum value was chosen to be the highest solubility of platinum in NiAl at 1000°C, which is the working temperature for turbine blades. Indeed, the presence of precipitates at this temperature is harmful for coating ductility [12], and (iv) production of fully dense nano-intermetallics by reactive sintering in an SPS machine; the simultaneous effect of an electrical field combined with an applied pressure during the combustion, using the field-activated pressure-assisted synthesis process, was found to be suitable to produce good quality dense intermetallic compounds in a "one step process" from mechanically activated powder mixtures [13]. The feasibility of this new approach for the simultaneous synthesis and densification of nanomaterials was demonstrated in the Fe/Al, $NbAl_3$, and Mo/2Si systems [13].

REFERENCES

[1] Bernard F, Gaffet E, Gramond M, Gailhanou M, Gachon JC. Simultaneous infrared and time-resolved X-ray diffraction measurements in order to study the self-sustained reactions. J Synchrotron Radiat 2000;7:27–33. Vrel D, Girodon N, Paris S, Mazue JF, Couqueberg E, Gailhanou M, et al. A new experimental setup for the time resolved X-ray diffraction study of self-propagating high-temperature synthesis. Rev Sci Instrum 2002;73(2):422–8.

[2] Charlot F, Bernard F, Gaffet E, Klein D, Niepce JC. In-situ time resolved diffraction coupled with thermal I.R. camera to study mechanically activated SHS reaction: case of Fe-Al binary system. Acta Mater 1999;47(2):619–29. Gras Ch, Charlot F, Gaffet E, Bernard F, Niepce JC. In-situ synchrotron characterization of mechanically activated self-propagating high-temperature synthesis applied in Mo/Si system. Acta Mater 1999;47(7):2113–23. Carole D, Fréty N, Paris S, Vrel D, Bernard F, Marin-Ayral R-M. Investigation of the S.H.S. mechanisms of titanium nitride by in situ time-resolved diffraction and infrared thermography. J Alloys Compd 2007;436:181–6. Gauthier V, Cochepin B,

Vrel D, Dubois S. Self-propagating high-temperature synthesis of Ti3SiC2: study of the reaction mechanisms by time-resolved X-ray diffraction and infrared thermography. J Am Ceram Soc 2006;89(9):2899. Gauthier V, Bernard F, Gaffet E, Vrel D, Gailhanou M, Larpin JP. Investigations of the formation mechanism of nanostructured NbAl3 via MASHS reaction. Intermetallics 2002;10:377–89.

[3] Gaffet E, Bernard F, Niepce JC, Charlot F, Gras C, Le Caer G, et al. Some recent developments in mechanical activation and mecanosynthesis. J Mater Chem 1999;9:305–14. Gaffet E, Bernard F. Mechanically activated powder metallurgy processing: a versatile way towards nanomaterials synthesis. Ann Chim Sci Mater 2002;27:45–59.

[4] Gras Ch, Vrel D, Gaffet E, Bernard F. Mechanical activation effect on self-sustaining combustion reaction in Mo-Si system. J Alloys Compd 2001;314:240–50. Gauthier V, Josse C, Bernard F, Gaffet E, Larpin JP. Study of niobium aluminides synthesis using MASHS and M2AP processes. Mater Sci Eng A 1999;265:117–28.

[5] Gras Ch, Zink N, Bernard F, Gaffet E. Assisted self-sustaining combustion reaction in the Fe-Si system: mechanical and chemical activation. Mater Sci Eng A 2007;456:270–7.

[6] Carole D, Fréty N, Etienne-Calas S, Merlet C, Marin-Ayral R-M. Microstructural and mechanical characterization of titanium nitride produced by S.H.S. Mater Sci Eng A 2006;419:365–71.

[7] Lemarchand A, Bonnet JP. Modeling silicon carbide formation during heating at constant rate. J Eur Ceram Soc 2006;26:2389–96. Lemarchand A, Bonnet JP. Modeling silicon carbide synthesis on a submicronic scale. J Phys Chem C 2007;111:10829–35.

[8] Baras F, Kondepudi D. A multilayer model for self-propagating high-temperature synthesis of intermetallic compounds. J Phys Chem B 2007;111:6457–68. Baras F, Kondepudi D, Bernard F. Combustion synthesis of MoSi2 and MoSi2-Mo5Si3 composites: multilayer modeling and control of the microstructure. J Alloys Compd 2010;505:43–53.

[9] Vrel D, Hendaoui A, Langlois P, Dubois S, Gauthier V, Cochepin B. SHS Reactions in the NiO–Al system: influence of stoichiometry. Int J Self-Propag High-Temp Synth 2007;16(2):62–9.

[10] Tingaud D, Stuppfler L, Paris S, Vrel D, Bernard F, Penot C, et al. Time-resolved X-ray diffraction study of SHS-produced NiAl and NiAl-ZrO2 composites. Int J SHS 2007;16(1):12–7.

[11] Dumont AL, Bonnet JP, Chartier T, Ferreira JMF. MoSi2/Al2O3 FGM: elaboration by tape casting and SHS. J Eur Ceram Soc 2001;21:2353–60.

[12] Pascal C, Marin-Ayral RM, Tedenac JC, Merlet C. Combustion synthesis: a new route for repair of gas turbine component—achievements and perspectives for development of SHS rebuilt. J Mater Process Technol 2003;135:91–100. De Jouvancourt H, Record MC, Marin-Ayral RM. The effects of platinum concentration on the combustion synthesis of NiAl. Application in repairing Ni-based super-alloys. Mater Sci Technol 2007;23(5):593–9.

[13] Bernard F, Charlot F, Gaffet E, Munir ZA. One-step synthesis and consolidation of nanoaluminides. J Am Ceram Soc 2001;84(5):910–4. Paris S, Bernard F, Gaffet E, Munir ZA. Spark plasma synthesis from mechanically activated powders: a versatile route for producing dense nano-aluminides. Scr Mater 2004;50(5):691–6. Cabouro G, Chevalier S, Gaffet E, Vrel D, Boudet N, Bernard F. In situ synchrotron investigation of MoSi2 formation mechanisms during current activated SHS sintering. Acta Mater 2007;55(18):6051–63.

SHS in Georgia

Giorgi Onashvilli

In the country of Georgia, the development of self-propagating high-temperature synthesis (SHS) technology began as a result of the scientific relationship between academicians F. Tavadze and A. Merzhanov. They subsequently established the SHS Laboratory at the Institute of Metallurgy and Materials Science at the Georgian Academy of Sciences, which enabled large-scale research efforts. For this research, the institute's staff members were twice awarded the Georgian State Prize in Science and Technology (1999 and 2010), and in 2013 they were conferred the award named for the renowned Georgian metallurgist, G. Nikoladze. In 1985 an all-union conference on the interconnectedness of self-propagating high-temperature synthesis and materials science was held in Kutaisi. Furthermore, during 2005 a workshop in Tbilisi was organized to discuss the problems of SHS development in the South Caucasus.

Important results from research carried out in Georgia using SHS technologies are as follows:

1. Research resulted in the development of a technology for receiving carbon-free and low-carbon ferromanganese from I and II quality concentrates of Chiatura manganese ores. This method for manganese extraction increased recovery from 75% to 89%, as compared to the aluminum thermal method. Continuation of this work includes obtaining low-carbon and low-phosphorus manganese-containing alloys from waste (from slags, enriched remains), thus providing high ecological efficiency. A technology for obtaining silicides from Chiatura-Sachkhere sands was also developed.
2. SHS centrifugal molding has led to achievements such as the production of special-purpose cast hard alloys based on tungsten carbide; ferroboron with a boron content of ~20% (mass percent); a wide assortment of ligatures based on recovered manganese waste; ferrovanadium; a nickel boron ferroaluminum; as well as various forms and sizes of composite ceramics and ceramic metals. At atmospheric pressure, a porous composite filtering material intended for purification of water and air from dust has been produced. In high-pressure reactors, ultradispersion powders of hard alloys, nitrides, carbides, borides, and silicides have been produced.
3. Specially designated functional gradient materials have been produced, as well as two-layer and multilayer materials that have potential armor applications. In addition,

Concise Encyclopedia of Self-Propagating High-Temperature Synthesis
http://dx.doi.org/10.1016/B978-0-12-804173-4.00064-8

these materials are used in mining, extraction of oil and gas, manufacturing of cutting blades, and on the surfaces of fast-wearing items.

4. A technology was developed for procuring single-phase titanium-aluminum nanocrystalline compounds. These materials are characterized by considerably higher physical-mechanical properties as compared to the properties of multiphase materials obtained by existing methods. A nanocrystalline material, Ti-Al-Ti-B, has been developed and is used for light armor and anti-radiation purposes.
5. Several methods have been developed for the manufacture of boron-isotope-containing (^{10}B and ^{11}B) antiradiation materials. Current technological questions focus on the production cost and the cost comparison to those for existing technologies.
6. A new class of specially designed, high-intensity, dynamic-loading, resistant ceramic materials, based on the Ti-B-N and Ti-B-C-N systems, as well as the technology for producing them via the SHS method, has been produced. The materials are used for manufacturing armor plates, casts, press forms, drawing dies, and more, which are used to protect individuals and vehicles.
7. A new energy-efficient technology for producing nonmetallic and intermetallic materials that combines electric resonance irradiation and SHS has been developed. Its first application focused on the microwave electric resonance effect on chasms. This effect increases the combustion velocity by at least one order of magnitude, while maintaining product homogeneity. The new method requires 20% to 25% less energy than other high-temperature approaches, such as laser and plasma synthesis. As a result, power requirements are minimal for the production of typically expensive products, such as construction ceramics, superconductors, abrasives, and armor materials.

SHS in India

Singanahally T. Aruna

Self-propagating high-temperature synthesis (SHS) or combustion synthesis activity in India is fairly recent. SHS activities in India can be broadly discussed under the topics of solid-state combustion and solution combustion. The solution combustion route pioneered by K.C. Patil is widely used in India and abroad. In this article, the focus is on the practice of solid-state SHS activities based on the technology developed in Russia [1]. This approach makes use of the solid-state combustion of metal powders with nonmetals, resulting in the formation of materials such as nitrides, carbides, borides, and silicides.

In India most of the SHS activities are carried out in government-funded organizations such as Defence Laboratories, the Council of Scientific and Industrial Research (CSIR), the and Department of Science and Technology (DST). The major centers for the activities are the CSIR-National Metallurgical Laboratory (NML), the CSIR-National Aerospace Laboratories (NAL), the Defense Metallurgical Research Laboratory (DMRL), and the International Advanced Research Centre for Powder Metallurgy and New Materials (ARCI).

With a view to indigenize the SHS technology, in 1994 ARCI initiated a project in collaboration with Samara State Technical University (Russia) and eventually developed the first SHS pilot-scale plant in India (Fig. 1). The SHS facility was optimized to conduct two kinds of reactions (i) SHS-azide where sodium azide becomes one of the reactants, and (ii) filtration combustion, which uses industrial waste as the raw material to produce value-added powders of certain industrial nitrides. In a more recent work, a nanostructured, hard, and wear-resistant composite coating of Al_2O_3, TiB_2, and TiN was formed on low-carbon steel substrate by combined SHS and laser surface alloying (LSA). A high-power laser beam is passed over the coating to remelt the coating top, which enhances microhardness and tribological performance through modification of the microstructure of the top portion of the coating [2].

The local expertise was successful in stabilizing the synthesis conditions for many varieties of compounds and established the SHS technology for the production of the following bulk materials: (i) Si_3N_4 and Si_3N_4-SiC (whiskers), (ii) AlN, TiN, TiCN (powders), (iii) Si_3N_4-TiN (powders + whiskers), (iv) TiN_x (powders) from Ti turning scrap, (v) ZrN_x (powders) from Zr turnings scrap, (vi) AlN powders (from Al turnings scrap), and (vii) $MoSi_2$ powders. A state-of-the-art facility for research and development in advanced materials using SHS was developed under the aegis of the Department of

Concise Encyclopedia of Self-Propagating High-Temperature Synthesis
http://dx.doi.org/10.1016/B978-0-12-804173-4.00074-0

Fig. 1 SHS reactor built at ARCI.

Science and Technology, Government of India, in collaboration with premier institutes in Russia, Belarus, Ukraine, and Moldova using the available material expertise of DMRL. An economical process for the synthesis of AlN from Al scrap was developed using the SHS route. $MoSi_2$-based materials were prepared in a single step by SHS, starting from elemental powders. Composites such as $MoSi_2$-WSi_2 alloys, $MoSi_2$-SiC, and $MoSi_2$-TiC and complex composites such as (Mo, W)Si_2-SiC and (Mo,W) Si_2-TiC were prepared by SHS.

Group leaded by Dr. S.K. Mishra from NML studied the preparation and properties of a variety of borides, carbides, and their composites with alumina in collaboration with ISMAN. The effect of processing parameters on the sintering characteristics and microstructure was investigated. Recently, a new technique, called DSHS (double SHS), was developed, in which the reaction product of the first SHS was mixed with calculated amounts of Mg and H_3BO_3 powder and subjected to a second SHS [3]. The ZrB_2 powder produced by the DSHS technique yielded increased conversion, and the NaCl used as a diluent during SHS facilitated a reduction in the average particle size of SHS ZrB_2 powder from a range of 75–25 nm to a range of 25–40 nm after DSHS.

The solid-state SHS has been practiced at the Indian Institute of Technology, Kharagpur, in collaboration with NML and ARCI. Recently, hard nanostructured Al_2O_3-TiB_2-TiN coatings were developed in collaboration with ARCI using a combination of SHS and LSA [4].

At NAL, the combustion synthesis of fully dense intermetallic-ceramic composites has been conducted. Techniques based on SHS were developed to produce centrifugal castings from metals, alloys, and ceramics that are useful for both coatings and component development. Intermetallics such as Ni-Al were prepared by centrifugal force-assisted

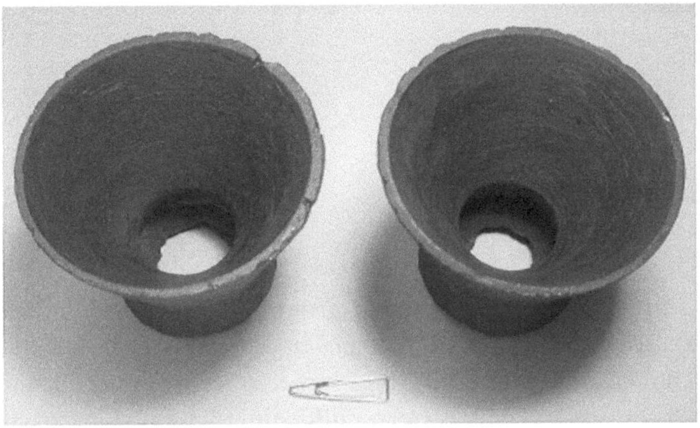

Fig. 2 Zirconia-based cermet convergent-divergent nozzles.

combustion synthesis (CFACS). The facility was created in-house. Significant progress was made in providing ceramic lining to axisymmetric metal components based on CFACS. The same route was also exploited to produce ceramics for high-temperature applications (>3000 K). Zirconia-based ceramic inserts were also successfully developed for rocket thrusters using combined thermite reactions under centrifugal force (Fig. 2) [5]. Thrusters were successfully tested, and the lining withstood temperatures as high as 3000 K.

A few conferences and workshops have been organized in India related to SHS process.

REFERENCES

[1] Merzhanov AG. 40 years of SHS: a lucky star of a scientific discovery a presentation with elements of a scientific lecture. Bentham eBooks; 2012.

[2] Chatterjee S, Shariff SM, Padmanabham G, Majumdar JD, Choudhury AR. Surf Coat Technol 2010;205:131–8.

[3] Khanra AK, Pathak LC, Godkhindi MM. J Mater Process Technol 2008;202(1):386–90.

[4] Chatterjee S, Shariff SM, Majumdar JD, Choudhury AR. Int J Adv Manuf Technol 2008;38:938–43.

[5] Arul Paligan A, Shubha V, Ramesh TG. Centrifugal thermit process—an overview. In: Proceedings of the international conference on aerospace science and technology, 26–28 June; 2008.

SHS in Italy

Giacomo Cao, Roberto Orrù

The scientific output of the seven research groups that have been active in Italy over the past 20 years cover diverse classical topics in SHS, that is, combustion theory and modeling, structural macrokinetics, synthesis of ceramics, intermetallics and composites, and nontraditional processes, which creates difficulties when taking trying to deal with them in a systematic manner. The space dedicated in this work to each institution is similar, and they have been ordered alphabetically. (It should be noted that we did not dedicate a section to the University of Rome because the SHS literature includes only one contribution from that institution.)

UNIVERSITY OF CAGLIARI (PROFESSORS G. CAO AND R. ORRÙ)

The SHS and field-activated combustion synthesis processes or their combination with ball milling or mechanical activation to synthesize various materials such as TiC/Fe, TiC-Ti, $BaTiO_3$, ultra-high temperature ceramics, $TiAl_3$, TiAl, Ti_3Al, and Fe-$SrTiO_3$ were conducted from 1997 to 2007. The collaborations of the Prof. G. Cao on the early stage of activity with Professor Z.A. Munir focused on the synthesis of titanium aluminides and with Professor A. Varma on the modeling of SHS have been important for the development of the research group.

UNIVERSITY OF GENOA (PROFESSOR M. FERRETTI)

This institution was involved in the combustion synthesis of various metal nitrides from 1998 to 2001. This research group now designs and uses SHS in a high-temperature, high-pressure stainless steel autoclave in a nitrogen atmosphere in the pressure range 5–100 MPa.

NATIONAL RESEARCH COUNCIL IENI, MILAN, ING. (C. ZANOTTI)

This group's research activity was devoted to SHS of Ti-Al, Ti-Ni and functionally graded materials (FGM) from 2002 to 2007. In particular, SHS was used to prepare boride-based FGM as a thermal barrier for space reentry vehicles. FGM comprised a

Concise Encyclopedia of Self-Propagating High-Temperature Synthesis
http://dx.doi.org/10.1016/B978-0-12-804173-4.00082-X

ceramic phase, that is, MB_2 (M = Ti, Zr, Hf), and an intermetallic phase, that is, NiAl, joined together by composite layers of the graded stoichiometries of the two components.

UNIVERSITY OF PAVIA (PROFESSOR U. ANSELMI TAMBURINI)

This group is active in SHS and related fields, and through several national and international collaborations. SHS of various systems such as Ni-Al foils, Y_2O_3-ZrO_2, Ni/YSZ, ZrO_2, Zr-Si, Fe- ZrO_2, Cr-Si, metastable solid solutions, Ta-Si, Ti-Si, Co-Al, Nb-Al, metal-metal oxides, Ti-Al, Ti-Ni, and FGM were investigated from 1989 to 2006. The synthesis of a nickel/yttria-stabilized zirconia (Ni/YSZ) cermet with tailored microstructural characteristics, which are generally obtained through a route based on the reduction, via hydrogen gas, of poorly sintered mixtures of YSZ with nickel oxide, deserves attention.

UNIVERSITY OF SASSARI (PROFESSORS G. COCCO AND A. MARIANI)

Two research groups are active on SHS and related subjects at the University of Sassari. Professor G. Cocco's group was involved in the investigation of combustion behavior dioxins, herbicides, and hexachlorobenzene from 1997 to 2004. Environmental SHS was also addressed in collaboration with the group operating at the University of Cagliari during 1999. Ignition mechanism of mechanically activated powders was investigated from 2000 to 2004.

The group, coordinated by Professor A. Mariani, was active in frontal polymerization (FP) of acrylic monomers, polyurethanes, dicyclopentadiene, epoxy resins from 2001 to 2007. Specifically, FP was successfully applied, for the first time, to obtain polymeric nanocomposites containing polyhedral oligomeric silsesquioxanes in an amine-cured epoxy matrix.

TURIN POLYTECHNIC UNIVERSITY (PROFESSOR I. AMATOE AND B. DEBENEDETTI)

SHS of several metal-ceramic or ceramic-ceramic composites was performed from 2003 to 2005. In addition, gel-combustion of tin oxide nanopowders was investigated in 2006. One of the most intriguing activities conducted was the use of metastable processing based on the SHS technology followed by quenching for the synthesis of a broad class of nanostructured materials.

SHS in Japan

Osamu Odawara

Japanese research activities on SHS process were started in 1979 by the R&D on ceramic-linings with thermite reactions performed inside a centrifuged hollow pipe (centrifugal-thermite process) conducted at the Advanced Industrial Science and Technology (AIST) in Sendai, Ministry of Economy, Trade and Industry (METI). This project aimed to develop geothermal casing pipes resistant to corrosion, abrasion, and heat as part of the national SUNSHINE project. The Japanese SHS-oriented investigations accelerated in 1982 under the influence of J.F. Crider's review paper published in Ceramic Engineering and Science Proceedings.

The Japan Research Association of Combustion Synthesis (JRACS) was established in 1987 with members from 5 Universities, 4 National laboratories, and 29 Industries. In addition, in 1984 the groups from the AIST in Sendai, Tohoku University, and the Kakuda Space Center discussed the concept of "functionally graded materials," and the Functionally Graded Materials Forum of Japan was established in 1988. At first the JRACS mainly focused on simultaneous synthesis and sintering of materials with the application of external forces:

- After applying high pressure in the SHS process, a high pressure combustion sintering technique was proposed by a group from Osaka University.
- An R&D on an energy-saving system for high-temperature inorganic materials was conducted by the AIST of METI in Sendai as the national MOONLIGHT project.
- As for the powder sintering techniques, not only HIP and CIP but also the spark isostatic pressing process, electrical discharge compaction, and plasma sintering were investigated.

The Japanese book Nensyou Gousei No Kagaku (Chemistry of Combustion Synthesis) was published by the JRACS in 1992. The book comprises 12 chapters:

(1) "Ceramic Powder Synthesis," (2) "Intermetallic Compounds Synthesis," (3) "Single Crystal Floating-Zone Formation With SHS Precursors," (4) "High Pressure Combustion Sintering Process," (5) "Centrifugal-Thermite Process," (6) "Combustion Impregnation Method," (7) "Reaction-Fusion-Adhesion Process," (8) "Pseudo-HIP Method," (9) "Nitride SHS Process," (10) "Thermite Reaction Under High Pressure Process," (11) "SHS-Bonding," and (12) "Shock Compaction Process."

Concise Encyclopedia of Self-Propagating High-Temperature Synthesis
http://dx.doi.org/10.1016/B978-0-12-804173-4.00083-1

The book played an effective role in providing knowledge to many researchers in Japan about the SHS process as a novel high-temperature material synthesis technology that works without high-temperature furnaces and complex processing equipment, and that is an energy-saving approach that allows rapid production of large quantities of ceramics.

Japanese R&Ds on the SHS process were widely developed in 2000s not only in short-time process controls for product formation and/or sintering by applying external forces to the SHS process [1] but also in other directions, including the following:

- Research on "nanosize diamond formation" was conducted with crystal diamond processing by Tomei Diamond Co. Ltd.; combustion-flame technology was developed by Tokai University and detonation diamond technology by Tokyo Institute of Technology.
- Combustion Synthesis Co. Ltd. in Shizuoka developed a manufacturing process of β-SiALON in collaboration with Hokkaido University by means of the SHS process using NaCl as a diluent. Their products are high-purity β-SiAlON-($Si_{6-z}Al_2O_2N_{8-z}$, z: 0.25–3) powders with submicron size [2]. The product's particle size decreases with the increase of NaCl, showing diameters of about 300 nm in particle size when $z=3$.
- "Hydride combustion synthesis" [3] was developed by Hokkaido University for the SHS of TiFe-based hydrogen storage alloy from Fe and TiO_2 using metallic calcium as a reducing agent and heat source. Finely powdered products were obtained without pulverization processes. This method created an innovative production route for TiFe.
- "Induction field activated combustion synthesis" [4] was successfully applied to SHS of Mg_2Ni with high purity and fine properties as a hydrogen storage material from elemental Mg and Ni powders by Ryukoku University.
- "Multilayered foil SHS" investigated by the Japan Aerospace Exploration Agency formulated the theory for heterogeneous flame propagation in the SHS process that proceeds in multilayer foils consisting of alternating layers of constituents.
- Porous TiC and TiC/Ti composites can be synthesized by SHS process. The heat of reaction between titanium and carbon is high enough to induce the self-sustaining reaction of TiC formation on condition that processing parameters of Ti/C ratio and porosity of the precursor were appropriately selected. The pore morphology is controlled by changing the Ti/C ratio and the preheat temperature [5].
- SHS processes have also been applied in Japan for frontier exploration of space, not only around the Earth's orbit environment, but also the orbits of the Moon, Mars, and other such environments. It is expected that resources such as metal oxides that exist on planets or satellites could be refined and utilized as a supply of heat energy, where the SHS processes, under high-gravitational or microgravity conditions will be an excellent approach for in situ resource utilizations.

REFERENCES

[1] Odawara O. Mass-forced SHS technology of ceramic materials. Adv Sci Tech 2010;63:302–11.

[2] Niu J, Yi X, Nakatsugawa I, Akiyama T. Salt-assisted combustion synthesis of β-SiAlON fine powders. Intermetallics 2010;35:53–9.

[3] Tsuchiya T, Yasuda N, Sasaki S, Okinaka N, Akiyama T. Combustion synthesis of TiFe-based hydrogen storage alloy from titanium oxide and iron. Int J Hydrogen Energy 2013;38:6681–6.

[4] Kodera Y, Yamasaki N, Yamamoto T, Kawasaki T, Ohyanagi M, Munir ZA. Hydrogen storage Mg_2Ni alloy produced by induction field activated combustion synthesis. J Alloys Compd 2007;446–447:138–41.

[5] Kobashi M, Ichioka D, Kanetake N. Combustion synthesis of porous TiC/Ti composite by a self-propagating mode. Materials 2010;2010(3):3939–47. http://dx.doi.org/10.3390/ma3073939.

SHS in Kazakhstan

Georgiy I. Ksandopulo, Zulkhair A. Mansurov, Nina N. Mofa, Sergey M. Fomenko, Roza G. Abdulkarimova

In the early 1980s, G.I. Ksandopulo became the first scientist in Kazakhstan to introduce the SHS process. At Kazakh State University (the present name is Al-Farabi Kazakh National University), the Chemical Kinetics and Combustion Chair was organized where research on combustion processes started and where, through the efforts of G.I. Ksandopulo and coworkers, the titanates with alkaline-earth elements were obtained.

Successful experimentation of SHS brought about the first symposium on SHS. Among further SHS research trends were layered reactions where the product of the first layer is a reagent for the next layer. But the most significant factor was the use of SHS for the production of ferro alloys of wolframite, molybdenum, titanium, nickel, and vanadium. At the Institute of Combustion Problems, a pilot-phase production of ferrous-based alloys as well as aluminum powder was created, which became the basis for the "Small metallurgy of Kazakhstan" enterprise with a productivity rate of 60 tons per month, and with ferrowolfram producing 50 tons per month and ferromolybdenum and aluminum powder producing 150 tons per month. In addition, the establishment of factories for the production of ferrovanadium (Kyzylorda) and ferrotitanium (Almaty) is planned.

The combustion of layered systems under the conditions of centrifugal acceleration was investigated theoretically and experimentally, and as a consequence, the composite gradient materials were obtained. With the participation of A.A. Sauhimov et al. at Kazakh National Research Technical University named after K.I. Satpayev, the tertiary oxides of the Y-Fe-O system were realized through SHS method. R.G. Abdulkarimova group developed the refractory materials and samples based on SiCAlON, AlCON, and corundum. Z.A. Mansurov group obtained the silicon nanopowders and silicon carbides. N.N. Mofa group are engaged in synthesis of SHS ceramic systems of different applications using mechanochemical treatments.

The most attractive aspect of SHS is the production of refractory materials that contain titanium carbides and silicon carbides for obtaining refractory carbides. Systematic investigations of solid-phase metallothermic combustion of oxide systems in order to develop a new production technology of refractory materials were carried out at the Institute of Combustion Problems [1]. The result of these studies was the creation of SHS technology for high-temperature refractory materials and welding compounds of

Concise Encyclopedia of Self-Propagating High-Temperature Synthesis
http://dx.doi.org/10.1016/B978-0-12-804173-4.00118-6

the "furnon" series, which today has wide practical application for the repair (fettling) of high-temperature thermal generating units such as furnaces for limestone and cement calcining, preheat furnaces for coke calcination and anode paste, vaults of electric steel furnaces, package boilers, and so on [2]. After successful tests at the factories in Germany, Spain, Great Britain, Cuba, and China, this technology was adopted by 120 large metallurgic and cement factories. With the help of the SHS method, materials such as ceramic electric heaters and chemically resistant coatings and SHS catalysts were produced. Equipment for processing lines and for the production of rare ferrous-based alloys (such as molybdenum, tungsten, vanadium) were created.

The use of previously activated raw materials for the production of ceramic systems is one of the most effective and rising areas in material science [3]. The mechanical effect on the development of chemical reactions allows expanding SHS possibilities. With the help of MA and SHS, the ceramic products of different applications were synthesized.

Silicon is the basic element necessary for the development of microelectronics, solar energy, and so on. For producing silicon, the SHS process has numerous advantages over many synthesis powder methods, especially the synthesis of high-melting compounds such as silicon carbide. A method for producing silicon nanopowders and silicon carbide in an SHS regime based on quartz of various sizes was developed [4].

Filtration combustion in revolving reactors has particular properties and a high energy capacity. Centrifugal acceleration reduces the lower concentration limit of combustion and provides a steady process-behavior regime even in highly diluted mixtures. The high speed of the combustion wave front leads to the transition of the entire system to a liquid-phase state, reduces melt viscosity, and encourages active separation of synthesis products in accordance with their density. The various composites, such as $Al_{20}B_4O_{36}$, WB_2, W_2B, WB, and ferrochrome-borium (FeB, Cr_2B, Cr_5B_3, CrB) have been synthesized [5–9].

The research area focusing on the production of borides, silicides, and carbides was based on the study of different reaction mixtures by centrifugal accelerated clusters of various metals, and it was successfully developed. The first materials obtained included radicals that contained extra stoichiometric oxygen as well as borides and carbides of various metals. It is expected that the materials synthesis has high-temperature conductivity properties, modification of crystallizable melting of metals, and an effectiveness in some biological processes. This part of the work is published and patented. There have been 69 applications for an invention, author certificates, and patents since Jan. 23, 1986.

REFERENCES

[1] Ksandopulo GI, Ismailov MB. SHS refractory materials "Furnon". In: Collection: problems of gaseous combustion and condensed systems, Alma-Ata, no. 5; 1993. p. 5.

[2] Mansurov ZA. SHS refractory materials "Furnon" and their practical implementations in Kazakhstan and Russia. Adv Sci Technol 2010;63:312–21.

[3] Mofa NN. Composite materials are obtained by SH-synthesis is pre-activated and modified by quartziferous material. In: Proceedings of All-Russian conference "Combustion processes and explosion in physicochemistry and technology of inorganic materials"; 2002. p. 278–82.
[4] Yermekova Z, Mansurov Z, Mukasyan A. Combustion synthesis of silicon nano powder. Int J SHS 2010;19(2):96–103.
[5] Mansurov ZA, editor. SHS-composite materials. In: Almaty: Қазақ университеті; 2015. 402 p.
[6] Ksandopulo GI. Non-chain autoacceleration of SHS-wave in condition of rotation. Int J SHS 2015;24(1):8–13.
[7] Ksandopulo G, Baydeldinova A, Kartkuzhakov S, Bairakova O, Gassan N, Xanthopoulou G. Preventation of radioactive and other contaminant leaching into soil by high temperature synthesis fixing. Euro-Asian J Sust Energy Dev Policy 2010;2(1):210–2.
[8] Ksandopulo GI, Baideldinova AN, Omarova KI, Ainabayev AM. Initiating potential of centrifugally accelerated metal particles in the inorganic synthesis reactions. Eurasian Chem Technol J 2014;16(l):49–53.
[9] Ismagilov ZR, Mansurov ZA, Shikina NV, Yashnik SA, Aldashukurova GB, Mironenko AV, et al. Nanosized Co-Ni/glass fiber catalysts prepared by "Solution-Combustion" method. Nanoscience and Nanotechnology 2013;3(1):1–9.

SHS in Liquid Nitrogen

Inna P. Borovinskaya

One of the unique methods for synthesizing metal and nonmetal nitrides is combustion of elements in liquid nitrogen. This method was carried out for the first time in ISMAN in the 1970s after scientists discovered that during annealing of the intermediate products of zirconium compact nitriding in gaseous nitrogen by "throwing" them into liquid nitrogen, the combustion process did not stop; it became more intensive. Then the combustion of zirconium samples was purposefully carried out in an open Dewar flask with liquid nitrogen at nitrogen excess [1]. A thin film formed around the sample; the combustion velocity was 1 cm/s. Later the experiments were organized in hermetically sealed metal reactors withstanding high pressures, higher than 4 kbar. Very high combustion temperatures were observed at the time ($\geq$2500°C). Nitrides of transition metals (Ti, Hf, Nb, Ta, etc.) and nonmetals (B, Al, Si) were synthesized by this method; previously, they had been obtained by combustion of elements in gaseous nitrogen. As a rule, the structure and chemical composition of the metal nitrides are the same irrespective of the method of their production including SHS in gaseous nitrogen. But niobium, tantalum, and hafnium nitrides can contain higher amounts of nitrogen, that is, they can be superstoichiometric; tantalum nitride forms a cubic modification which was unknown before [2]. In the X-ray diffraction patterns of aluminum nitride, some additional lines were observed that belonged to an unknown phase of AlN. The most interesting results were obtained at boron powder combustion in liquid nitrogen [3]. Combustion of boron powders or compacts is carried out in special hermetically sealed cryogenic reactors. Theoretical calculations prove that the pressure in such reactors can rise up to $\geq$2000 MPa at liquid nitrogen evaporation. Both elemental boron and its mixtures with magnesium borides and other compounds (Li, LiN, Mg, Sn, etc.), which act as catalysts in formation of cubic β-BN (borazon), are used in the experiments. For the first time, β-BN (wurtzite-like form) was discovered at combustion of boron with metal magnesium. In such experiments, hexagonal α-BN is obtained as thin (<10 nm) transparent laminar crystals or "ingots" of 0.5 mm in size, which is equal to β-BN in chemical stability. This product is suggested to appear because of the reverse transition β-BN → α-BN during the reactive mass cooling. After the reactive mass treatment with $NaOH + Na_2CO_3$, the crystals of cubic BN that precipitated in bromoform are 100–150 μm in size.

Concise Encyclopedia of Self-Propagating High-Temperature Synthesis
http://dx.doi.org/10.1016/B978-0-12-804173-4.00126-5

But it's necessary to note that experiments in combustion of boron and other elements in liquid nitrogen in hermetically sealed reactors are very dangerous. In some experiments, the pressure jumps exceeded nitrogen pressure by 20 kbar, and even combustion changeover to detonation up to the installation breakdown occurred. Treatment of the annealed products with acids can be followed by popping with fire. Probably, the intermediate or by-products at fast cooling are explosive azides.

Because of the available literature, the process is investigated in Japan. Some experiments in nitriding of titanium wire in liquid nitrogen are described [4]. The wire is heated and ignites a titanium compact inside the hermetically sealed reactor. An increase in nitrogen pressure >60 MPa and temperature is observed. The sample is densified to 90%. The product is $TiN_{0.87}$ without any titanium impurities. Changes of combustion temperature and nitrogen pressure during the process have been studied; characteristics of the obtained products are demonstrated. Some other results have been published, too.

REFERENCES

[1] Merzhanov AG, Borovinskaya IP. A new class of combustion processes. Combust Sci Technol 1975;10:195–201.

[2] Merzhanov AG, Borovinskay IP, Butakov AA, Rabinkin AG, Shehtman VSh. Method for the production of tantalum mononitride. Patent SU 264365; 1970. Bull. Invent; 1970.

[3] Borovinskaya IP, Loryan VE, Zakorzhevsky VV. Combustion synthesis of nitrides for development of ceramic materials of new generation. In: Gromov AA, Chukhlomina LN, editors. Nitride ceramics: combustion synthesis, properties, and applications. Weinheim: Wiley-VCH Verlag GmbH & Co. KGaA; 2014. [chapter 1], ISBN 978-3-527-33755-2 [chapter 1].

[4] Shiuya M, Despres JF, Odawara O. TiN_x combustion synthesis with liquid nitrogen filled in a closed vessel. Int J SHS 1999;8(3):315–20.

SHS in Poland

Jerzy Lis

BRIEF HISTORY

SHS was introduced to Poland in the 1970s by Professor Roman Pampuch and his coworkers at the AGH-University of Science and Technology, Faculty of Materials Science and Ceramics, Kraków (AGH-UST). They initially studied combustion phenomena which occur during processing of SiC-Si composites [1]. The majority of later work confirmed the applicability of SHS to preparation of sinterable non-oxide ceramic powders. The aim of the AGH-UST work was preparation of single-phase sintered polycrystals, as well as multiphase ceramic composite materials from SHS-derived powders. Dense polycrystals with interesting properties were synthesized for materials in numerous explored systems: Si-C-B, Si-C-N, Si-Al-O-N, Si-Al-N-C, Al-O-N, Ti-B-C, Ti-Si-C, Ti-Si, Ti-Al-C, Ti-Si-C-N, and Ti-Al-C-N [2–5]. The main researchers working in the SHS AGH-UST group were R. Pampuch, L. Stobierski, J. Lis, D. Kata, M. Bućko, L. Chlubny, D. Zientara, A. Gubernat, M. Rączka, T. Rudnik, and M. Łopaciński.

A second SHS research effort took place at the Centre of High Pressures UNIPRESS Polish Academy of Sciences in Warsaw. These investigations, performed by Dr. Małgorzata Skibska in 1980s, focused on the preparation of silicon nitride powders by SHS.

A new research line opened at the Laboratory of Nanomaterials Physics and Chemistry, Department of Chemistry, Warsaw University (WU) through a group headed by Professor Andrzej Huczko. There, they have made extensive studies on SHS and solution combustion synthesis (SCS) applications, including the discovery of SHS of SiC nanofibers [6].

PRESENT SHS ACTIVITY

AGH-UST

The following selected systems are investigated in terms of their preparation of ceramic powders and materials:

(i) Studies on solid combustion in the Si-C-N system were carried out by Professors D. Kata and J. Lis. The goal was to prepare composite $SiC\text{-}Si_3N_4$ and pure SiC nano-powders [7]. Conventional SHS in the Si-C system provided products in the form

Concise Encyclopedia of Self-Propagating High-Temperature Synthesis
http://dx.doi.org/10.1016/B978-0-12-804173-4.00109-5

of strongly agglomerated SiC coarse grains required long milling times to obtain the desired nanopowders. However, under controlled nitrogen pressure, the Si+C reaction pathway provided a much more efficient method for preparation of SiC and composite SiC-Si_3N_4 nanopowders. Obtained powders had high sinterability and could be used in a wide range of applications including: refractory materials, armor material, cutting tools, and different parts of chemical equipment.

(ii) The group headed by Professor Mirosław Bućko with J. Lis, D. Zientara, and PhD students, worked on SHS of powders in the Al_2O_3-AlN system. This system allows preparation of precursor powders for single-phase γ-AlON materials [8] or composites with nitride inclusions contained within a γ-AlON matrix. In this method, a mixture of aluminum, aluminum oxide, and selected metal (B, Ti, Ta, Nb, or Cr) powders were reacted in a nitrogen atmosphere. Hot pressing or sintering of powders produced composites of γ-AlON + nitride that possessed fracture toughness, hardness, and wear resistance significantly higher than pure γ-AlON materials. In the case of hexagonal boron nitride inclusions, composites were textured and demonstrated strong anisotropy of thermal conductivity.

(iii) The group headed by Professor Jerzy Lis and Dr. Leszek Chlubny have extensive experience in the manufacturing of MAX phases by the SHS method. Their work concerned the preparation of Ti_3SiC_2 in which the authors were the first to synthesize active and sinterable Ti_3SiC_2 powders by SHS and to obtain bulk materials by sintering. Many different applications of Ti_3SiC_2-based ceramics such as ceramic armor, cutting tools, FGM materials, and many others, were tested and reported. The purpose of the work was to produce dense, single, or near-single phase ternary MAX phases in the system of Ti-Al-C-N system during hot-pressing or pressure-less sintering of SHS-derived active precursor powders. Dense MAX materials (Fig. 1) possess unique structures and functional properties [9].

Warsaw University

The synthesis of SiC nanofibers proceeded according to the scheme presented in Fig. 2. The process was optimized and SiC nanofibers produced were used for preparation of nanocomposites [10]. In the past decade, many other compounds (TaC, C_3N_4, YAG: Ce^{+3} garnet, and graphene) have been synthesized, and the synthesis has developed online via optoelectronic diagnostics.

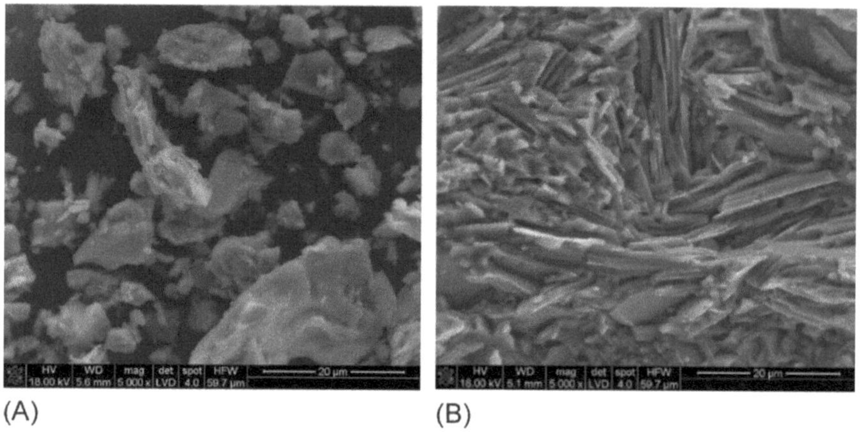

Fig. 1 Ti_3AlC_2 MAX phase from SHS powders: (A) SHS powder; (B) sintered dense polycrystal.

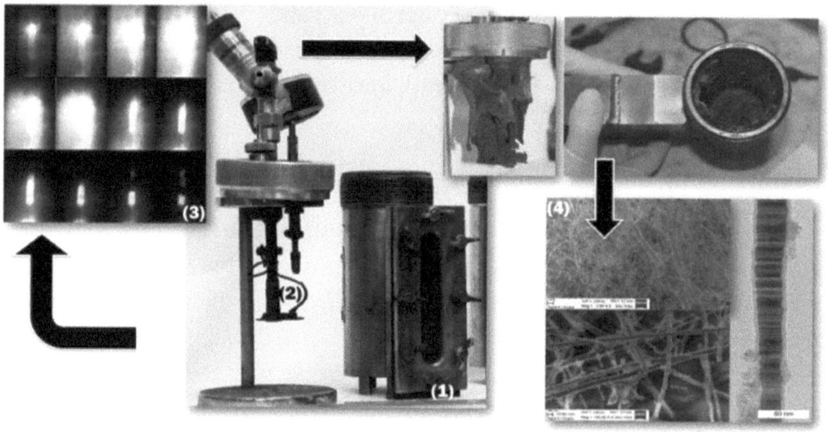

Fig. 2 SHS production of 1D SiC. (1,2): experimental setup; (3): optical diagnostics of combustion; (4): reaction products.

REFERENCES

[1] Pamuch R, Białoskórski J, Walasek E. Ceram Int 1987;13(1):63–7.
[2] Stobierski L, Ermer E, Pampuch R, Lis J. Ceram Int 1993;19:231–4.
[3] Pampuch R, Stobierski L. J Int J SHS 1993;2(3):159–64.
[4] Lis J, Miyamoto Y, Pampuch R, Tanihata A. Mat Let 1995;22:163–8.
[5] Lis J, Kata D, Chlubny L, Łopaciński M, Zientara D. Ann Chim Sci Mat 2003;28:S115–22.
[6] Huczko A, Lange H, Chojecki G, Cudziłło S, et al. J Phys Chem B 2003;107:2519–24.
[7] Kata D, Lis J. Glass Phys Chem 2005;31(3):364–9.
[8] Zientara D, Bućko MM, Lis J. J Eur Ceram Soc 2007;27:775–9.
[9] Chlubny L, Lis J, Chabior K, Chachlowska P. Arch Metall Mat 2015;60(2A):859–63.
[10] Huczko A, et al. J Crys Growth 2014;401:469–73.

SHS in Russia

Alexander S. Rogachev

Russia takes a special place in the history of SHS: the *Discovery of SHS* was made in Russia. The First experimental results and products of SHS were obtained by A.G. Merzhanov, I. P. Borovinskaya, and V.M. Shkiro in the Branch of the Institute of Chemical Physics, Russian Academy of Sciences in Chernogolovka (a scientific center near Moscow) in 1967. Later (1987), a new institute, *ISMAN*, was founded for research and development in the field. Up to now, *ISMAN* keeps leading positions in theoretical, experimental, and technological aspects of SHS [1,2].

New scientific directions required human resource development, first of all, training of high-educated researches and engineers. A leading university that solved this problem in Russia is the National Research Technical University, Moscow Institute of Steel and Alloys (MISiS) NITU. The Department of Powder Metallurgy and Functional Coatings, as well as the Scientific-Education Center of SHS (MISiS-ISMAN; head of the department and director of the center is Professor E.A. Levashov), teach bachelor, master and PhD candidates in the field of SHS. Curricula for foreign students (in English) are also available. The research activity of MISiS involves a wide variety of SHS materials and technologies: cemented carbides, STIM, superhard diamond-containing materials, electrodes and targets for surface engineering, biocompatible alloys, super-refractory compounds, intermetallics, multifunctional coatings, and so on. A research center of Structural Ceramic Nanomaterials at MISiS (director of the center is Professor A.S. Mukasyan), equipped with advanced laboratory and technological installations, develops methods of SHS + SPS (spark plasma sintering), mechanically activated SHS, *Solution Combustion Synthesis*, and more for production of nanocrystalline ceramics of SiC, B_4C, oxide, and metal nanopowders joining materials.

Among other Russian *SHS Centers*, the Department of Structural Macrokinetics at the Tomsk Scientific Center, Siberian Branch of the Russian Academy of Sciences (founded in 1974 by Professor Yu.M. Maksimov) in collaboration with Tomsk Polytechnic University and Tomsk State University demonstrate a wide spectrum of research and development. New physical effects, such as emission of electrons, ultraviolet and X-ray, and regimes of SHS were discovered in this department.

The Engineering Center of SHS in Samara and the Department of Physical Metallurgy, Powder Metallurgy and Nanomaterials of the Samara State Technical

Concise Encyclopedia of Self-Propagating High-Temperature Synthesis
http://dx.doi.org/10.1016/B978-0-12-804173-4.00120-4

University (headed by Professor A. P. Amosov) mostly work with SHS using azides and halides for the production of Si_3N_4, SiC, nanocomposites Al-AlN, and many other materials [3–5].

Mechanical activation of SHS is the focus of the Institute of Solid State Chemistry and Mechanochemistry, Siberian Branch of the Russian Academy of Sciences in Novosibirsk (director academician is N.Z. Lyakhov). This institute is a worldwide leader of this research direction, it also produces unique equipment for mechanical activation. Outstanding results on the mechanisms of structure formation during SHS have been obtained by scientists of this institute in cooperation with the Budker Institute of Nuclear Physics and V.V. Voevodsky Institute of Chemical Kinetics and Combustion, Siberian Branch of the RAS, with participation of ISMAN. They perform a series of research based on time-resolved synchrotron rays diffraction of different SHS processes (academician is V.V. Boldyrev, Professors B.P. Tolochko, M.A. Korchagin, A.S. Rogachev, and others). The results shed light on the mechanisms of SHS reactions; temporal resolution better than 1 ms was achieved [6].

Significant scientific and methodological contributions were made by Kirensky Institute of Physics, Russian Academy of Sciences, Krasnoyarsk (Dr. S. Zharkov, TRTEM, thin films), Polzunov Altai State Technical University (group of Professor V.V. Evstigneev (1937–2008), high-speed micro-pyrometry), and many other research groups, laboratories, and institutes [7].

REFERENCES

[1] Merzhanov AG. 40 Years of SHS: a lucky star of a Scientific Discovery (A presentation with elements of a scientific lecture). Bentham Science Publishers; 2012. 104 p.

[2] Merzhanov AG. It is better to be required than to be free. Chernogolovka: Territoria; 2005. 255 p. [Russ.].

[3] Maksimov YuM, Chuklomina LN, Braverman BS, Smirnov LA. Self-propagating high-temperature synthesis of nitrogen-containing alloys for metallurgy. Novosibirsk: Nauka; 2014. 231 p.

[4] Amosov AP, Borovinskaya IP, Merzhanov AG. Powder technology of self-propagating high-temperature synthesis of materials, vol. 1. Moscow: Mashinostroenie; 2007 [Russ.].

[5] Levashov EA, Rogachev AS, Kurbatkina VV, Maksimov YuM, Yukhvid VI. Advanced materials and technology of self-propagating high-temperature synthesis. Moscow: Izd. MISiS; 2011. 378 p. [Russ.].

[6] Grigorieva T, Korchagin M, Lyakhov N. Combination of SHS and mechanochemical synthesis for nanopowder technologies. KONA 2002;20:144–58.

[7] Rogachev AS, Mukasyan AS. Combustion for material synthesis. Boca Raton, London, New York: CRC Press Taylor & Francis Group; 2015. 397 p.

SHS in South Korea

Hayk Nersisyan, Sr., Hyong Il Won, Chang Whan Won, Jong Hyeon Lee

SHS processes were introduced in Korea in early 1995. A number of universities and organizations were involved in research activity for the production of various compact and porous inorganic materials. A list of organizations involved in SHS activity and their main field of interest is as follows:

Chungnam National University: Nanostructured materials and LED phosphor powders
Korea Institute of Machinery & Materials: Porous $MoSi_2$
Korean Atomic Energy Research Institute: TiC, SiC, and ZrC powders and C/SiC/C multilayer composite
Chonbuk National University: $MoSi_2$-SiC composite
Ulsan University: Porous TiNi alloy
Hong Ik University: $MoSi_2$ and AlN
Pusan National University: AlN and ZrB_2
Korea Institute of Geoscience and Mineral Resources: AlN powder
Changwon National University: SiAlON
Hanbat University: $MoSi_2$-SiO_2
Kyungnam University: TiC, SiC, WC, and composites
Sun-Moon University: TiC nanofibers
Dong-A University: Ba-Zn ferrites

Among the preceding organizations listed, a research group in RASOM (Chungnam National University, CNU), which involved Professsor Chang Whan Won, Professsor Jong-Hyeon Lee, Dr. Hayk Nersisyan, and Dr. Hyung-Il, carried out fundamental investigations in the field of SHS of 1D, 2D, and 3D nanostructures and LED phosphor powders. The methodology included the reaction of oxides (WO_3, TiO_2, B_2O_3) and organic polymers (e.g., $C_2H_3Cl_n$ and C_2F_{4n}) with inorganic reduction agents (e.g., Mg, Zn, NaN_3, $NaBH_4$) and use of special additives for controlling the shape and size of product particles. The combustion processes that operated under optimized conditions resulted in the fabrication of technologically useful inorganic nanomaterials [1,2]. TEM images for some of these nanoparticles are shown in Fig. 1. Typical examples include (A) carbon nanoballs, (B) hollow carbon nanospheres, (C) honeycomb-structured carbon, (D) a few atomic layer BN nanosheets, (E) B_4C nanosheets, and (F) MoS_2 nanoparticles.

Concise Encyclopedia of Self-Propagating High-Temperature Synthesis
http://dx.doi.org/10.1016/B978-0-12-804173-4.00138-1

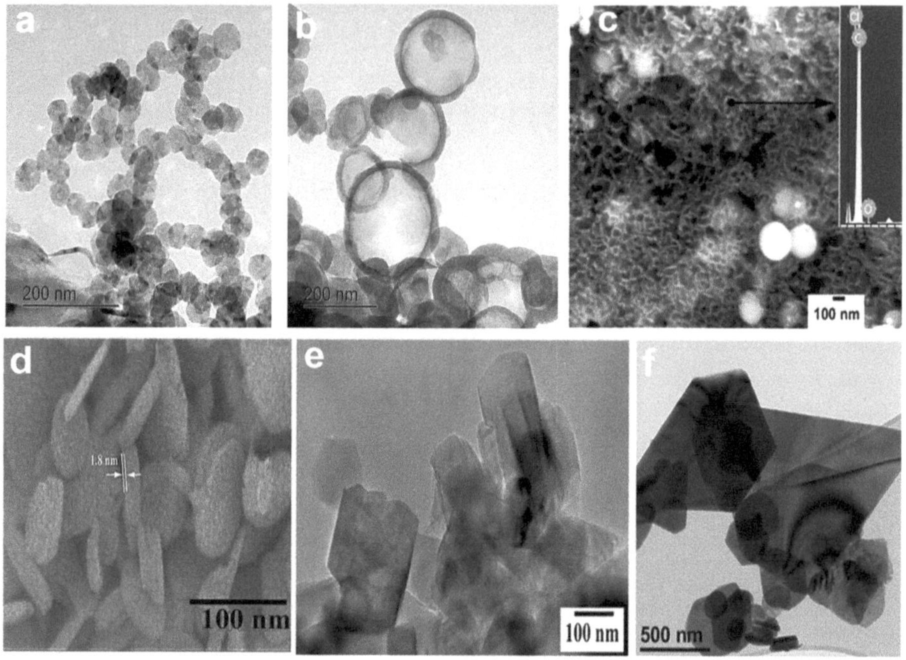

Fig. 1 TEM morphology of combustion derived nanomaterials: (A) carbon nanoballs, (B) hollow carbon nanospheres, (C) honeycomb-structured carbon, (D) a few atomic layer BN nanosheets, (E) B_4C nanosheets, and (F) MoS_2 nanosheets.

Impressive results were achieved in the field of the synthesis of phosphor powders for LED. A new combustion synthesis technology was developed to produce high-quality $Me_2Si_5N_8{:}Eu^{2+}$ (Me is Sr, Ca and Ba) and $CaAlSiN_3{:}Eu^{2+}$–based phosphors for emitting red and green lights under 450 nm blue-light excitation. The morphology, particle size, and photoluminescence characteristics (PL) of combustion synthesized phosphors are shown in Fig. 2.

Based on achievements in the field of the nitride phosphor powders, in 2011 a LED phosphor manufacturing company, Ellim Advanced Materials, Korea, was founded under the supervision of Professor C.W. Won [3,4]. Ellim Advanced Materials manufactures and sells high-performance LED phosphors to domestic and overseas markets. Ellim's major customers are LG Innotek and LG Display.

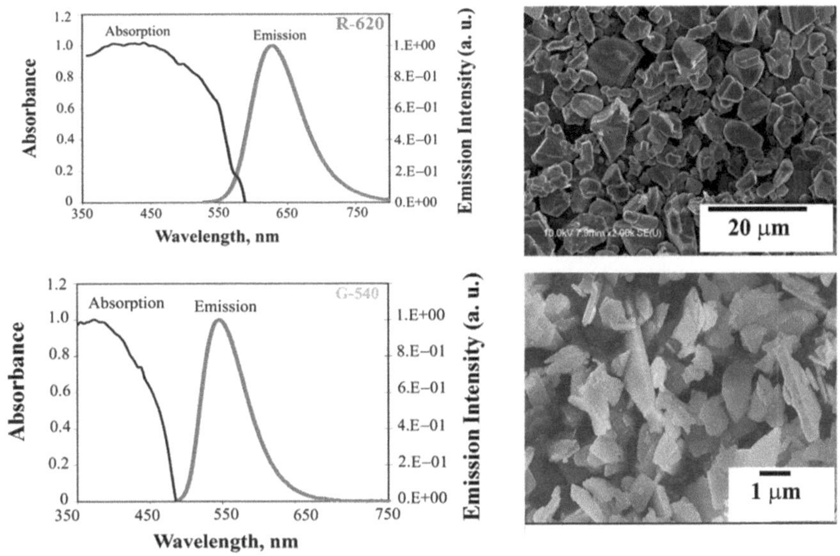

Lot. number	R-628	G-540
Apparent density	1–1.5	1–1.5
Particle size	10.7 μm	6.5 μm
Luminescence (%)	130	110
Color coordinate	X=0.642 Y=0.357	X=0.357 Y=0.605
Emission peak	628 nm	542 nm

Fig. 2 The morphology, particle size, and photoluminescence characteristics (PL) characteristics of $Me_2Si_5N_8:Eu^{2+}$ phosphor.

REFERENCES

[1] Won CW, Nersisyan HH, Won HI, Lee JH. Refractory metal nanopowders: synthesis and characterization, Curr Opin Solid State Mater Sci 2010;14:53–68.

[2] Nersisyan HH, Lee TH, Lee KH, An YS, Lee J-S, Lee JH. Few-atomic-layer boron nitride nanosheets synthesized in solid thermal waves. RSC Adv 2015;5:8579–84.

[3] Nersisyan H, Won HI, Won CW. Highly effective synthesis and photoluminescence of $Sr_2Si_5N_8:Eu^{2+}$ red-emitting phosphor for LEDs. Chem Commun 2011;47:11897–9.

[4] Won CW, Won HI, Won HS, Nersisyan H. Nitride phosphor, reaction mixture, and method production and light emitting device comprising such a phosphor, US2012/0228550 A1; 2012.

SHS in Spain

Miguel A. Rodríguez

Activities on SHS began in Spain in 1991 with a joint project between Institute of Structural Macrokinetics (ISMAN, USSR Academy of Sciences), United Technologies Corp. (UTC), and Tecnología del Grupo INI (TGI, INI group). As support centers, the Institute of Ceramics and Glass (ICV-CSIC) and the INASMET Technology Centre collaborated in this project. The aim was to produce a pilot plant for production of powders by SHS and evaluate their quality and production costs. All these data would serve to address the industrialization of raw materials synthesis by SHS.

The primary people involved at the beginning of the activity in Spain include Professor Alexander G. Merzhanov, Professor Inna P. Borovinskaya from ISMAN, Jose M. Bernardo from TGI, and Miguel A. Hernán from UTC.

These early works included the construction of a pilot plant (SHS Spain AIE) to develop powders of Si_3N_4, AlN, TiC, and Cr_2C. In approximately 2 years, a team was formed by people from ISMAN, TGI, Empresa Nacional Santa Barbara, and UTC. During this time, ICV-CSIC and INASMET provided financial support to this group. One of the scientific challenges was the synthesis of large quantities of α-Si_3N_4 powder. The goal was successfully accomplished and led to the creation of SHS Ceramicas, S. A., the first Spanish company to obtain raw SHS materials. Primary partners in these activities included ENUSA (Spain) and ISMAN (Russia) (Fig. 1). The chief activity of this company was the production of α-Si_3N_4 and BN powders. Unfortunately, the economic crisis prevented the further development of SHS Ceramicas S.A. and finally it was closed.

However, above works promoted scientific activity in Spain regarding combustion synthesis, initially in collaboration with the researchers from ISMAN and later independently. These Spanish centers have been part of multinational teams that have developed collaborative projects within the European Union and other countries. Among the scientific achievements, especially noteworthy is the synthesis of single crystal fibers of Si_3N_4 [1], AlN [2], and SiAlON [3] using the SHS method. These achievements were possible thanks to the work in the pilot plant SHS Spain and its collaboration with the other research centers. From this beginning came work at the ICV-CSIC in nitrides [4] (Fig. 2), carbides, and intermetallics, mostly through in situ investigation of the kinetics of chemical reactions using synchrotron radiation [5,6]. The other primary focus was

Concise Encyclopedia of Self-Propagating High-Temperature Synthesis
http://dx.doi.org/10.1016/B978-0-12-804173-4.00139-3

Fig. 1 SHS reactor to produce BN and Si_3N_4 ceramics.

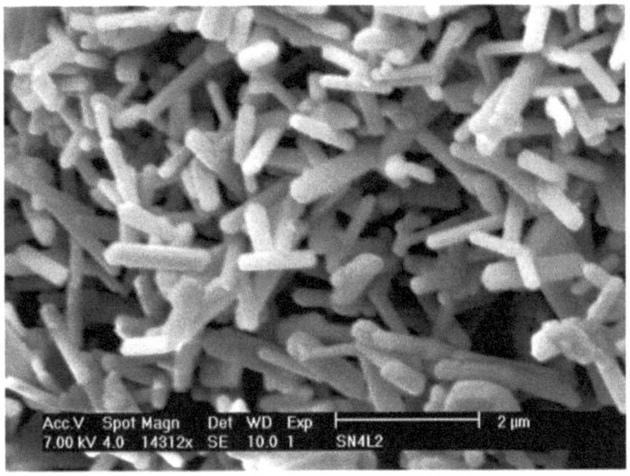

Fig. 2 Si_3N_4 whiskers obtained by SHS at the ICV.

devoted to SHS of materials with functional properties, such as catalysts for fuel cell applications [7].

It is worth noting that the activity of INASMET in the fields of metals and intermetallic alloys were executed in collaboration with ISMAN (Russia) and with the Armenian Academy of Sciences [8].

Currently, the work on combustion synthesis is ongoing in several new different directions, including field-assisted combustion synthesis [9] and solution combustion synthesis [10]. The materials being developed are aimed at different fields, such as structural applications, functional applications, and biomaterials.

REFERENCES

[1] Rodríguez MA, Makhonin NS, Escriña JA, Borovinskaya IP, Osendi MI, Barba MF, et al. Moya Adv Mater 1995;7(8):745–7.
[2] Moya JS, Iglesias JE, Limpo FJ, Escriña JA, Makhonin NS, Rodríguez MA. Acta Mater 1997;45(8): 3089–94.
[3] Rodríguez MA, Makhonin NS, Escriña JA, Limpo FJ, Osendi MI, Barba MF, et al. Scripta Mater 1997;37(4):405–10.
[4] Cano IG, Rodríguez MA. Scripta Mater 2004;50:383–6.
[5] Contreras L, Turrillas X, Vaughan GBM, Kvick Å, Rodríguez MA. Acta Mater 2004;52:4783–90.
[6] Curfs C, Turrillas X, Vaughan GBM, Terry AE, Kvick Å, Rodríguez MA. Intermetallics 2007;15:1163–71.
[7] Chinarro E, Segadães AM, Jurado JRJ. Solid State Electrochem 2004;8:606–10.
[8] Agote I, Coleto J, Gutierrez M, Sargsyan A, Garcıa de Cortazar M, Lagos MA, et al. Intermetallics 2008;16:1310–6.
[9] Mas-Guindal MJ, Benko E, Rodríguez MA. J Alloys Comp 2008;454:352–8.
[10] Rojas RE, Rodriguez MA, Fernández JF. RSC Adv 2015;3104–12.

SHS in Turkey

Onuralp Yücel

The beginning of metallothermic reduction-based studies in Turkey was a result of studies on the carbothermal synthesis of advanced ceramics (e.g., B_4C) and ferroalloys (e.g., FeB) at Istanbul Technical University (ITU), Metallurgical Engineering Department in the late 1980s [1,2] (Fig. 1). By the 1990s, the first metallothermic reduction experiments were carried out to develop alternative production routes for ferroalloys. Çinar et al. reported a study on the aluminothermic production of FeV in 1992. Therefore the publication in question was the first manuscript that focused on metallothermic reduction processes in Turkey [3]. The first research article in Turkey on metallothermic reduction was published in 1996 by Yucel et al. [4]. Although the number of studies, concerning metallothermic reduction (W, Mn, Cr, and their alloys, AlTiB, etc.), rapidly increased after that time, studies on SHS began by the 2000s in Turkey. In the SHS field, the first study originating in Turkey was reported by Demircan et al. in 2004. The study was about the determination of optimum SHS parameters for the synthesis of TiB_2 [5].

Many SHS studies have been conducted at ITU concerning the SHS of borides and carbides, such as TiB_2, B_4C, W_xB_y, and ZrB_2; and the metallothermic reduction of iron and its alloys, such as carbon free Cr-Ni-Mo-Ti-containing alloys and shape memory alloys [6]. Middle East Technical University (METU) in Turkey also contributed to the literature on volume combustion synthesis (VCS) [7].

The seventeenth International Symposium on Boron, Borides, and Related Materials (ISBB 2011) was held in September 2011 in Istanbul, Turkey, and one of the symposium sessions was dedicated to SHS [8]. The thirteenth International SHS Symposium (SHS XIII) was held in Antalya, Turkey, between Oct. 12, 2015 and Oct. 15, 2015. Seventy-four participants from 17 countries attended the SHS XIII (Fig. 2) [9].

Turkish companies use metallothermic reduction processes in all their processes or in some steps of their processes, such as primary magnesium production (Eczacıbaşı Esan: 15,000 t/y capacity; Kümaş A.Ş.: pilot-scale plant), production of steel deoxidation agents from Al drosses (Mesek Metal A.Ş.), and production of low carbon ferrochromium (Eti Elektrometalurji A.Ş.: 10,000 t/y capacity).

Concise Encyclopedia of Self-Propagating High-Temperature Synthesis
http://dx.doi.org/10.1016/B978-0-12-804173-4.00163-0

Fig. 1 Research group that conducted the first studies on the carbothermal synthesis of advanced ceramics in Turkey (1989). Adnan Tekin stands in front left side, Chiranjib Kumar Gupta (visiting scientist) stands in the middle, Okan Addemir stands on the far right, Onuralp Yucel sits in front left side.

Fig. 2 Participants of the thirteenth International Symposium on Self-Propagating High-Temperature Synthesis (SHS XIII).

REFERENCES

[1] Yücel O, Göller G, Addemir O, Tekin A. Production of boron carbide. In: Proceedings of the 4th International Ceramics Technical Congress, Istanbul, Turkey; 1990. p. 176–85.
[2] Yücel O, Addemir O, Tekin A. The optimization of parameters for the carbothermic production of ferroboron. In: Proceedings of the 6th International Ferroalloys Congress, Cape Town, South Africa; 1992. p. 285–9.
[3] Çınar F, Sesigür H, Yücel O, Addemir O, Tekin A. Production of ferrovanadium from vanadium sludge. In: Proceedings of the International Conference on The Recycling of Metals, Dusseldorf, Germany; 1992. p. 209–17.
[4] Yücel O, Çınar F, Addemir O, Tekin A. High Temp Mater Proc 1996;15(1–2):103–9.
[5] Demircan U, Derin B, Yücel O. Synthesis of TiB_2 powder by pyro-hydrometallurgical process. In: Proceedings of the XII International Materials Research Congress, Cancun, Mexico; 2004.
[6] Turan A, Alkan M, Tavsanoglu T, Sahin FC, Goller G, Yucel O. The journey from macro to nano via SHS. In: 12th International symposium on self-propagating high-temperature synthesis (SHS XII), South Padre Island, Texas, USA; 2013.
[7] Çamurlu HE, Sevinç N, Topkaya Y, Bilgi E, Akgün B. Mater Res Bull 2008;43:873–81.
[8] Program of 17th international symposium on boron, borides and related materials (ISBB 2011), Istanbul, Turkey, 2011.
[9] Yücel O. Conclusions and closing remarks. In: 13th International Symposium on Self-propagating High-temperature Synthesis (SHS XIII), Antalya, Turkey; 2015.

SHS in United States of America

Jan A. Puszynski

In 1967 Merzhanov, Skhiro, and Borovinskaya [1] published the first comprehensive paper describing the self-sustaining character of reactions in a condensed phase, which could be utilized for synthesis of many ceramic and intermetallic materials. In this paper, the authors demonstrate the principle of "solid flame" using reactions between transition metals and boron, carbon, or nitrogen.

In the United States, the first reported research utilizing self-sustaining character of thermite condensed-phase reactions was conducted by Walton and Poulos [2] in the mid- and late-1950s. These authors explored thermite reactions to make refractory coatings. Mixtures of aluminum and/or magnesium with oxides of iron, cobalt, and vanadium were used to produce different cermets. In 1973 Hardt and Phung [3] published a very important paper on propagation of gasless reactions in solids, which further alerted US scientists about importance of the combustion synthesis. After sporadic activities in the Western world in the 1950s and 1960s, more significant research efforts were made in the United States starting in the early 1980s. In 1982 McCauley et al. [4] and Holt and Kingman [5] published new results in the area of combustion synthesis, which generated interest at several universities and US government laboratories. A pioneering work by Holt mainly focused on combustion synthesis of ceramic powders and refractory materials in general, which was more aligned with the research activities conducted in the former Soviet Union's laboratories.

A turning point in the US efforts in SHS was catalyzed by a major contract from the Defense Research Projects Agency (DARPA) that was carried out during 1984 and 1986. The overall contract manager was J.W. McCauley and the program manager was J. Birch Holt at the Lawrence Livermore National Laboratory with subcontracts at University of California, Davis, Ceramatec, Los Alamos National Laboratory, and Rice University.

The key universities that conducted research in combustion synthesis included University of California, Davis; Georgia Institute of Technology; State University of New York, Buffalo; Northwestern University; Colorado School of Mines; University of Notre Dame, and University of Houston. These research activities were supported by the National Science Foundation, the Department of Energy (Los Alamos National Laboratory and Sandia National Laboratory), and the US Army. Also, some research in the area of combustion synthesis was conducted in US government laboratories, especially the Department of Energy, US Army, and US Navy.

Concise Encyclopedia of Self-Propagating High-Temperature Synthesis
http://dx.doi.org/10.1016/B978-0-12-804173-4.00166-6

Table 1 summarizes the research activities in academia, government laboratories, and industry in the United States. The experimental research conducted in the United States' national laboratories and universities resulted in many accomplishments, which led not only to significant contributions in the fields of physics, materials science, ceramic engineering, and reaction engineering but also to the development of several technologies that were commercialized [6].

Table 1 SHS R&D groups in the United States

Organization	Investigators	Focus
Department of Defense		
Army Materials Technology Lab.	Croft, Marzik, McCauley	Powder characterization sintering
Army Ballistic Research Lab.	Niller, Kottke	Dynamic compaction; modeling
Army Research Office	Crowson	Coordination and management
Department of Energy		
Los Alamos National Lab.	Behrens	High-temp. chem; laser ignition
Lawrence Livermore National Lab.	Holt, Halverson, Chow	SHS; bulk materials, models
Sandia National Laboratory	Margolis	Modeling
Academia		
Alfred University	Sprigs	Materials process, ceramics
Colorado School of Mines	Moore	Intermetallics, gravity influence
Georgia Tech Research Institute	Logan	Material process., borides
John Hopkins University	Wheis	Reactive foils, welding
New Jersey Institute of Technology	Dreizin	Mechanical activation
New Mexico Inst. of Mining &Tech.	Thadani	Mechanical stimulation
Northwestern University	Matkovwsky, Volpert	Modeling, instability
Oregon State University	Kanury	Modeling, kinetics
Purdue University	Varma, Son	Catalysis, mechanical stimulation
Rice University	Margrave	High-temp. mass spectrometry
State University of NY-Buffalo	Hlavacek, Puszynski	SHS powders, process, modeling
University of California, Davis	Munir	Fundamentals, elect. field SHS
University of California, San Diego	Meyers	Dynamic compaction

Table 1 SHS R&D groups in the United States—cont'd

Organization	Investigators	Focus
University of Florida	Clark, Dalton	Microwave processing
University of Houston	Luss, Martirosyan	Nanopowder, SHS & elect. fields
University of Notre Dame	Varma, Mukasyan	Fundamentals, solution SHS
University of Texas, Rio Grande V.	Martirosyan	Nanomaterials, biomaterials
Washington State University	Wojcicki	Materials process, eutectics
Industry		
Research Triangle Institute	Mullins	Fibers and metal matrix composites
CERAMETEC	Cutler	SHS; powders; thermite
General Sciences Inc.	Zavistanos	SHS densification
System Planning Corp.	Frankhouser	Reviews and analyses
Lockheed Corp.	Hardt	SHS, sintering; phase equilibria
Corning Glass Works	DeAngelis	Reactive hot pressing
W.R. Grace	Rice	Materials processing
Advanced Refractory Technologies	Blakely	SHS powders; whiskers
Innovative Materials, Inc.	Puszynski and Hlavacek	SHS, nitride, boride, and carbide
Benchmark Structural Ceramics		Ceramics and intermetallics
Powder Technologies, Inc.	Hida	SHS powders and whiskers
Synergetic Materials, Inc., Kiser	Logan	SHS powders and bulk materials
Research, Inc.	Halverson	Advanced materials
	Kiser	Soviet SHS technologies

REFERENCES

[1] Merzhanov AG, Shkiro VM, Borovinskaya IP. USSR inventor certificate, 255221; 1967.
[2] Walton JD, Poulos NE. J Am Ceram Soc 1959;42(1):40–9.
[3] Hardt AP, Phung PV. Combust Flame 1973;21(1):77–89.
[4] McCauley JW, Corbin ND, Resetas TM, Wong P. Ceram Eng Sci Proc 1982;3(9–10):538–54.
[5] Holt JB, Kingman DD. In: Proceedings of university conference on ceramic science, November 8–10. Raleigh, NC: North Carolina State University; 1982.
[6] McCauley JW, Puszynski JA. Int J SHS 2008;17(1):58–75.

SHS in Vacuum

Valery I. Uvarov

Being an exothermic process that does not require a gaseous oxidizer, SHS in vacuum is favorable for obtaining nanoporous materials to be used for making catalytically active membranes, since the porosity increases naturally because of the loosening effect of impurity gas pressure and that of liquid-phase vapors of the green mixture. The pressure of saturated vapors of the liquid phase, which is determined by Langmuir and Knudsen techniques, and impurity gases extends the reactive liquid mass and forms the product porosity and pore size. The terms of SHS in vacuum govern the formation mechanism of catalytically active porous materials with a high rate of open porosity (up to 90%). The melting point of the materials in vacuum is known to be practically equal to the boiling temperature of the initial products at synthesis. For instance, the boiling temperature of cobalt at $P_{Ar}=1$ atm is 3100°C and that of copper is 2360°C. At $P=0.1$ mm Hg, the boiling point of cobalt is 1645°C and that of copper is 1280°C [1,2]. In addition, during membrane synthesis in vacuum, significant evaporation and green mixture material transfer is observed. The velocity of substance evaporation in vacuum is two to three orders of magnitude higher than that at the atmospheric pressure of the inert gas. For instance, when the membranes are synthesized from the powders of alloy ((95%Ni–5%Al) + Co_3O_4), it results in surface and volume segregation of nanocomponents of the green mixture. Therefore nanoparticles (20–40 nm) of nickel-cobalt alloy are formed in the surface layers. The alloy segregation can also be observed because alloys are crystallized at the temperature range but not at one exact point. The composition of the crystals formed at the beginning of the process can differ from that at the end. The wider the temperature range, the more significant the segregation. At segregation, the alloys can disintegrate with the transition from the liquid to the solid state of their components or can separate compounds with different melting points. Nanoparticles of Ni and Co (10–20 nm) can be formed in the surface layers of the membrane. During the experiments, the open pore size of the ceramic membranes was measured by Hg porosimeter. The open pore size, which is equivalent to the hydraulic diameter of the channel where a gas or liquid flows, was 2–3.5 μm. The sample open porosity analyzed by hydrostatic weighing was 42%. The phase analysis proved the existence of Ni, Co, and Al_2O_3 phases. The morphology, structure, composition, and particle size distribution of the synthesized porous ceramic membranes were studied by SEM (Fig. 1).

Concise Encyclopedia of Self-Propagating High-Temperature Synthesis
http://dx.doi.org/10.1016/B978-0-12-804173-4.00127-7

Fig. 1 Ni-Co metal particle on γ-Al_2O_3 surface.

REFERENCES

[1] Uvarov VI, Borovinskaya IP, Tsodikov MV, Teplyakov VV. Kinetika I Kataliz 2006;47(1):29–39 [Russia].

[2] Tsodikov MV, Teplyakov VV, Fedotov AS, Uvarov VI, Ruazar D, Kinneman A, et al. Izvestiya Akademii Nauk Seriya Khimicheskaya 2006;12:128–55 [Russia].

SHS Materials in Medicine

Dmitry V. Shtansky

The application of bulk materials and coatings either obtained directly by the SHS method or fabricated by various deposition techniques using targets and electrodes produced by SHS is an exciting and rapidly growing area. The SHS method was employed to fabricate porous bulk biomaterials. One of the benefits of using porous structures is the possibility of bone ingrowth through the porous space accompanied with the transport of body fluids, thus ensuring strong implant integration into the bone. Another advantage of a porous material is a reduction in stiffness, which minimizes the risk of stress shielding and provides better transfer of an applied load to the bone. The SHS of porous NiTi alloys, as well as of TiB_2 ceramics and TiC/Ti cermets, for medical applications was reported. These materials were prepared through the following exothermal reactions:

$$Ni + Ti \rightarrow NiTi$$

$$3Ti + B_4C \rightarrow 2TiB_2 + TiC$$

$$Ti + C \rightarrow TiC$$

Elahinia et al. [1] discussed challenges and opportunities with using NiTi in implants. Nitinol is known for its biocompatibility and shape memory properties and has been approved by the US Food and Drug Administration (FDA) for clinical applications worldwide. Biorthex (Montreal, Canada) has been producing porous NiTi implants with about 65% porosity by SHS method. Porous NiZr-based biomaterials for orthopedic applications obtained via combustion reaction of Ni with Zr in the presence of additives (ZrH_2, $CaCO_3$, NH_4F, and Teflon) were reported. Combustion synthesis was also employed to fabricate nanopowders. For instance, Si-doped hydroxyapatite nanobioceramics suitable for biomedical applications were produced by solution combustion synthesis.

Physical vapor deposition (PVD) is an effective tool to modify surface characteristics of various biomaterials (e.g., metals and polymers), while retaining their bulk properties. The SHS method allows the production of a wide spectrum of highly dense composite materials with uniform structure and the mechanical, thermal, and electrical properties needed for composite targets used in PVD technology. Various multicomponent films can be deposited using a single SHS-fabricated composite target [2], which enables the use of a simple sputtering unit. During the past decade, various composite targets

Concise Encyclopedia of Self-Propagating High-Temperature Synthesis
http://dx.doi.org/10.1016/B978-0-12-804173-4.00128-9

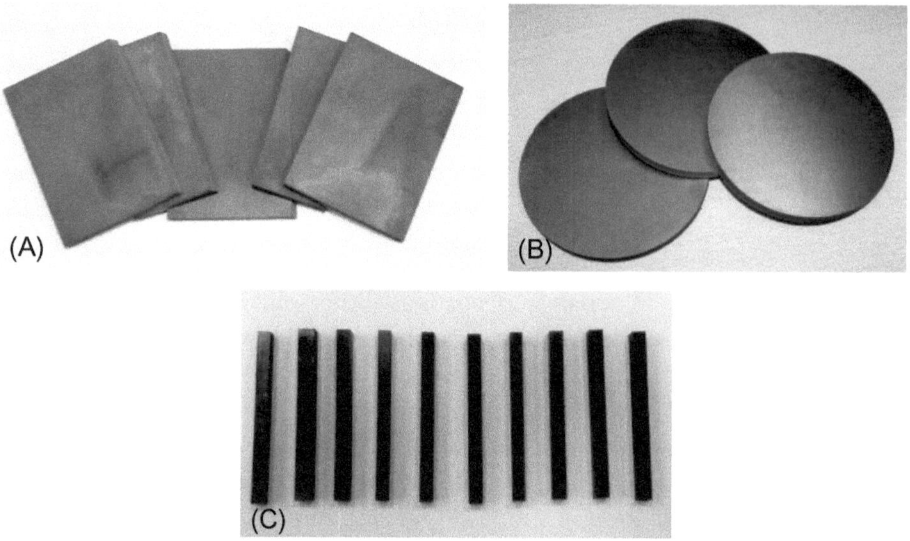

Fig. 1 (A, B) SHS targets and (C) electrodes.

produced by SHS have been developed and synthesized for the deposition of multicomponent bioactive nanostructured films, such as $TiC_{0.5}+CaO$, $TiC_{0.5}+ZrO_2$, $TiC_{0.5}+CaO+TiO_2$, $TiC_{0.5}+Ca_{10}(PO_4)_6(OH)_2$, and (Ti,Ta)C + CaO, (Ti,Ta)C-Ti_3PO_x-CaO, TiC-Ti_3PO_x-CaO [3,4]. Bioactive yet antibacterial films can be fabricated by two consecutive processes: either by deposition of bioactive TiCaPCON films via magnetron sputtering of composite target TiC-Ti_3PO_x-CaO followed by Ag^+ ion implantation or by means of cosputtering of SHS-produced composite TiC-Ti_3PO_x-CaO and metallic (Ag) targets [5]. Recently, Ag-doped bioactive TiCaPCO films fabricated by pulsed electrospark deposition using SHS-produced composite TiC-Ti_3PO_x-CaO-Ag electrode were also reported (Fig. 1).

TiCaCON and TiCaPCON films deposited on the surface of metallic implants for cranial-jaw-face surgery, spine surgery, and hip joint surgery have successfully passed clinical studies at the P.A. Hertsen Moscow Research Oncological Institute and the Central Institute of Traumatology and Orthopedics, Moscow, after which they were approved for medical applications in the territory of the Russian Federation.

REFERENCES

[1] Elahinia MH, Hashemi M, Tabesh M, Bhaduri SB. Prog Mater Sci 2012;57:911–46.
[2] Shtansky DV, Levashov EA, Batenina IV, Gloushankova NA, Anisimova NY, Kiselevsky MV, et al. RSC Adv 2013;3:11107–15.
[3] Potanin AYu, Levashov EA, Pogozhev YuS, Shvindina NV, Kovalev DYu. Ceram Int 2015;41: 8177–85.

[4] Shtansky DV, Kiryukhantsev-Korneev PhV, Bashkova IA, Sheveiko AN, Levashov EA. Int J Refract Met Hard Mater 2010;28:32–9.
[5] Sukhorukova IV, Sheveyko AN, Zhitnyak IY, Gloushankova NA, Denisenko EA, Filipovich SYu, et al. Colloid Surf B 2015;135:158–65.

SHS of Heat-Treated Mixtures

Nikolay G. Kasatsky, Yury S. Nayborodenko, Galina V. Lavrenchuk

The effect of vacuum heat treatment (VHT) was studied during the combustion of the $Ti\text{-}2B_{am}$, $Ti\text{-}2B_{cr}$, $Zr\text{-}2B_{am}$, Ti-C, Ni-Al, and Zr-3Al systems. Titanium (grade PTM "A," commercial and divided into narrow fractions in the centrifugal classifier), amorphous boron (TU 6-02-923-74), crystalline boron (TU 6-08-296-74), carbon (soot PM-15), zirconium (M-41), nickel (PNK-1BL-7), and aluminum powders (ASD-4) were used. The powder mixtures were mixed in alcohol, followed by drying in a vacuum oven. The heat treatment of the pressed samples (10 m in diameter, $18\text{–}20 \times 10^{-2}$ m in height, and porosity of 40%) was conducted in a vacuum assembly ($P \approx 133 \times 10^{-3}$ Pa) at temperatures between 423 and 973 K for 5 hours. The combustion rate reached saturation for such duration of annealing time. After heat treatment, the samples were burned in a constant-pressure bomb in an inert atmosphere ($P_{Ar} = 2 \times 10^5$ Pa), and the combustion rate was measured by the photorecording method.

As shown in Fig. 1, there is a significant (two- to threefold) increase in the combustion rate after vacuum treatment at a temperature of 873 K in the metal-metalloid systems with the amorphous components such as $Ti\text{-}2B_{am}$ (1), $Zr\text{-}2B_{am}$ (2), and Ti-C (3). In the mixtures with crystalline components such as $Ti\text{-}2B_{cr}$ (4), Ni-Al (5), and Zr-3Al (6), the VHT effect is insignificant.

The combustion rates of the annealed samples practically coincided with the combustion rates of the initial samples. The length of the samples remained unchanged after combustion with and without annealing.

For the unannealed systems with amorphous components, there was a significant (more than twofold) increase in the initial length after combustion. The linear length changes of the $Ti\text{-}2B_{am}$ and Ti-C burned samples were eliminated after vacuum annealing at 873–973 K [1]. In all cases, the pressure in the bomb was changed less during the combustion of the annealed samples compared to the unannealed samples.

The increase in the length observed during the combustion of the unannealed samples was caused by the release of adsorbed and dissolved gases in the heating zone. Therefore, in this case, the combustion front moved along the moving substance, which led to the decrease in the combustion rate measured. To find the true combustion rate, we propose

Concise Encyclopedia of Self-Propagating High-Temperature Synthesis
http://dx.doi.org/10.1016/B978-0-12-804173-4.00042-9

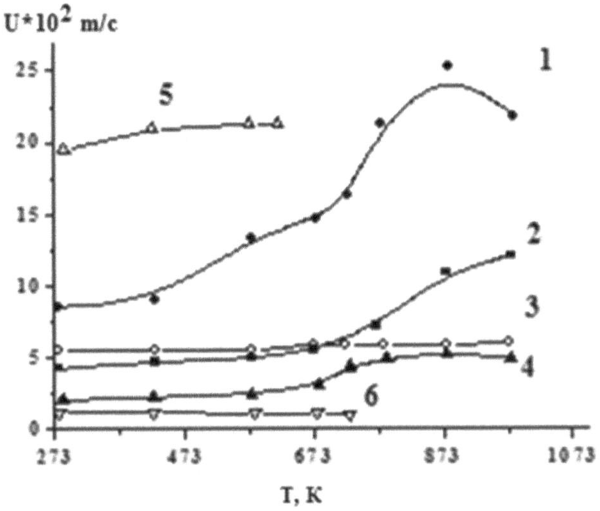

Fig. 1 Combustion rate of the mixtures vs the temperature of the vacuum treatment: (1) Ti-2B$_{am}$, (2) Zr-2B$_{am}$, (3) Ti-C, (4) Ti-2B$_{cr}$, (5) Ni-Al, (6) Zr-3Al.

$$U = U_0(1 + \Delta l/l)$$

where U is the true combustion rate, U_0 is the observed combustion rate, and $\Delta l/l$ is the elongation of the sample.

After the heat treatment of the system with amorphous components, there were practically no linear length changes of the samples after combustion, because the samples start "gassing" and the gaseous adsorbed impurities were removed from the powders during vacuum annealing, beginning with the temperatures between 423 and 477 K. The composition of the impurities and adsorbed gases included H_2O, CO, CO_2, N_2, O_2, and H_2 according to the mass spectral analysis [1,2]. Comparing the results obtained for the Ti-2B system with amorphous and crystalline boron, it can be concluded that the gas release of the amorphous component is much higher. Removing the adsorbed gases during VHT caused the combustion front to be moved along the motionless substance during the subsequent combustion. In this case, the measured combustion rates must coincide with the true rates. For the Ti-2B$_{am}$ system, the combustion rate measured after VHT was equal to 25×10^{-2} m/s at 873 K, which was 2.5 times higher than the combustion rate observed for the unannealed samples ($U_0 = 8\text{–}10 \times 10^{-3}$ m/s). The high temperature annealing was accompanied by the isothermal interaction of the components and led to the decrease in the combustion rate.

VHT of the systems with amorphous components led to combustion of the annealed samples in the gas-free mode at a pressure of inert gas less than the atmospheric one. The combustion rate of the Ti-2B_{am} and Ti-C systems was qualitatively changed for the annealed samples, depending on the particle size of titanium. Continuous increase in the combustion rate was observed with the decrease in the particle size of titanium [1].

REFERENCES

[1] Nayborodenko YuS, Kasatsky NG, Lavrenchuk GV, et al. Effect of vacuum heat treatment on combustion of gas-free systems. Combustion of condensed and heterogeneous systems. In: Proceedings of the VI all-union symposium on combustion and explosion. Alma-ata, 23-26 Sep., Chernogolovka; 1980. p. 74–77.
[2] Itin VI, Nayborodenko YuS. High-temperature synthesis of intermetallic compounds, 1989; Tomsk: Tomsk State University, p. 36–41.

SHS-Rolling in Vacuum

Evgeny A. Levashov

Among SHS technologies currently used in order to prepare articles made of synthetic hard materials, an important role is occupied by processes based on the combination of SHS with consolidations such as pressing, extrusion, and rolling. The most widely studied processes are SHS-pressing, which is in practical use, and SHS-extrusion. Little data exist for studying and assimilating the process of SHS-rolling, which is of interest from the point of view of producing items for different purposes in the form of strips, sheet, bars of different cross section, pipes, and in individual cases, finished articles [1–3]. In order to remove impurities and protect hot SHS products from oxidation, rolling in a vacuum has been considerably effective. A United States patent [4] to bring this process to fruition was proposed. However, in the opinion of the author [1–3], this method proposed has a number of disadvantages: does not provide sufficient productivity, has difficult to implement under industrial conditions, and exhibits limited production possibilities. Under the leadership of the author [1–3] over a number of years, work has been carried out on studying hot rolling in a vacuum and the creation of special equipment. The experience accumulated has made it possible to perform experiments for studying and for the practical application of the SHS-rolling method in a vacuum.

Various exothermic mixtures were selected as study materials: Ti-C, Ti-Al, and Ti-C-Ni, calculated for conditions for preparing nonstoichiometric titanium carbide, γ-TiAl, and STIM alloys based on titanium carbide. Studies were carried out in the vacuum rolling mill-155. The mill of a roll-chamber type has vacuum chambers connected successively: loading, heating (continuous type with heater unit), working (with rolls 155 mm in diameter), cooling, and unloading. The mill is fitted with a transportation system. During experiments, the temperature in the furnace, the change in degree of vacuum in the system, and the heating rate and deformation degree were optimized to control residual porosity and microstructure of product. Results point to the possibility of removing gas admixtures by special treatment of the original charge in vacuum; therefore a significant portion of the studies were devoted to developing the optimum heat-treatment regimens for charge mixtures. The test temperature range for treatment was limited to below 500°C, above which desorption is markedly activated and self-ignition of the mixture occurs. Optimization of thermovacuum treatment parameters is required in order to prevent breakage of combustion products and to retain their correct geometric shape.

Concise Encyclopedia of Self-Propagating High-Temperature Synthesis
http://dx.doi.org/10.1016/B978-0-12-804173-4.00131-9

The subsequent experiments showed that charge mixtures, apart from purity with respect to gas admixtures, should satisfy the following requirements: there should be a sufficiently strong briquette of correct geometric shape with uniform distribution of porosity throughout the volume, and the maximum combustion rate should be provided [2,3]. An important result from this work [1,2] was the creation of a technology for preparing a rolled product in a vacuum of high-quality briquette charges with a relative density of 0.5–0.95. A study of the effect of the relative density of briquettes on the combustion rate showed that for the Ti-Al system, the dependence has the form of a straight line with a very high rate of ascent [2]. An increase in the relative density of briquettes from 0.4 to 0.95 leads to an almost tenfold increase in the combustion rate. For multicomponent systems based on Ti-C-Ni, this dependence takes the form of a curve with a maximum corresponding to 66% of the theoretical density.

A considerable effect on the combustion rate of the preliminary heating temperature was revealed: for a multicomponent system, an increase in temperature from 600°C to 700°C leads to an almost twofold increase in the combustion rate. These studies were the basis for experiments. As a result, with the SHS-rolling method in a vacuum test, specimens were prepared with a porosity ranging from 5% to 60% in the form of strips 2–5 mm thick, 30–50 mm wide, and 150 mm long.

Experiments have confirmed the possibility of preparing items by the SHS-rolling method in a vacuum. In recent years, a number of special vacuum rolling mills have been created. They include a mill with a multiroll pass for rolling bars of different cross sections; an automatic mill for the production of components, i.e., stepped bodies of rotation by transverse-wedge rolling; and one of the world's largest vacuum rolling mills, the 600, which is intended for producing sheets with a thickness of 1–10 mm, with a width up to 500 mm and a length up to 3000 mm. The existence of special equipment will make it possible to prepare high-quality materials comprising synthetic hard materials in the form of strips, large sheets, bars with different cross sections, pipes, and components in the form of stepped bodies of rotation [3]. Significant contribution to development of SHS-rolling was made by Professor V.N. Chernyshov and Dr. E.E. Osipov.

REFERENCES

[1] Osipov EE, Levashov EA, Chernyshov VN, Merzhanov AG, Borovinskaya IP. Int J SHS 1992;1(2):241–6.
[2] Chernyshov VN, Osipov EE, Levashov EA, Merzhanov AG, Biyachi L. Int J SHS 1993;2(3):315–21.
[3] Chernyshov VN. Refract Ind Ceram 2007;48(5):359–61.
[4] Rice RW. Hot rolling of self-propagating synthesis. US Patent No. 4642218, 1987.

Sialon

Konstantin L. Smirnov

High hardness, strength, wear and corrosion resistance characterize sialon ceramics (compounds of Si-Al-O-N system). Sialons are widely applied in various fields of engineering, including refractory items, bearing parts, and tools for machining [1]. Here are some of their characteristics and applications:

- Solid solutions based on α-modification of Si_3N_4 (α-sialons). Their composition corresponds to the formula $\alpha\text{-}Me_mSi_{12-(pm+n)}Al_{pm+n}O_nN_{16-n}$, where m and n are substitution parameters, p is Me^{p+} ionic valency, and Me^{p+} is an ion of Li, Mg, Ca, Y, and other rare-earth metals introduced into $\alpha\text{-}Si_3N_4$.
- Solid solutions based on β-modification of Si_3N_4 (β-sialons). Their composition is $\beta\text{-}Si_{6-z}Al_zO_zN_{8-z}$, and the value of the substitution degree z changes constantly from 0 to 4.2.

Ceramics based on α-sialon are distinguished by higher hardness and lower creep under high temperatures. β-Sialon ceramics are characterized by higher fracture toughness.

SHS of sialons is based on filtration combustion of reactive mixtures containing aluminum, silicon powders, and corresponding oxides in gaseous nitrogen or air. In such systems, the gaseous reagent flows into the reaction zone through the pores of the condensed phase because of the pressure gradient forming between the zone and its environment. Filtration combustion is a complicated macrokinetic process, and its behavior depends on a number of parameters of the reaction system: reactive gas pressure and composition, geometry of initial samples, their piezo-thermal conductivity, and kinetic characteristics of the chemical process. SHS can be used for both obtaining powders as raw materials for conventional sintering of sialon ceramics and direct production of ceramics in one stage during filtration combustion.

At powder synthesis, nitrogen pressure is 2–10 MPa. In this case, the reactive gas content in the pores is not higher than 50% to the complete conversion of aluminum and silicon powders. However, gas permeability of initial billets with 0.2–0.4 relative porosity provides infiltration of the required amount of nitrogen from the external environment. A refractory diluent is introduced into the initial mixture in order to eliminate the restrictions in nitriding degree connected with the product high-temperature dissociation and initial reagent sintering in the burning wave. Silicon and aluminum nitrides as well as sialon with the elemental composition of the target product are used for this purpose. Under such terms, SHS provides high velocities of phase formation without mass losses

Concise Encyclopedia of Self-Propagating High-Temperature Synthesis
http://dx.doi.org/10.1016/B978-0-12-804173-4.00132-0

and allows obtaining powders of α- and β-sialons with a high degree of homogeneity. Moreover, variation of maximum temperature in the filtration combustion front and composition of initial mixture as well as its mechanical activation allow to change the dispersion degree (from some tenths of a micron to several hundred microns) and morphology (isomorphous, prismatical, fibrous) of the synthesized powders [2–4].

In the case of direct SHS of ceramic materials based on sialons, initial powder compacts with a relative density of 0.60–0.65 are used. Such a value restricts both the content of the gas reagent in the billet pores and its following inflow from the external environment. Therefore nitriding of these systems at the filtration combustion mode can be possible only at a minimum nitrogen pressure of 30 MPa. But at 30 MPa, filtration combustion can occur either in nonstationary or surface regimes, and it results in a remarkable inhomogeneity of the phase product in its volume. When the initial nitrogen pressure is increased up to 100 MPa filtration combustion changes into the layer-by-layer stationary mode, which is more favorable for synthesizing materials with homogeneous structures. The maximum temperature in the front and its propagation rate achieve the highest values (2400–2600 K and 0.9–1.6 mm/s, correspondingly). The important features in structure formation of sialon ceramics at filtration combustion are the deformation shrinkage determined by the existence of oxide melts in the reaction zone and the development of the reactive gas-pressure gradient near this zone. This factor allows obtaining ceramic materials based on sialons, including composites that contain other refractory compounds (SiC, BN, TiN, TiB_2) with a relative density of 0.8–0.95 [5]. They are characterized by high resistance to thermal shock, oxidation, and corrosion action of metallurgical melts. Synthesized composites are readily machinable with WC-based cutting instruments (Fig. 1).

Fig. 1 Items made of SHS sialon-based composites: nozzles for casting metal and lining plates.

REFERENCES

[1] Ekstrom T, Nygren M. J Am Ceram Soc 1992;75(2):259–76.
[2] Zhao YS, et al. Int J SHS 2009;18(2):87–91.
[3] Liu G, et al. Mater Lett 2005;59:3955–8.
[4] Rodriguez MA, et al. Scr Mater 1997;37(4):405–10.
[5] Borovinskaya IP, Smirnov KL. Refract Ind Ceram 1999;40(3–4):94–7.

Silicon Carbide

Alexander S. Mukasyan

Silicon carbide (SiC) ceramics have a set of unique physical-chemical properties, such as high hardness and mechanical stability at high temperatures, excellent thermal conductivity and low coefficient of thermal expansion, high resistance to corrosion and oxidation, wide bandgap, and others [1]. This low density ceramic has a variety of applications including as an abrasive material, wide bandgap semiconductors in high-power electronics, and structural ceramics in the automotive and aerospace industries. The most widely used approach for powder synthesis is known as the Acheson process [2]: namely, a reduction of silica by carbon at an elevated temperature. This method allows the production of powders with different grades of purity and particle sizes $>5\ \mu m$. Pure submicrometer SiC particles can be synthesized by CVD methods in silane/hydrocarbons systems, sol-gel processes, plasma, and microwave radiation [1]. However, the major disadvantage of all above-mentioned methods is their low efficiency, i.e., a long process duration and relatively large energy consumption, which limits their application for large-scale production.

Combustion synthesis provides an energy-saving method for production of SiC powders [3]. In general, there are two main routes to fabricate SiC compound in the self-sustained CS mode: (i) from elements and (ii) by reduction of silicon oxide. The gasless combustion reaction from elemental powders in the Si-C system is characterized by the following equation:

$$Si + C = SiC;\ \Delta H_{273} = -73\,kJ/mol \tag{1}$$

The reaction (1) has a moderate enthalpy of product formation, and thus is characterized by a relatively low adiabatic combustion temperature ($T_{ad} = 1873$). This feature implies that it is not easy to accomplish a self-sustained CS process in this system. Several approaches have been developed to enhance the reactivity of the Si-C system [4]: (i) CS with preheating of the reactive media [5]; (ii) CS with electrical field [6]; (iii) chemical activation of the CS process [7]; (iv) SHS synthesis in air/nitrogen [8]; and (v) mechanical activation of the initial mixture [9]. The alternative CS method for obtaining SiC is to use the reduction route [10,11]. The overall combustion reaction for reduction of SiO_2, when magnesium (Mg) is used as a reducing element, can be written as follows:

Concise Encyclopedia of Self-Propagating High-Temperature Synthesis
http://dx.doi.org/10.1016/B978-0-12-804173-4.00133-2

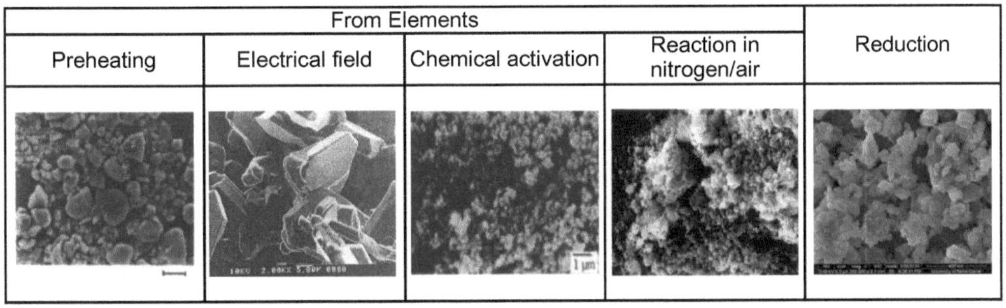

Fig. 1 Microstructures of SiC powders produced by different SHS routes.

$$SiO_2 + 2Mg + C = SiC + 2MgO;\ \Delta H_{273} = -122\,kJ/mol \quad (2)$$

The absolute value of T_{ad} for reaction (2) exceeds 2000 K, which is above the melting and boiling points of magnesium (922 K and 1363 K), as well as the mp of silicon (1683 K) and silicon oxide (1923 K), but well below the mp of MgO (3073 K) and carbon (4093 K) [10].

Typical microstructures of the synthesized SiC powders by different CS approaches are shown in Fig. 1. The BET surface area of the β-SiC obtained by using the preheating approach was about 6 m^2/g^5. The field-assisted SHS β-SiC powder is produced with well-crystallized morphology in the form of plates having a thickness ~2 μm and a size of ~20 μm [6]. By using PTFE as an activator, cube-shape SiC particles with sizes on the order of 10 μm can be produced, while for the other compositions, long ~1 mm thin (0.5 μm) fibers were fabricated [7]. Powder with a surface area of 10 m^2/g that contains predominately (>90%) SiC phase and a balance of Si_3N_4 was synthesized by the combustion reaction of Si + C mixture in an air/nitrogen atmosphere [8]. The morphology, surface area, and particle size of the SiC powder synthesized by the reduction combustion reaction in the SiO_2 + C + Mg system depend on the microstructure of the initial silica. It was shown that SiC powders with a BET surface area up to 100 m^2/g can be fabricated by this CS route [11]. Submicron β-SiC powders may be synthesized from elements by using high-energy ball milling of the initial (Si + C) mixture to fabricate reactive composite Si/C particles [9]. The morphology and size of the CS-SiC particles proved to be identical with those for the mechanically induced reactive nanocomposite.

REFERENCES

[1] Abderrazak H, Hmida E. In: Gerhardt R, editor. Properties and applications of silicon carbide. Rijeka, Croatia: InTech; 2011. p. 361–88.
[2] Krstic VD, Am J. Ceram Soc 1992;75(1):170–4.
[3] Mukasyan A. In: Gerhardt R, editor. Properties and applications of silicon carbide. Rijeka, Croatia: InTech; 2011. p. 389–409.
[4] Pampuch R, Stobierski L, Liz J. J Am Ceram Soc 1989;72(8):1434–5.

[5] Yamada O, Miyamoto Y, Koizumi M. Amer Cer Soc Bul 1985;64(2):319–21.
[6] Feng A, Munir ZA. Metall Mater Trans B 1995;26(3):587–93.
[7] Nersisyan GA, Nikogosov VN, Kharatyan SL, Merzhanov AG. Comb Explos Shock Wave 1991;27(6). 720–724.
[8] Agrafiotis Ch, Lis J, Puszynski JA, Hlavacek V. *J Am Ceram Soc* **72** (9), 1990, 1735–1738.
[9] Mukasyan AS, Lin YC, Rogachev AS, Moskovskikh DO. J Am Ceram Soc 2013;96:111–7.
[10] Merzhanov AG, Borovinskaia IP, Mamian SS, et.al. Patent no.4445557/26; 1994, Russia.
[11] Ermekova ZS, Mansurov ZA, Mukasyan AS. Ceram Inter 2010;36(8):2297–305.

Silicon Nitride

Vladimir V. Zakorzhevsky

Silicon nitride was originally synthesized in 1844 through silicon annealing in nitrogen atmosphere. Si_3N_4-based ceramics are used to manufacture units and parts operating under severe thermal and mechanical loading, e.g., cutting plates, ball-bearings, turbine blades, armor elements.

There are two modifications of silicon nitride, α and β, and they differ in lattice structure. Low-temperature α-Si_3N_4 is used to make high-quality ceramics. It is a metastable compound and transitions to stable β-Si_3N_4 at $T > 1450°C$ under specific conditions (treatment period, catalytic additives).

Silicon nitride synthesis in the combustion mode is based on combustion of silicon powder in nitrogen with regulating additives such as silicon nitride powder as an inert diluent, salts (NH_4Cl, NH_4F, Na_2SiF_6, $(NH_4)_2$ SiF_6), and so on, as illustrated here:

$$3Si + 2N_2 \xrightarrow{\text{combustion, } Si_3N_4} Si_3N_4$$

$$3Si + Si_3N_4 + NH_4Cl + 2N_2 \rightarrow 2Si_3N_4 + HCl + 0.5N_2 + 1.5H_2$$

The first papers on SHS of silicon nitride were published in the 1980s; such gasifying additives as ammonium chloride and fluoride were used to synthesize α-phase for the first time [1].

Solid nitriding components such as alkali metal azides were used as a source of nitrogen [2]. The synthesis is carried out by the following scheme:

$$14Si + 6NaN_3 + (NH_4)_2SiF_6 + N_2 \rightarrow 5Si_3N_4 + 6NaF + 4H_2$$

It is also possible to use NH_4Cl, NH_4F, Na_2SiF_6, and $(NH_4)_2SiF_6$ as salt additives to the green mixture.

When Si burns in nitrogen, Si_3N_4 is formed by the mechanism of a vapor-liquid-crystal process. The limiting stage of the process is silicon evaporation [3], which

Concise Encyclopedia of Self-Propagating High-Temperature Synthesis
http://dx.doi.org/10.1016/B978-0-12-804173-4.00134-4

means that the temperature in the synthesis zone must exceed the melting point of silicon (1420°C). The minimum temperature of silicon combustion in the $Si + Si_3N_4 + N_2$ system is about 1750°C, because lower temperatures do not allow evaporation of silicon in an amount sufficient to maintain SHS [4].

Introduction of salt additives is necessary for controlling the synthesis temperature regime; it is especially important for the synthesis of α-Si_3N_4. Salt decomposition products react with silicon particles and provide silicon transition to the gas phase, which allows carrying out the low-temperature regime by means of an intermediate product formation.

The α-phase of silicon nitride was obtained in the $Si + Si_3N_4 + N_2$ system without gasifying additives, but with submicron silicon powder ($d_{50} = 200$ nm, $S_{sp} = 12$ m^2/g). The temperature of the green mixture combustion was noticed to be 1346°C, which is lower than the Si melting point (1420°C). This result was explained by the existence of the nanofraction of the silicon powder [5].

The particle morphology of silicon nitride synthesized in the combustion mode depends on the synthesis conditions. At the low-temperature mode ($T < 1650$°C) with participation of gasifying additives, α-Si_3N_4 formed as whiskers (Fig. 1). Particles of high-temperature β-phase formed as hexagonal columnar crystals (Fig. 2).

Fig. 1 SHS silicon nitride crystals obtained by low temperature burning.

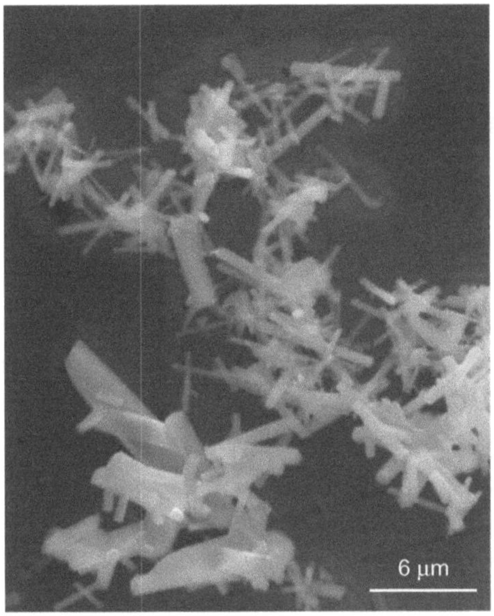

Fig. 2 SHS silicon nitride crystals obtained by high temperature burning.

REFERENCES

[1] Merzhanov AG, Borovinskaya IP, Martynenko VM. Production method of silicon nitride, Auth. Cert. No. 1533215, USSR; 1983.
[2] Kosolapov VT, et al. Production method of carbonitrides, Auth. Cert. No 738242, USSR; 1978.
[3] Mukasyan AS, Stepanov BV, Galchenko YuA, Borovinskaya IP. Fizika goreniya i vzryva 1990;1:45–52.
[4] Grachev VV, Shatalov BN, Borovinskaya IP. All-Russian conference "combustion and explosion processes in physical chemistry and technology of inorganic materials," Moscow, 2002, 89-93; 2002.
[5] Zakorzhevsky VV, Borovinskaya IP. Powder Metall Metal Ceram 2009;48(7/8):375–80.

Solid Flame

Alexander S. Mukasyan

In 1967 a group of researchers from the Academy of Sciences of USSR, headed by Professor A.G. Merzhanov discovered a new phenomenon called "wave localization for self-retarding solid state reactions [1]," or in modern terms the *solid flame* [2]. The fundamentals for this phenomenon are as follows: (i) in some exothermic reactive systems (e.g., Ta-C and Mo-B), the adiabatic combustion temperature, while being high, is still well below the melting points of the precursors, any intermediates, and final products; (ii) the amount of equilibrium gas-phase products in these systems are negligibly small; and (iii) solid precursors react along the surface contact area and form a layer of solid products, which separates the reagents thus retarding the diffusion rate. However, it was experimentally shown that such self-retarding solid-state reactions, when localized in the narrow zone, may rapidly self-propagate in the form of a combustion wave.

From the standpoint of conventional combustion, a solid flame is enigmatic, because it is hard to believe that solely solid-state diffusion can define the self-sustaining nature of the combustion process. It is a prevalent opinion that self-propagating combustion processes, such as metallothermic reactions, exist because of relatively fast mass transport in the liquid phase (diffusion and convection); therefore melting of at least one component in the system is considered as a necessary condition for combustion. This requirement was even announced as a "new criterion" for sustainability of the combustion reaction: "adiabatic temperature … must be high enough to melt the lower melting point component [3]." On the other hand, the regularities of combustion in Ta + C mixtures were studied by Dr. V.M. Shkiro and coauthors assuming that the reaction proceeds completely in the solid state [4]. It was shown that the combustion wave in the Ta-C system can propagate in the pulsating mode at temperatures below the lowest melting or eutectic point.

More careful experimental investigation of the combustion mechanism revealed, however, a gas-transport stage in the combustion reaction for this system [5]. In this article, which was based on a detailed investigation of the structural transformations taking place in the combustion front, it was stated that impurities enhance mass transport in the Ta-C system. Investigation of the microstructural transformations in the quenched combustion wave revealed that the tantalum carbide crystals that form in the leading zone of the reaction front have columnar morphology and uniformly cover the particle

Concise Encyclopedia of Self-Propagating High-Temperature Synthesis
http://dx.doi.org/10.1016/B978-0-12-804173-4.00135-6

surface, despite the fact that in the initial mixture, the reagents have only point contacts. These structural transformations led to the hypothesis that the existence of a gas phase and melt leads to enhanced reagent transport, and consequent product formation in this system. These phases form because of the impurities present in the initial reagents, such as oxygen, in the form of oxide film on the surface of the metal particles, and iron introduced into the system during the mixing process. In 2001 the terminology "catalytically assisted combustion synthesis of tantalum carbide" was introduced by Kim and Wooldridge [6]. They indicated that the presence of iodine vapor and carbon dioxide significantly enhances the combustion synthesis process, leading to higher conversion efficiencies and influencing the product's microstructure.

In 2016 [7] a preparation of highly pure and pore-free Ta/C composite particles, which were used for experimental validation of the solid-flame, was reported. Preparation of these composite particles involves short-term, high-energy ball milling of tantalum and carbon powders. High-resolution microscopy coupled with three-dimensional reconstruction techniques were used to characterize the volume nanostructure of mechanically fabricated composite particles. It was quantitatively shown that the particles have nanoscale mixing of the reagents and possess a high contact surface area between tantalum and carbon. Experiments revealed that the ignition temperature of as-fabricated composite particles was 1243 ± 15 K, and the maximum combustion temperature was shown to be 2487 ± 50 K, which is well below possible solid-liquid transitions. It was also revealed that in such nanostructured, pore-free media, a carbon diffusion through the tantalum grain boundaries and subsequent formation of a Ta(C) solid solution defines low-temperature ignition of mechanically fabricated particles. The high surface area contact between the Ta and C nanoscale reagents allows the reaction to propagate in a self-sustained manner, owing solely to a solid-state diffusion mechanism. Thus it was unequivocally shown that Mother Nature indeed "allows" the solid flame phenomenon.

REFERENCES

[1] Merzhanov AG, Borovinskaya IP, Shkiro VM. Phenomenon of wave localization for self—retarding solid state reactions. Diploma no 287, filed 05.07.1967, Vestnik USSR Acad Sci 1984;10 [in Russian].
[2] Merzhanov AG, Mukasyan AS. Solid-flame combustion. Moscow: Torus Press; 2007 [in Russian].
[3] Su X, Fu F, Yan Y, Zheng G, Liang T, Zhang Q, et al. Nat Commun 2014;5:1–7.
[4] Shkiro VM, Nersisian GA, Borovinskaya IP. Combust Explos Shock Waves 1978;14(4):455–60.
[5] Merzhanov AG, Rogachev AS, Mukasyan AS, Khusid BM, Borovinskaya IP, Khina BB. J Eng Phys Thermophys 1990;59(1):809–16.
[6] Kim T, Wooldridge MS. J Am Ceram Soc 2001;84(5):976–82.
[7] Shuck CE, Manukyan KV, Rouvimov S, Rogachev AS, Mukasyan AS. Comb Flame 2016;163:487–93.

Solution Combustion Synthesis

Singanahally T. Aruna

In the last decade, solution combustion synthesis (SCS) has been one of the most widely used methods for the preparation of a plethora of oxide materials. This method was pioneered and developed by K.C. Patil et al. in the 1980s in India [1] and was introduced to the SHS community around the world by A.G. Merzhanov in 1993 [2]. For the first time, α-alumina was synthesized, though accidently when a mixture of $Al(NO_3)_3.9H_2O$ and urea was rapidly heated around 500°C in a preheated muffle furnace [3]. The reaction mixture foamed and ignited to burn with an incandescence flame yielding a voluminous white product that was identified as α-Al_2O_3. In order to understand the highly exothermic nature of the combustion reaction, concepts used in propellant chemistry were employed [3]. In the solution combustion calculations, the valence of the oxidizing elements was modified to be considered as negative, and the reducing elements as positive, similar to the oxidation number concept familiar to chemists. Accordingly, the elemental valency of C, Al, and H was +4, +3, and +1, respectively, while for oxygen it is considered to be −2 and zero for nitrogen. Based on this concept, the oxidation number of aluminum nitrate and the reducing valency of urea are $Al(NO_3)_3 = -15$; ($Al = +3$, $3N = 0$, $9O = 9x{-}2 = -18$) and $CH_4N_2O = +6$; ($C = +4$, $4H = 4x + 1 = +4$, $2N = 0$, $O = 1x{-}2 = -2$). Thus, for the complete combustion of the (aluminum nitrate/urea) mixture, the molar ratio becomes $15/6 = 2.5$. The stoichiometric equation for this reaction can be presented as follows:

$$2Al(NO_3)_3(aq) + 5\underset{\text{Urea}}{CH_4N_2O}(aq) \xrightarrow{500°C} \alpha - Al_2O_3(s) + 5CO_2(g) + 8N_2(g) + 8H_2O(g)$$

During the preceding reaction, the energy released is at maximum, and the combustion is complete with no carbon residue. It is feasible to dope the desired amount of metal ions in the host oxide matrix to form pigments, phosphors, and catalysts. By this method, 1% of Cr was doped uniformly in the alumina matrix by adding chromium nitrate to the redox mixture of aluminum nitrate and urea to get pink alumina (Fig. 1).

Later a wide range of oxides were prepared by using hydrazides instead of urea. Hydrazides such as carbohydrazide, oxalyldihydrazide, tetraformal trisazine, maleic hydrazide, and malonic acid dihydrazide were used by Patil's group [4,5]. Other researchers explored glycine, alanine, hexamine, citric acid, and so on as fuels [6]. Urea

Concise Encyclopedia of Self-Propagating High-Temperature Synthesis
http://dx.doi.org/10.1016/B978-0-12-804173-4.00136-8

Fig. 1 SCS-derived voluminous pink alumina powder.

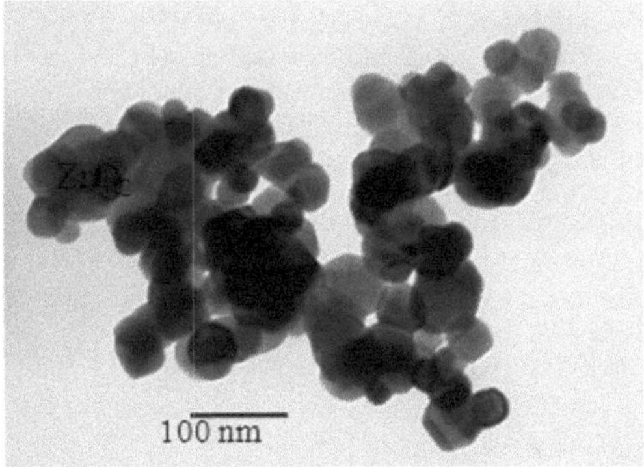

Fig. 2 TEM image of monoclinic ZrO_2 nanopowder.

has been used as a fuel for the synthesis of sulfides by SCS. A mixture of fuels resulted in the synthesis of nanosize oxides (see Fig. 2). More recently, plant extracts were used as fuels to prepare different morphologies of nanostructured materials [7].

The particle size of solid combustion products were controlled by controlling the nature of combustion and by using suitable fuels that result in (i) a slow burning rate, (ii) a reduction in exothermicity (making the reaction smolder rather than flam), and (iii) an increase in the number of gaseous products (H_2O, N_2,CO_2, etc., which dissipate the heat). Linear combustion appears to yield nanosize oxide powders unlike volume combustion. It was also observed that metal acetates favored the formation of nano scale oxides.

The salient advantages of SCS process are (i) that it is a fast process that uses relatively simple equipment and is energetically attractive as it makes use of the exothermicity of the redox reaction rather than external heat source; (ii) that it permits control over the composition, structure, homogeneity, and stoichiometry of the products; (iii) that it leads to the formation of high purity products; (iv) that it allows incorporation of the desired quantity of impurity ions or dopants in the oxide hosts to prepare industrially useful materials; (v) that it permits stabilization of metastable phases (γ-Fe_2O_3, t-ZrO_2, anatase TiO_2, etc.); and (vi) that it allows the formation of virtually any size (micron to nano) and shape (spherical to hexagonal shape) products. The materials include, among others, magnetic oxides, pigments, phosphors, high T_c cuprates, solid oxide fuel cell materials, M/oxide catalysts, and tough materials (e.g., t-ZrO_2/Al_2O_3).

The SCS-derived powders are very effective for environmental remediation. The possibility of depositing nanocrystalline ionic catalysts on ceramic cordierite honeycombs by a single-step combustion method has demonstrated [8]. There are a few reports on scaling up this process; however, this process has not been adopted by industries.

REFERENCES

[1] Patil KS, Gajapathy D, Verneker VRP. Mater Sci Lett 1983;2(6):272–4.
[2] Merzhanov AG. Int J Self-Propag High-Temp Syn 1993;2:113–58.
[3] Patil KC, Hegde MS, Rattan T, Aruna ST. Chemistry of nanocrystalline oxide materials: combustion synthesis, properties and applications. Singapore: World Scientific; 2008.
[4] Patil KC, Aruna ST, Mimani T. Curr Opin Solid State Mater Sci 2002;6:507–12.
[5] Aruna ST, Mukasyan AS. Curr Opin Solid State Mater Sci 2008;12:44–50.
[6] Udayabhanu, Nethravathi PC, Pavan Kumar MA, Suresh D, Lingaraju K, Rajanaika H, et al. Mater Sci Semicond Process 2005;33:81–8.
[7] Aruna ST. Solution combustion synthesis: novel routes to novel materials. In: Lackner M, editor. Sharjah: Bentham Publishers; 2010. p. 206–21.
[8] Hegde MS, Madras G, Patil KC. Acc Chem Res 2009;(42):704–12.

Solution Combustion Synthesis of Catalysts

Khachatur V. Manukyan

The preparation, characterization, and use of catalysts are recognized as areas of major interest in chemical engineering and the production of chemicals. Industrial catalysts exist in various forms, and their preparation involves different methods. Solution combustion synthesis (SCS) is a relatively new method for catalyst preparation [1,2]. It uses elements of conventional preparation approaches and involves its own unique techniques. SCS-derived catalysts can be categorized into two major groups: bulk and supported.

In the first group, combustion synthesized porous foam-like aggregates or powders can be used directly as catalysts, while in the second group, the active phase is deposited onto a solid support using a rapid high-temperature reaction. In the first group, SCS catalysts include oxides, metals, and alloys [3,4]. Depending on the specific catalytic reaction, prepared materials can be subjected to postcombustion treatment. An example of the bulk catalyst prepared by SCS is Ce- and Ni-substituted $LaFeO_3$ perovskites, which showed high activity and excellent stability for autothermal reforming of hydrocarbon mixtures [3]. Other examples of bulk metallic catalysts synthesized by solution combustion are Ni (Fig. 1A) and Cu or their alloys for ethanol reforming reactions [4,5].

As an example of supported catalysts utilizing SCS, manganese oxides can be synthesized onto pore-free magnesium aluminum silicate (cordierte) monolith filters [6]. This synthesis formulation includes impregnating the filters into the reactive solution of manganese nitrate and glycine and then placing them in a preheated furnace to initiate combustion. High surface area $Ni/\gamma\text{-}Al_2O_3$ catalysts were also prepared by immersing the $\gamma\text{-}Al_2O_3$ pellets into reactive solutions containing nickel nitrate and glycine [7]. The dried pellets were then placed on the preheated hot plate to initiate combustion.

Direct synthesis of high surface area (155 m^2/g) Ni (or NiO) catalysts supported on SiO_2 was also reported [8]. In this case, a reactive solution of nickel nitrate, glycine, and ammonium nitrate was mixed with silica powder. Drying of the resultant slurry yielded a gel, which was then placed in a reactor with gas fed under air or argon. After ignition, a combustion front propagated through the sample. Combustion of gels in air resulted in highly dispersed nickel nanoparticles (5 nm), which instantly oxidized in air (Fig. 1B). This catalyst showed high activity and selectivity for ethanol reforming toward hydrogen generation.

Concise Encyclopedia of Self-Propagating High-Temperature Synthesis
http://dx.doi.org/10.1016/B978-0-12-804173-4.00137-X

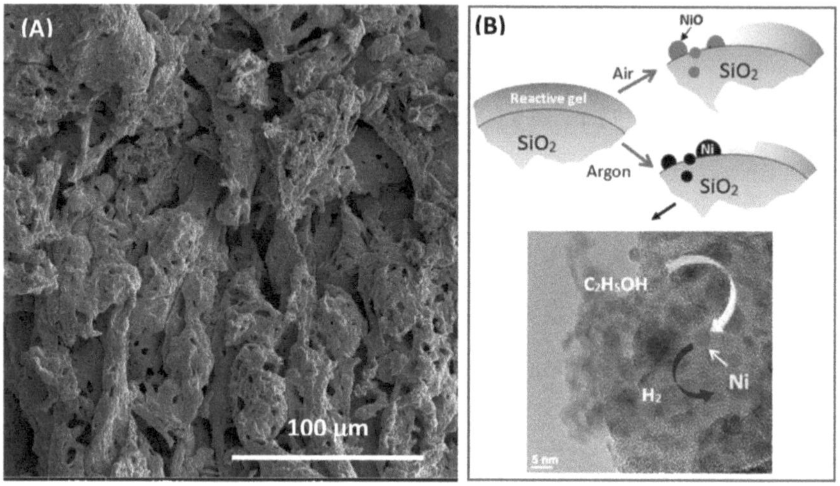

Fig. 1 SEM/TEM image of porous Ni (A) and Ni (NiO) supported (B) catalysts.

The impregnated active layer combustion method was developed for the preparation of catalysts without using a high-temperature calcination step [9]. This approach is beneficial for the synthesis of metal oxide or metals with low caloric nitrate fuel systems. The additional heat formed through burning of the cellulose makes the oxide formation reaction self-sustaining.

The evident advantages of combustion reactions to prepare solid catalysts include utilization of straightforward and rapid techniques and the ability to control the catalyst microstructure by changing the combustion conditions. One of the most significant advantages relates to the preparation of doped complex catalysts with the highest possible structural homogeneity. Combustion synthesized catalysts exhibit high stability toward deactivation.

REFERENCES

[1] Hegde MS, Madras G, Patil KC. Acc Chem Res 2009;42:704–12.
[2] González-Cortés LS, Imbert FE. Appl Catal A 2013;452:117–31.
[3] Erri P, Dinka P, Varma A. Chem Eng Sci 2006;61:5328–33.
[4] Kumar A, Miller JT, Mukasyan AS, Wolf EE. Appl Catal A Gen 2013;467:593–603.
[5] Kumar A, Cross A, Manukyan K, Bhosale RR, van den Broeke LJP, Miller JT, et al. Chem Eng J 2015;278:46–54.
[6] Piumetti M, Fino D, Russo N. Appl Catal B 2015;163:277–87.
[7] Cross A, Kumar A, Wolf EE, Mukasyan AS. Ind Eng Chem Res 2012;51:12004–8.
[8] Cross A, Roslyakov S, Manukyan KV, Rouvimov S, Rogachev AS, Kovalev D, et al. Phys Chem C 2014;118:26191–8.
[9] Kumar A, Mukasyan AS, Wolf EE. Appl Catal A Gen 2010;372:175–83.

Spark Plasma Sintering of SHS Powders

Roberto Orrù, Giacomo Cao

The electric current sintering technique, often referred to as spark plasma sintering (SPS), is an efficient method for consolidation of powders [1]. As schematically represented in Fig. 1A, the graphite die containing the processing powders and the powders themselves (when electrically conductive) are directly crossed by an electric current while a mechanical pressure is concurrently applied during SPS. Consequently, powders undergoing sintering are very rapidly heated by the Joule effect so that dense materials with relatively fine microstructure can be produced under less severe sintering conditions than with conventional hot pressing. Indeed, lower dwell temperatures and shorter processing times allow inhibiting the grains growth phenomena during the consolidation of nanopowders, making it possible to obtain bulk products with submicrometer- or nanometer-size grains correspondingly by SPS [1]. The presence of additional phenomena (electro-migration, plasma formation among particles, etc.) has also been postulated to take place during SPS [1]. The previously mentioned advantages also hold true when SPS apparatuses are utilized to perform the reaction synthesis and densification in a single process (reactive SPS or RSPS), starting with the suitable reactants. However, when highly exothermic systems are processed by RSPS, the possible occurrence of the synthesis reaction under the combustion regime is accompanied by safety problems and inhomogeneities in the end product [2]. These factors are crucial for possible scale-up of reactive SPS processes.

Therefore the availability of suitable synthesis routes able to yield powders to be more efficiently consolidated by SPS is very important, particularly when processing difficult-to-sinter ultra-refractory ceramics. In this regard, the capability of the SHS technique to produce powders characterized by higher sintering ability, with respect to products synthesized by conventional routes (furnace, solution methods, etc.), was recently proved [3–5]. For instance, when ZrB_2 powders prepared by replacement reaction or carbothermal reduction were processed for 1 h at 1800°C by pressureless sintering, 86%–87% dense samples were obtained [3]. On the other hand, relative densities up to 93% were achieved when the same sintering condition was applied to ZrB_2 powders synthesized by SHS. The improved sinterability of SHS powders can be ascribed to their significantly higher defect concentration, on the order of 10^{12} cm^{-2}, compared with 10^{8} cm^{-2} observed in powders synthesized by replacement reaction and carbothermal reduction

Concise Encyclopedia of Self-Propagating High-Temperature Synthesis
http://dx.doi.org/10.1016/B978-0-12-804173-4.00140-X

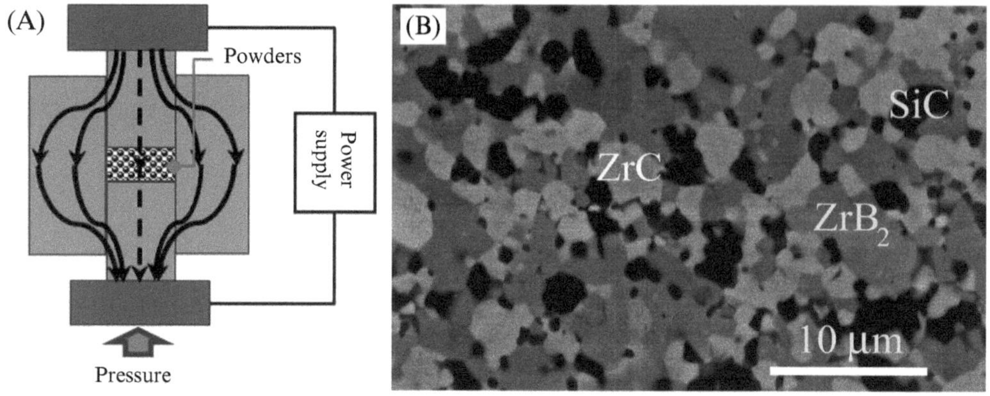

Fig. 1 (A) Schematic representation of the SPS technique: the electric current flows through the die (solid lines with arrows) and the processing powder when it is electrically conductive (dotted lines); (B) SEM back-scattered micrograph of ZrB_2-40 vol.%ZrC-12 vol.%SiC sintered product obtained from SHS powders.

methods. This feature may be a consequence of the extreme heating and cooling rates, up to 2000–200,000 K min^{-1} established during combustion front propagation [3].

In order to exploit the advantages displayed by both the SHS and SPS techniques, the two methods have been recently coupled for the fabrication of several monolithic (HfB_2, La_x-$Sr_{1-x}TiO_3$, $La_{0.7}Sr_{0.3}MnO_3$, $MoSi_2$, NiAl, β-Si_3N_4, $Sr_{1-x}R_xTiO_3$, etc.) and composite (Al_2O_3-TiC, Cr_2AlC matrix composites, HfB_2-$HfSi_2$, HfB_2-SiC, HfB_2-HfC-SiC, NbC-NbB_2, TaB_2-SiC, TaB_2-TaC-SiC, Ti-Al_2O_3-TiC, $TiC_{0.7}$-TiB_2, TiC-Si_3N_4, ZrB_2-SiC, ZrB_2-ZrC-ZrC, ZrB_2-ZrC, etc.) systems in the bulk form [5]. The benefits arising from their combination are confirmed particularly when considering ceramic composites. For instance, an investigation involving the fabrication of bulk ZrB_2-40 vol.%ZrC-12 vol.%SiC materials showed that, under appropriate SPS conditions (1800°C, 10 min, 20 MPa), a fully dense product (Fig. 1B) can be obtained when using composite powders (particle size in the range of 0.04–60.00 μm) synthesized by SHS according to the following reaction:

$$8Zr + 2B_4C + 1.5Si + 3.5C \rightarrow 4ZrB_2 + 4ZrC + 1.5SiC \qquad (1)$$

On the other hand, when a relatively fine commercial powder mixture (particle size in the range 0.04–20.00 μm) with identical nominal composition was processed under the same SPS conditions, the resulting product was only 91% dense (Fig. 2A). The importance of synthesizing by SHS composite mixtures in one step was further shown in a recent study on the influence of $HfSi_2$ addition on the consolidation of HfB_2 powders by SPS (1800°C, 20 min, 50 MPa). Higher density samples can be produced when using

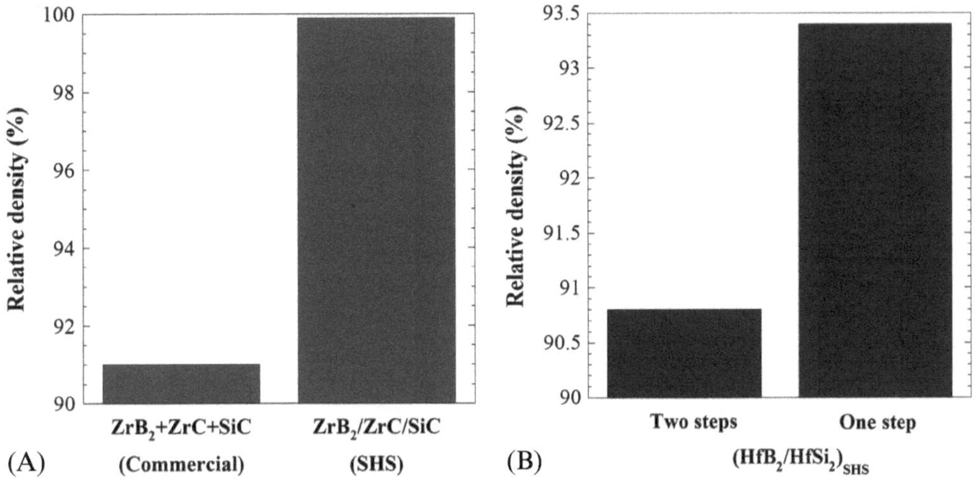

Fig. 2 Comparison between density of (A) ZrB_2-40 vol.%ZrC-12 vol.%SiC sintered products from commercial and SHS powders, and (B) HfB_2-10 vol.% $HfSi_2$ material from SHS powders synthesized in one or two steps.

powders consisting of HfB_2 and $HfSi_2$ simultaneously synthesized by SHS (one-step synthesis) instead of a mixture of the ceramic products obtained separately (two-step synthesis) (Fig. 2B). It is then apparent that the concurrent synthesis of composite constituents clearly promotes the formation of intimate contacts among grains and, consequently, the reduction of diffusion distances so that powders sintering can be significantly facilitated.

REFERENCES

[1] Orrù R, Licheri R, Locci AM, Cincotti A, Cao G. Mater Sci Eng R 2009;63:127–287.
[2] Licheri R, Musa C, Orrù R, Cao G. J Eur Ceram Soc 2015;35(4):1129–37.
[3] Mishra SK, Das S, Pathak LC. Mater Sci Eng A 2004;364:249–55.
[4] Musa C, Licheri R, Orrù R, Cao G. Ind Eng Chem Res 2014;53(22):9101–8.
[5] Orrù R, Licheri R, Musa C, Cao G. Adv Sci Tech 2014;88:111–20.

Spectroscopy of SHS

Alexander S. Rogachev

Hot condensed combustion products emit a continuous spectrum of gray-body radiation that can be used for temperature measurement by optical pyrometry [1]. The emission spectra of hot gases normally consist of individual lines and bands. In the first experiments [2], the absorption bands of gases were recorded in the emission from a burning sample (Table 1). Most of these bands were attributed to molecular hydrogen, sometimes even with the vibrational structure.

Minor contamination of samples with sodium (most likely Na salts) gave intense absorption and emission lines of atomic Na. It should be noted that the emission line around 420 nm observed for the Ti + C system was identified as the emission from Ti atoms [3]. However, the intense line (or band) at about 422 nm was also observed in the systems that did not contain titanium [2]. This can be attributed in all likelihood to the H_2 (420.5 nm band) and O_2 (421.5 nm band) impurity gases. Apparently, under normal (without disturbances) combustion of SHS systems, the spectra of metal vapor cannot be registered because of their low intensity. The emission spectra were recorded for the Ti + C mixture in conditions of laser ignition [3]. Laser irradiation can be expected to intensify the evaporation of Ti and thus strengthen the intensity of emission/absorption bands.

Table 1 Absorption lines of the optical spectra for various SHS systems [2]

Wave-length (nm)	Relative intensity of the absorption peak for the system (%) (Intensity of gas laser line 632.4 nm is taken as 100%)						Probable source of the line
	Ti + 2B	Zr + 2B	Ta + 2B	Ti + C	Zr + C	Ta + C	
402.2	–	–	12	14	12	–	
402.4	–	11	10	11	–	–	H_2, O_2
402.6	–	–	12	12	11	–	
422.1	12	14	33	19	33	–	H_2, O_2, (Na)
461.5	–	–	11	24	12	–	H_2
588.8	16	17	20	17	17	17	(Na)
589.4	12	16	17	20	11	19	(Na)
670.2	–	–	14	10	–	–	H_2

Concise Encyclopedia of Self-Propagating High-Temperature Synthesis
http://dx.doi.org/10.1016/B978-0-12-804173-4.00141-1

In a second series of experiments [2], only the emission of the gas phase was recorded. For this purpose, the burning sample was placed behind an opaque screen that cut off the glow emission of condensed products. It allowed observing not only the absorption but also emission spectra of gases. In these experiments, the emission line at 656.25 nm was observed and was attributed to atomic hydrogen (Hα of the Balmer series). Thus it was shown that the hydrogen released during combustion is partly in its atomic state.

REFERENCES

[1] Andreev VA, Maltsev VM, Seleznev VA. Combustion of Hf–B mixtures as studied by optical pyrometry. Combust Explos Shock Waves 1980;16(4):374–8.
[2] Merzhanov AG, Rogachev AS, Umarov LM, Kir'yakov NV. Experimental study of the gas phase formed in the processes of self-propagating high-temperature synthesis. Combust Explos Shock Waves 1997;33(4):439–47.
[3] Hansen GP, Fredin L, Margrave JL, Behres RG. Characterization of vapors evolved during high-temperature synthesis. Defense advanced research project agency (DARPA) report; 1986. p. 63–73.

Spin Combustion of Gas and Gas-Free Systems

Inna P. Borovinskaya, Yury M. Maksimov

Spin combustion can be described as follows: when a sample is ignited, a burning spot appears on its side wall and moves in a spiral by the unburnt part leaving a bright trace; the trace merges gradually with the one from the previous spiral. This process was first discovered when combustion of zirconium porous samples in gaseous nitrogen diluted with argon was studied. The aim of those experiments was to shift the kinetic mode of filtration gas admission to the combustion front to a diffusion mode. Fig. 1 shows the dependence of the zirconium combustion rate on the degree of nitrogen dilution with argon. A shift of gas supply mode to the diffusion one results in abrupt deceleration of the combustion process and spin appearance. Even slight amounts of argon decrease the combustion rate five times, and argon introduction at the rate of $\sim$0.26 (parts) leads to complete combustion decay. Spin combustion occurs at the combustion threshold and appears to be one of the types of unsteadiness at stationary surface combustion [1]. This type is called spin combustion, similarly to the well-known phenomenon of spin detonation. The thorough investigation of the spin combustion mechanism is described in papers [2,3]. Spin combustion is shown to influence such parameters as a sample density, its diameter, and the nitrogen-argon ratio in the gas mixture. The direction of the hot spot motion (to the right or left) is spontaneous; sometimes some points moving in the same or opposite direction are observed. According to the analysis of the longitudinal cross-section of the burnt sample, the thickness of the surface layer covered with spin combustion does not exceed 2 mm. Fig. 2A shows photoregistration and a rapid movie picture for one turn in the case of hafnium combustion in a nitrogen-argon mixture ($P_{N_2} = 6.0$MPa, $P_{Ar} = 0.1$ MPa). The combustion rate is $U \cong 1$ cm/s, and the rotating frequency is $V \cong 0.3\ s^{-1}$. All the parameters are decreasing by the end of combustion.

Paper [3] studies the influence of parameters such as combustion temperature at nitrogen dilution with argon and helium, sample density, heat losses (sample diameter change), and titanium combustion rate at nitrogen dilution with inert gases ($P_{\frac{Ar}{N_2}} = 0.24$, $\rho = 1.85$ g/cm^3) on spin combustion limits and characteristics (rate and rotating frequency) (Fig. 2B). The obtained experimental results prove that spin

Concise Encyclopedia of Self-Propagating High-Temperature Synthesis
http://dx.doi.org/10.1016/B978-0-12-804173-4.00142-3

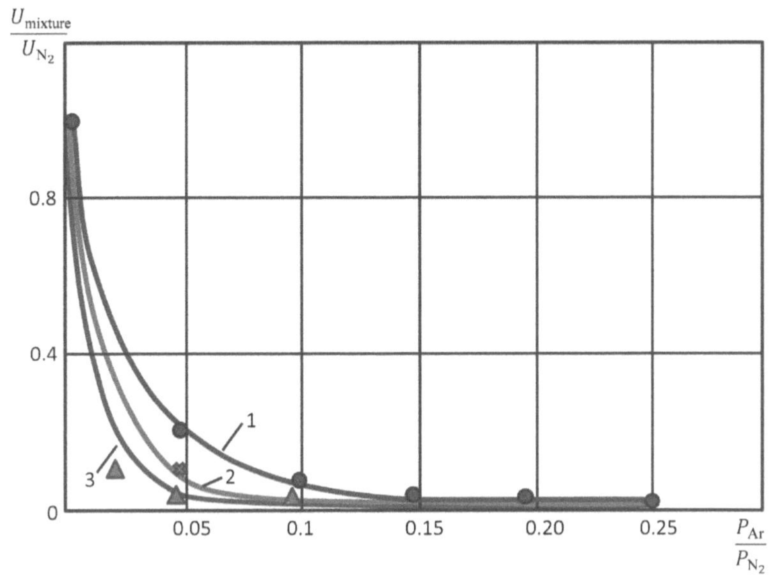

Fig. 1 Dependence of zirconium combustion rate on nitrogen dilution with argon: (1) $P_{total} = 1.0$ MPa; (2) $P_{total} = 3.0$ MPa; (3) $P_{total} = 7.5$ MPa.

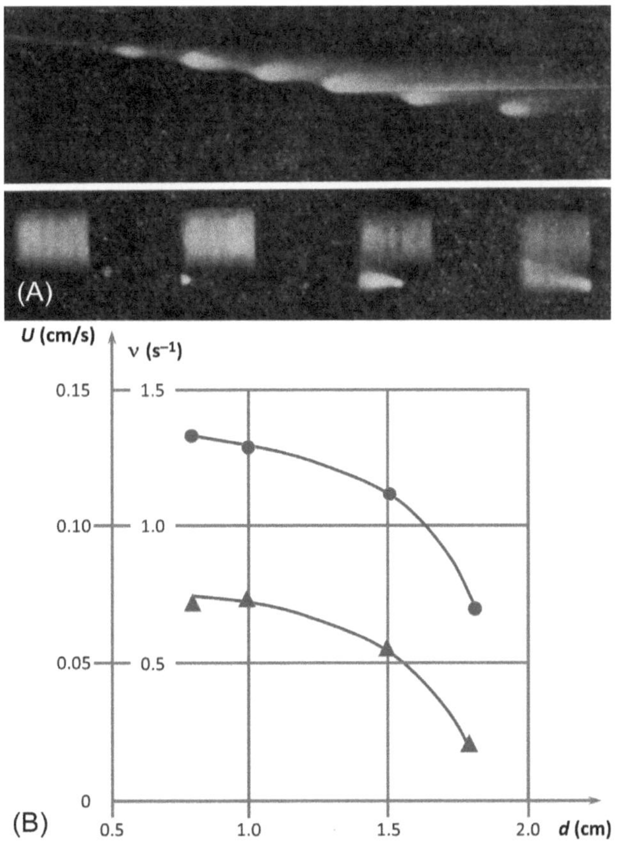

Fig. 2 (A) Photoregistration and rapid movie picture of spin combustion; (B) Dependence of titanium combustion rate and rotating frequency on sample diameter, $P_{total} = 2.5$ MPa.

Fig. 3 A disk burnt in spin mode.

combustion is a transition process from stationary combustion to combustion extinction. The extinction occurs in the case of heat removal dominance over heat release in the combustion front. The picture of gasless system combustion does not differ from that of metal-nitrogen spin combustion. Similar to metal combustion in gases, a lightening spot appears on the surface of a sample made of solid reagents mixture, for example Ti-B-αCu; it moves in a spiral with a bright trace after it. But unlike combustion in gases, a spiral line with a constant lead shows up well on the samples (disks); it is a trace of the chemical reaction [4] (Fig. 3). In this case, the reaction does not occur on the sample surface; it takes place in the sample volume. During the disks' combustion, the center of the reaction after its initiation moves on the flat surface leaving a two-dimensional spiral trace. With an increase in the disk radius, the motion dynamics become more complicated. In the case of small diameters, only one spin is observed. With the increase in diameter, two- or three-headed spin appears. With further increase in the diameter, the number of spins grows.

The surface of a plate burned at spin mode looks like a zebra with dark and light stripes corresponding to the places of the chemical reaction. The reaction center originates at the angular point and moves along the warmed-up angle sides. Point-like centers constantly appear, and the bent flat combustion front straightens. The reaction centers are often generated by the plate edges, which are ways of significant heat flow. In this case, the center motion is of the ordered character (Fig. 4).

Spin combustion of solid systems can be considered as a display of unsteadiness to lateral disturbance of one-dimensional auto-oscillatory mode. From a physical viewpoint, this phenomenon is caused by enthalpy excess in the combustion front because for gas-free

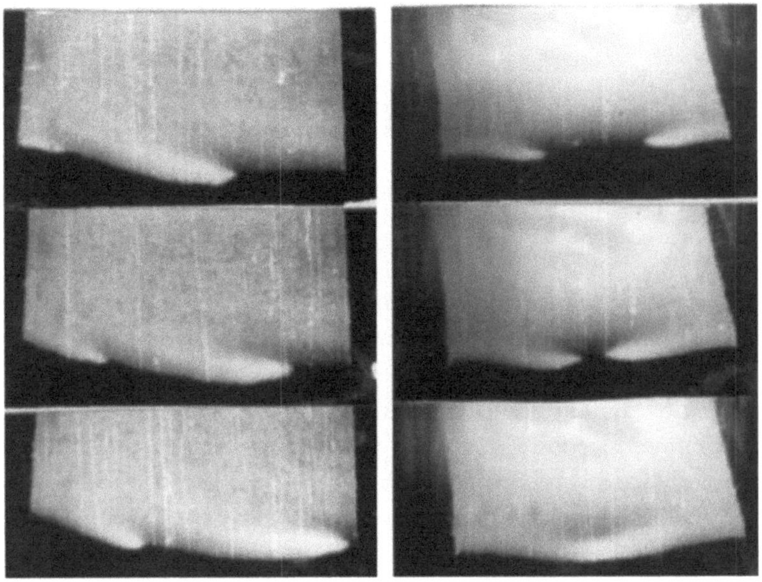

(A) (B)

Fig. 4 Movie pictures of spin combustion of a plate. Heat center generation by one (A) and two (B) sides.

systems, the area of a strong temperature gradient is much smaller than the size of the zone where chemical interaction is essential [5]. In this case, the auto-oscillatory mode that is characterized by pulsations, that is, consecutive alternation of flashes and depressions, is a consequence of two major factors: heat excess in the combustion front and heat flow from the reaction products to the conversion zone [5]. Under such conditions, a thermal center (spin) can arise and start moving by the most convenient thermodynamic path—along the heated layer. Spin combustion appears to be of great interest to theoreticians of combustion science who thoroughly study various types of this process [6].

REFERENCES

[1] Merzhanov AG, Borovinskaya IP. A new class of combustion processes. Combust Sci Technol 1975;10 (5–6):195–201.
[2] Merzhanov AG, Filonenko AK, Borovinskaya IP. New phenomena at condensed systems combustion. Dokl Akad Nauk SSSR 1973;208(4):892–4.
[3] Filonenko AK, Vershinnikov VI. Regularities of titanium spin combustion in nitrogen. Fiz Goren Vzryva 1975;3:353–62.
[4] Maksimov MYu, Lapshin OV. Peculiarities of unsteady combustion of flat systems of $Ti + 2B + \alpha Cu$. Khim Fiz 2015;34(11):50–4.
[5] Ivleva TP, Merzhanov AG, Shkadinsky KG. Mathematical model of spin combustion. Dokl Akad Nauk SSSR 1978;239(5):1086–8.
[6] Merzhanov AG, Mukasyan AS. Solid flame combustion. Moscow: Torus-Press; 2007.

Spinel-Based Pigments

Vladimir I. Vereshchagin, Nina I. Radishevskaya

SHS is applicable to the production of pigments that belong to different classes of inorganic compounds: colored spinelides, corundums, sulfides, perovskites, etc. However, the vast majority of these pigments are oxide pigments with a corundum-spinel structure that capable of withstanding high temperature, UV-irradiation, and chemical erosion. SHS of oxide pigments requires oxide powders and transition metal salts. Aluminum powder is primarily used as a fuel, although it is also possible to use magnesium. Metal oxidation and metallothermic reactions are the parallel reactive driving forces which result in self heating of the mixture to the synthesis temperature of the target spinel, which forms in an additional exothermic step (Table 1).

The synthesis of colored corundums occurs only through aluminum oxidation reactions. The last stage of pigment formation corresponds to the burning-out zone, where the following processes occur: oxidation of metals; secondary physicochemical transformations, which result in structural transformation; and growth of crystals and other processes that do not affect the combustion front propagation rate. Table 2 presents the colors of pigments produced by SHS which consist of colored spinels and various solid solutions [1–2] (Fig. 1).

The advantage of SHS-produced corundum-spinel pigments is their finely dispersed state, a direct result of their formation in the combustion wave.

Table 1 SHS reactions for a corundum-spinel structure

SHS stages	**SHS reactions**	
I	Metal (Me[a]) oxidation	$Al + 0.75O_2 \rightarrow 0.5Al_2O_3 + 1675\,kJ$ (1)
		$2Mg + O_2 \rightarrow 2MgO + 1204\,kJ$ (2)
	Metallothermic reactions	$MeO + 2Al + O_2 \rightarrow Al_2O_3 + Me$ (3)
		$2Me_2O_3 + 8Al + 3O_2 \rightarrow 4Me + 4Al_2O_3$ (4)
II	Spinel synthesis	$MeO + Al_2O_3 \rightarrow MeAl_2O_4$ (5)
III	Afterburning processes	$2Me + O_2 \rightarrow 2MeO$ (6)
		$MeO + Al_2O_3 \rightarrow MeAl_2O_4$ (7)
		$\gamma,\theta\text{-}Al_2O_3 \rightarrow \alpha\text{-}Al_2O_3$ (8)

[a] Me = Co, Ni, Fe, Cr, Zn, Mg, etc.

Concise Encyclopedia of Self-Propagating High-Temperature Synthesis
http://dx.doi.org/10.1016/B978-0-12-804173-4.00144-7

Table 2 Composition of coloring phases of SHS pigments from corundum-spinel structure

No.	System	Color	Main coloring phases
I group: pigments on the basis of white spinels			
1	MgO-Al_2O_3	White	$MgAl_2O_4$
2	MgO-CoO-Al_2O_3	Light-blue	$MgAl_2O_4$, $CoAl_2O_4$
3	MgO-Cr_2O_3-Al_2O_3	Pink	$Al_{1.772}Cr_{0.03}Mg_{1.208}O_4$, $MgAl_2O_4$
4	ZnO-CoO-SnO_2-Al_2O_3	Brilliant blue	$Zn_xCo_{1-x}Al_2O_4$, $Zn_xCo_{2-x}SnO_4$
II group: pigments on the basis of colored spinels			
5	CoO-Al_2O_3	Blue	$CoAl_2O_4$, $AlCo_2O_4$
6	ZnO-MgO-CoO-Al_2O_3	Ultramarine	$Zn_xMg_{1-x}Co_{1-x-y}Al_2O_4$
7	ZnO-NiO-Cr_2O_3-Al_2O_3	Green	$Zn_xNi_{1-x}Al_2O_4$ $Zn_xNi_{1-x}Cr_2O_4$
8	ZnO-MgO-CoO-Cr_2O_3-Al_2O_3	Cyan	$Zn_xMg_{1-x}Co_{1-x-y}Al_{2-z}Cr_zO_4$
9	FeO-CoO-Al_2O_3	Black	$Fe_xCo_{1-x}Al_2O_4$ ($x=0.4$–0.5), Fe_3O_4, $CoFe_2O_4$
10	ZnO-CoO-Fe_2O_3-Cr_2O_3-Al_2O_3	Brown	$Zn_xFe_yCo_{1-x-y}Al_2O_4$, $Zn_xFe_yCo_{1-x-y}Cr_2O_4$, Fe_3O_4, $CoFe_2O_4$
III group: pigments on the basis of corundums			
11	Al_2O_3-Cr_2O_3	Pink	α-Al_2O_3

Fig. 1 Blue cobalt-aluminate (dark gray in print version) spinel produced by layer-by-layer combustion (longitudinal section of a specimen with the diameter of 80 mm).

REFERENCES

[1] Gladun GG, Iskalova AZ, Ksandopulo GI. Mixture for ceramic pigment, RF patent 2029745; 1995.
[2] Radishevskaya NI, Kasatsky NG, Shulpekov AM, Chapskaya AYu, Vereshchagin VI, Naiborodenko YuS, et al. Method for production of ultramarine ceramic pigment, RF patent 2305075; 2007.

Stationary Combustion Wave

Oleg V. Lapshin

One of the main problems in combustion theory is study of the exothermic reaction wave front that is initiated through local ignition of reactive substances [1]. Stationary combustion modes are studied in detail regarding the case when the velocity of the flame front remains practically constant. The profiles describing the kinetics of the process (temperature, completeness of chemical transformations, function of the heat release rate) are also unchanged.

A similar problem was first considered and solved theoretically by Ya.B. Zel'dovich and D.A. Frank-Kamenetsky for combustion in gaseous systems [2]. Based on their approach, B.V. Novozhilov solved the problem of the condensed system combustion [3]. The formulas for calculating the velocity of a stationary combustion wave are (1) and (2) [2,3].

For combustion in gas:

$$u=\rho^{-\frac{3}{2}}\sqrt{\frac{2\lambda}{cQ}\left(\frac{RT_f^2}{E}\right)^{n+1}\left(T_f-T_0\right)^n n!W\left(T_f\right)} \tag{1}$$

For combustion in condensed systems:

$$u=\sqrt{n\frac{\lambda}{\rho Q}\frac{RT_f^2}{E}k_0\exp\left(-\frac{E}{RT_f}\right)} \tag{2}$$

where λ is the thermal-conductivity coefficient; c is the heat capacity; ρ is the density; R is the universal gas constant; k_0 is the preexponential factor; E is the activation energy of a chemical reaction; n is the numerical factor that depends on the reaction order ($n=1$, the first order reaction, $n=2$, the zero order reaction); Q is the heat of reaction; T_0 is the temperature of the reactive substance in the initial condition; and $W(T_f)$ is the reaction rate at the combustion temperature T_f. If the reaction is initiated under adiabatic conditions, then the expression for the combustion temperature T_f corresponding to the maximum value in the front of the exothermic reaction is given by

$$T_f=T_0+Q/c \tag{3}$$

The typical structure of the combustion wave propagation is shown in Fig. 1. As can be seen from the figure, in heating zone I, the initial substance has the initial temperature and the

Concise Encyclopedia of Self-Propagating High-Temperature Synthesis
http://dx.doi.org/10.1016/B978-0-12-804173-4.00145-9

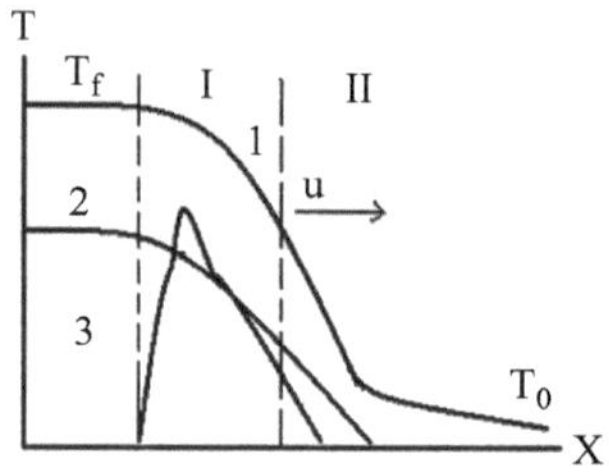

Fig. 1 The typical structure of the combustion wave: (1) temperature, (2) completeness of the chemical conversion, and (3) function of the heat release rate.

rate of the chemical reaction is very low. In the area of high temperatures (reaction zone II), there is intensive heat release as a result of the very strong (exponential) temperature dependence of the chemical transformation rate. The resulting temperature gradient leads to heat transfer from the high temperature regions to the nearby cooler layers, which then heat up and accelerate the reaction. The combustion wave starts propagating through the reactive substance.

For convenience, the stationary combustion wave is defined in the x' coordinate system which moves along with the wave front. The real coordinate system x and the relative coordinate system x' can be related by

$$x' = x + ut \tag{4}$$

where t is the time and u is the rate of propagation of the combustion wave. In Eq. (4) u propagates in the direction opposite to the real spatial coordinate x. Substituting Eq. (4) can greatly simplify the equations of heat conductivity and chemical transformations describing the combustion process.

It should be noted that although to a certain extent the concept of a stationary combustion wave is an approximation, it has repeatedly proven its importance to the development of combustion theory and practical purposes.

REFERENCES

[1] Merzhanov AG. Heat waves in chemistry. Heat and mass transfer in combustion processes. Chernogolovka: BICP AS USSR; 1980. p. 36–58.
[2] Zel'dovich YaB, Frank-Kamenetsky DA. Phys Chem 1938;12(1):100–5.
[3] Novozhilov BV. Proc USSR Acad Sci 1961;141:151.

Steady Propagation of Combustion Wave

Peter M. Krishenik

Theoretical analysis of SHS reactions is oriented toward elucidating the conditions at which frontal combustion of heterogeneous systems can occur in a steady wave propagation mode, which is most convenient for practical synthesis of desired condensed products. The unsteady processes of ignition and onset of the steady combustion wave both require a transient period, which precedes the steady propagation of frontal combustion. For transient periods that are too long, the onset of frontal combustion becomes unlikely because of partial chemical transformation within the starting reactive mixture. A steady propagation mode can be established for an optimal duration of ignition pulse such that the influence of previous inflammation becomes insignificant, while the extent of thermal reaction in the fresh mixture remains negligible.

In a steady propagation mode, burning velocity u is practically constant, with a constant temperature profile and extent of conversion in the combustion wave.

Burning velocity and temperature profile in a steady wave are independent of the ignition procedure (except for a few systems, e.g., Ti-C-H_2, where two different modes of combustion can be realized depending on ignition). The ignition method can affect the transient period only between ignition and the onset of steady wave propagation. The necessary conditions for establishing a steady mode are a sufficiently exothermic reaction and a strong temperature dependence of the reaction rate (i.e., very slow at initial temperatures and exceedingly high at temperatures attained in the combustion wave).

Concise Encyclopedia of Self-Propagating High-Temperature Synthesis
http://dx.doi.org/10.1016/B978-0-12-804173-4.00146-0

STIM Alloys

Inna P. Borovinskaya, Evgeny A. Levashov

Forced SHS consolidation [1], which combines the SHS process and further hot deformation of the synthesis product, is one of the many promising SHS technologies. This technology makes it possible to obtain compact low-porosity and porous-free materials and items based on simple and complex carbides, borides, silicides, carbonitrides, and other refractory compounds, most often with binding metallic components and alloying additives. Obtaining tungsten-free synthetic hard instrumental materials (STIM), disposable cutting plates based on carbides and borides with binding components (Co, Ni, Mo, etc.), and alloying additives (Sc, Cr, Re, Cu, Al, CuAl, etc.) [2] in particular, are a few of the areas developed because of to the capabilities of forced SHS consolidation. These materials are all referred to as "STIM alloys." Depending on the wear-resistant constituent, the STIM alloys are roughly divided into three groups: carbide-borides, carbides, and borides. The carbide-boride group comprises alloys with the wear-resistant constituent TiC-TiB_2 (STIM-1, STIM-1A, STIM-1B/1, STIM-1B/2, STIM-1B/3). The carbide group contains alloys with the wear-resistant constituent TiC, the solid solution TiC-Cr_3C_2 (STIM-2, STIM-2A, STIM-3A, STIM-3B), and the carbonitride TiMoCN with Ni-Al-Mo-Nb-Co-Cr binder (STIM-5) [3,4]. Alloys with a titanium boride TiB wear-resistant constituent (STIM-4, etc.) belong to the third group. A range of modernized STIM alloys: carbide and boride-oxide, carbonitride, and other alloys with mainly structural properties have been developed. Table 1 presents characteristics of several STIM alloys differing in composition.

Mechanical properties of the STIM alloys depend on the composition of the wear-resistant constituent and the metallic binder, as well as on the alloy structure. STIM alloys are materials with both high hardness and high strength. Some of them have unique properties. For example, STIM-5 is a new class of dispersion-hardened alloys with the effect of simultaneously dispersedly strengthening the carbonitride grains due to nanosized precipitations of molybdenum carbide phase from the oversaturated TiMoCN solid solution, as well as to the strengthening of the Ni-Al-Mo-Nb-Co-Cr metallic binder as a result of the precipitations of (Ni_3Al) γ'-phase [3,4]. For machining steels (Steel 40, Steel 45, KhNM Steel, etc., at cutting rates from 120 to 80 m/min), the cutting stability capacity of STIM-5 is 1.7–3.0 times higher than that of its industrial analogs T30K4, T15K6, and KNT16.

Concise Encyclopedia of Self-Propagating High-Temperature Synthesis
http://dx.doi.org/10.1016/B978-0-12-804173-4.00122-8

Table 1 Tungsten-free hard metals of STIM grade

Alloy grade	Composition	Density (g/cm^3)	Average particle size (μm)	HRA	Bending strength (MPa)	Application
STIM-1B/3	($TiC-TiB_2$) + Cu	4.94	5–7	94	700–800	Cutting tools
STIM-2	TiC + Ni	5.50	5–7	90	1000–1100	Armored plates
STIM-2A	TiC + (Ni-Mo)	6.40	1–2	87	1600–1800	Stamp instrument
STIM-3B	($TiC-Cr_3C_2$) + Ni	5.37	3–4	93	800–1000	Cutting tools
STIM-3C	($TiC-Cr_3C_2$) + steel	5.40	2–4	93	700–800	Nonscaling items
STIM-4	TiB + Ti	4.20	1–2	86	1200	Items resistant to thermal shock
STIM-5	TiMoCN + (Ni-Al-Mo-Nb-Co-Cr)	5.80	1–2	92	1200–1400	Cutting tools

The STIM-1B/3 alloy has a high cutting capacity; it functions on the level of the best ceramic disposable plates during final treatment at high cutting rates. When synthesizing STIM-3C, the complex carbide $Cr_{16}Fe_7C_6$ is formed, making this alloy the record holder for scale resistance. STIM-3C is applied in stamping accessories that operate at temperatures higher than 1100°C. The scale resistance of these accessories is five times higher than that of its analog made from ZhS6U. STIM-4 has a high thermal stability and corrosion resistance. Based on these properties, a wide range of large-size items for various purposes are produced using the forced SHS compaction, for example, rolls for forming nonferrous metals, hard alloy disks with a diameter of 370 mm for manufacturing special mirrors, and new types of macrocomposites with a metallic matrix that combine properties of hard alloys and ductile metals (Ti, Fe, steel). The large-scale SHS-made items are virtually unrivaled throughout powder metallurgy [5].

REFERENCES

[1] Merzhanov AG. Ceramics: toward the 21st century. In: Soga N, Kato A, editors. Proc. of symp. on ceram. commemorating the centennial of the Ceram. Soc. of Japan, 16–18 Oct., 1991, Yokohama, Japan. Tokyo: Ceramic Society of Japan; 1991. p. 378–403.

[2] Merzhanov AG. Int J SHS 2001;10(2):237–57.

[3] Levashov EA, Shtansky DV, Lobov AL, Borovinskaya IP. Int J SHS 1993;2(2):165–73.

[4] Levashov EA, Shtansky DV, Lobov AL, Merzhanov AG. Phys Met Metallogr 1994;77(2):118–24.

[5] Merzhanov AG. Russ Chem Bull 1997;46(1):1–27.

Structural Macrokinetics

Alexander S. Rogachev, Alexander S. Mukasyan

The concept of structural macrokinetics (SM) was suggested by A.G. Merzhanov and his school near the end of the 1980s [1,2]. At that time, development of SHS focused on the synthesis of powders with subsequent processing for the production of the final materials and net-shape articles in a single step by using the combustion process. This approach greatly increased the requirements for methods of controlling the structure of products, which determine the properties of the materials. Therefore A.G. Merzhanov suggested searching for a solution through by studying the patterns and mechanisms of a product's structure formation in SHS waves. This task merged with an investigation of the microstructure of combustion waves, which is one of the fundamental aspects of the combustion theory. A reciprocal influence of combustion and structure formation occurs in SHS: propagation of the combustion wave in a heterogeneous environment leads to the destruction of the original structure and the formation of a new structure of the media; simultaneously, these structural changes have a decisive influence on the propagation mechanism of the reaction front. The SM combines kinetics of chemical reactions, the theory of heat and mass transfer, and the kinetics of structural transformations in the heterogeneous media. The following formulas were suggested by A.G. Merzhanov as a simplified definition of the SM:

$$\text{Macrokinetics} = \text{Chemical kinetics} + \text{Heat} - \text{and mass} - \text{transfer}$$

$$\text{Structural macrokinetics} = \text{Macrokinetics} + \text{Structural transformations}$$

Therefore the main task of SM research is the study of the evolution of the structure of the medium during chemical transformations, taking into account the processes of heat and mass transfer. The SHS process with its extremely high temperatures, temperature gradients, and rates of physical-chemical and structural transformations has provided an experimental basis for the development of this new research.

The term macrokinetics was introduced by D.A. Frank-Kamenetskii [3]. The prefix "macro" means large or long (from the Greek "makros"). In the case of macrokinetics, it means that in addition to the elementary processes of interaction between two or more atoms or molecules, the study of macrokinetics processes includes a statistically large number of particles, which participate in different processes, such as diffusion, heat transfer, convection, and so on. However, a statistically large number of atoms can be

Concise Encyclopedia of Self-Propagating High-Temperature Synthesis
http://dx.doi.org/10.1016/B978-0-12-804173-4.00147-2

contained in a very small micrometer or nanometer-scale regions. This concept may be confusing, because in materials science, these scales are usually attributed to the microstructure of the media. Nevertheless, expressions such as "macrokinetics of reactions in a microheterogeneous medium" or "macrokinetics of interactions of the nanoparticles" do not contain contradictions; they just include terms from different fields of study in which concepts of micro and macro are defined differently [4].

In spatially inhomogeneous processes, in addition to the concepts of concentration and temperature gradients, a crucial role is played by the concept of the structure of the medium, involving macrostructure (scale, visible to the naked eye, e.g., cracks and large pores), microstructure (particles, grains, pores with a size from several nanometers to tens of microns, which can be seen in optical or electron microscope), and atomic/crystal structure (packing of atoms in crystals, short-range order in the amorphous phases and melts, etc.).

All these structural levels may influence the rate of heterogeneous chemical reactions. For example, consider the reaction of a solid particle with the melt, which yields a solid product. The reaction route depends on whether the solid product forms as a thin continuous layer on the surface of the particle or as separate grains on the surface or in the form of grains in the melt. In the first case, the reaction requires the diffusion of reactants through a layer of the solid product; in the second case, the mechanisms of grain boundary and surface diffusion govern the process; and in the third case, the processes of dissolution, diffusion in the melt, crystallization, and grain growth need to be considered. Thus the structure of the medium influences the macrokinetics of the chemical interaction. There is also an opposite effect. For example, depending on the reaction rate, the system can be heated up rapidly or slowly, and the heating rate has a decisive effect on the structure and phase composition of the products. Thus the study of chemical reactions taking into account the reaction kinetics, the processes of heat and mass transfer, as well as the dynamics of structural transformations of the medium is the subject of study for SM.

REFERENCES

[1] Merzhanov AG. Self-propagating high-temperature synthesis: twenty years of research and findings. Combustion and plasma synthesis of high-temperature materials. New York, NY: VCH Publishers Inc.; 1990. p. 1–53.

[2] Merzhanov AG, Rogachev AS. Structural macrokinetics of SHS processes. Pure Appl Chem 1992;64(7):941–53.

[3] Frank-Kamenetskii DA. Diffusion and heat transfer in chemical kinetics. Princeton, NJ: Plenum Press; 1967.

[4] Rogachev AS, Mukasyan AS. Combustion for material synthesis. Boca Raton, FL: CRC Press Taylor & Francis Group; 2015. 397 p.

Supported Catalysts

Vyacheslav N. Borshch

Many catalysts that are in mass production and wide use in industry are systems of an active phase applied to a support [1]. First of all, the use of the support generally sharply (by an order of magnitude or more) decreases the consumption of the more expensive active phase. Support determines the physical-mechanical properties (strength, abrasion resistance, heat resistance) of the catalyst, maintains the active phase's highly dispersed state, and improves the stability of active-phase particles by preventing their destruction or excessive growth because of migration of active-phase atoms. Support often increases the activity and selectivity of the active phase both by influence on its electronic structure and the morphology of its particles through chemical bonds and by a number of auxiliary reactions on a support.

The most widely used catalyst supports for many processes, especially in oil refining and petrochemistry, are alumina and silica and also silica-alumina, both amorphous and crystalline (zeolites). A new (for catalysis) class of supports are sialons, produced for the first time by SHS in ISMAN [2,3]. The structure of sialons, in particular, β-sialons $Si_{6-Z}Al_ZO_ZN_{8-Z}$ (Z = 1, 2, 3), is a combination of two lattices constructed from tetrahedrons $[SiN_4]$ and $[AlO_4]$ with bond lengths $R(\text{Si—N}) = 1.72$ Å and $R(\text{Al—O}) = 1.75$ Å, and partial substituted $[AlO_{4-k}N_k]$ and $[SiN_{4-k}O_k]$ (k = 1–3). Consequently, sialons can be considered as N-substituted analogs of silica-alumina consisting of $[AlO_4]$ and $[SiO_4]$. The physical-mechanical characteristics of sialons are much better, though their specific surface reach only a few m^2/g. But catalysts containing transition metal oxides on activated sialons demonstrated very high activity in deep oxidation of carbon monoxide and hydrocarbons (Fig. 1) and partial oxidation of methane (Table 1) [4,5].

Concise Encyclopedia of Self-Propagating High-Temperature Synthesis
http://dx.doi.org/10.1016/B978-0-12-804173-4.00148-4

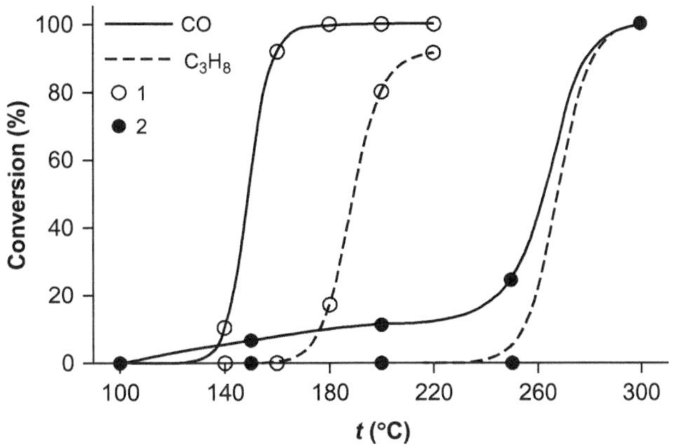

Fig. 1 Deep oxidation of CO and propane on 1 cm^3 of 0.1–0.2-mm granules of: (1) sialon with $Z=3$, activated in $Fe(NO_3)_3$ solution, 6.8% Co_3O_4 applied, and (2) 0.1%Pt/γ-Al_2O_3. Oxygen excess $\alpha=1.1$, GHSV = 120,000 h^{-1}.

Table 1 Oxidation coupling of methane

Catalyst	Gas mixture	t (°C)	GHSV (h^{-1})	% C_2H_4[a]	% C_2H_6[a]	Selectivity, $C_2H_4/\Sigma C_2$ (%)
Ca-sialon, Co, Mn, K, Cl	$CH_4+O_2+N_2$	800	1500	4.8	4.7	50.5
Ca-sialon, Co, Mn, K, Cl	CH_4+O_2+Ar	800	1500	15.3	9.0	63.0
Sialon $Z=1$, Co	$CH_4+O_2+N_2$	860	3000	15.8	4.7	77.0
Sialon $Z=1$, Mn, K	$CH_4+O_2+N_2$	890	3000	16.4	6.2	72.6
Sialon $Z=1$, Co, Nd	CH_4+O_2+Ar	800	1600	17.8	5.4	76.7

[a]Per methane passed.

REFERENCES

[1] Krylov OV. Geterogennyi kataliz. Moscow: Akademkniga; 2004.
[2] RF patent no. 1 626 601; 1995.
[3] Borovinskaya IP, Smirnov KL. Nauka Proizvod 1998;8:70.
[4] Borshch VN, Zhuk SYa, Vakin NA, Smirnov KL, Borovinskaya IP, Merzhanov AG. Dokl Phys Chem 2008;420:121.
[5] Borshch VN, Zhuk SYa, Vakin NA, Smirnov KL, Borovinskaya IP, Merzhanov AG. Int J SHS 2009;18:38.

Tantalum Nitride

Inna P. Borovinskaya

The tantalum-nitrogen system is characterized by a complicated state diagram including nitrogen solid solution in tantalum and some nitride phases. One of the phases is the well-known hexagonal ε-TaN. The cubic phase of δ-TaN was first obtained by the SHS (self-propagating high-temperature synthesis) method in regard to combustion of powder tantalum in nitrogen [1]. Then the technology of δ-TaN production in high-pressure gaseous nitrogen was developed, and the mechanism of its formation was studied [2]. Later δ-TaN was synthesized by other methods [3]. TaN cubic phase obtained by SHS has a wide range of homogeneity, and the nitrogen content in it is higher than that in the equiatomic composition ($TaN_{1.2}$). Superstoichiometric concentration of nitrogen grows with an increase in gas pressure from 50 to 300 MPa at nitride combustion synthesis in gaseous nitrogen. Neutron diffraction studies show that stoichiometric deviation occurs because of vacancies in metal sublattices of the type NaCl. Therefore the chemical composition is written as Ta_yN ($y<1$), not as TaN_x ($x>1$) [4]. This tantalum nitride is characterized by the highest microhardness (3200 kgf/mm^2) among well-known transition metal nitrides. It exists only in equilibrium at high temperatures and pressures. However, at low temperatures (<1000°C), this phase is stable, obviously because of insufficient atom diffusion mobility. Instability of TaN cubic modification is connected with a large number of vacancies in the metal sublattice. After annealing at $T>1000$°C, a slight decrease in nitrogen concentration and some phase transformations are observed. Annealing of $TaN_{1.05}$-$TaN_{1.13}$ at 1250°C results in decomposition of the cubic phase to two or more phases (ε-$TaN_{hex.}$, $Ta_5N_{6\ hex.}$, x-phase). In X-ray patterns of the samples after annealing, x-phase is seen as strong diffraction images that are not characteristic of any known Ta-N systems. After annealing at 1250°C, $TaN_{1.15}$-$TaN_{1.17}$ nitrides develop into two-phase state δ-TaN + Ta_5N_6. The quantity of the second phase grows with an increase in the annealing time, so at the end of the process, these nitrides form only Ta_5N_6. X-ray. Neutron structure analyses show that Ta_5N_6 has a hexagonal structure with lattice parameters $a=5.16$ Å, $c=10.27$ Å, spatial group $P6_3$ mcm; 10 atoms of tantalum are located in 4(d) and 6(g) at $x=0.330$; 12 atoms of nitrogen are located in 12(K) at $x=0.337$, $z=0.621$. Previously, this phase was observed only in thin films, but it had different atom coordinates in the same elementary cell. Ta_5N_6 appeared to be the most thermally resistant compound and can be decomposed in vacuum at

Concise Encyclopedia of Self-Propagating High-Temperature Synthesis
http://dx.doi.org/10.1016/B978-0-12-804173-4.00150-2

2500°C with a following formation of Ta_2N. This compound is characterized by superconductivity at $T < 7$ K. Annealing of δ-TaN in ammonium at 800°C during 12 h resulted in nitride formation of a Ta_3N_5 composition of the characteristic red color. This compound is poorly studied because of the difficulty of separating the δ-TaN_{cub}, and ε-$TaN_{hex.}$ impurity phases.

REFERENCES

[1] Merzhanov AG, Borovinskay IP, Butakov AA, Rabinkin AG, Shehtman VSh. Method for the production of tantalum mononitride. Patent SU 264365. Bull Invent 1970;1970.
[2] Borovinskaya IP, Merzhanov AG, Pitylin AN. Self-propagating high-temperature synthesis of tantalum nitrides. In: Combustion processes in chemical technology and metallurgy, Chernogolovka. 1975. p. 113–8.
[3] Boyko PG, Popova SV. JETP Lett 1970;12:101 [in Russian].
[4] Petrunin VF, Sorokin NI, Borovinskaya IP, Pityulin AN. Stability of cubic tantalum nitrides at thermal treatment. Poroshk Metall 1980;3:62–3 [in Russian].

Technological Routes

Alexander G. Merzhanov, Inna P. Borovinskaya

SHS technology is a large group of new technological processes united by a common feature—optimum use of heat evolved by exothermic reactions and organization of their spontaneous propagation. These processes differ from each other because of their secondary characteristics (chemical classes of raw materials and products, reaction conditions, combustion and structure formation modes, types and intensity of external effects, etc.). The main peculiarities of the SHS technology are as follows [1,2]:

- Technological self-propagating high-temperature synthesis (SHS) processes do not include heating devices because the process itself heats the reagents to the required high temperatures. This simplifies the construction of SHS reactors.
- There is no significant difference between processes occurring in small and large volumes. It allows prediction of SHS process performance in large installations, which is critical for the production of large-scale items.
- For technological processes of the discontinuous type, it is not necessary to use follow-up arrangements because the process occurs rather fast as is consistent with the predictions based on laboratory experiments.
- The distinction between the investigators studying the SHS regularities and mechanism and the technologists realizing the processes under the production terms is eliminated. There are several instances in which a researcher has become a developer and then a technologist.

At the present time, six technological types (TT) of SHS processes have been developed [1–3]:

- *TT-1*: A technology of simple burning of SHS systems that obtains an SHS semifinal product (including azide-based technologies, solution combustion) of binary (carbides, borides, nitrides, silicides, chalcogenides, hydrides, intermetallides), ternary, and multicomponent systems as a cake or an ingot that is further processed into powders
- *TT-2*: A gasostat technology for porous and low-porous binary (nitrides, oxides, hydrides), ternary (e.g., nitrides, carbonitrides, AlON), multicomponent ceramics (e.g., sialon) systems and items
- *TT-3*: Forced SHS compaction (densification of a hot combustion product by pressing, extrusion, rolling) of dense composite materials, tools (including synthetic hard instrumental material (STIM), diamond containing materials, etc.) and large-scale items

Concise Encyclopedia of Self-Propagating High-Temperature Synthesis
http://dx.doi.org/10.1016/B978-0-12-804173-4.00151-4

- *TT-4*: A technology for high-temperature SHS melts (high-caloric, high-temperature SHS processes forming melt products) or SHS metallurgy (centrifugal casting, deposition by welding)
- *TT-5*: SHS welding
- *TT-6*: Gas-transport technologies of fine powders and coatings

Specific equipment has been created for realization of TT-1 to TT-6:

- SHS reactors of various volumes and application purposes (vacuum, open-type, pressure chambers)
- Thermo-vacuum chambers, high-pressure vessels, special molds, extruders, mills, centrifugal reactors, welding devices, etc.

REFERENCES

[1] Merzhanov AG. Advanced SHS ceramics: today and tomorrow morning. In: Soga N, Kato A, editors. Ceramics: toward the 21st century. Proceedings of symp. on ceram. commemorating the centennial of the ceram. soc. of Japan, Yokohama, Japan, 16–18 Oct. 1991. Tokyo: Ceramic Society of Japan; 1991. p. 378–403.
[2] Merzhanov AG. SHS on the pathway to industrialization. Int J SHS 2001;10(2):237–56.
[3] Merzhanov AG. Fundamentals, achievements, and perspectives for development of solid-flame combustion. Russ Chem Bull 1997;46(1):1–27.

Temperature Measurements

Alexander S. Rogachev, Alexander S. Mukasyan

The most accurate method for measuring the sample temperature during combustion is the use of thermocouples. Because of the high temperatures developed in the combustion process, it is necessary to use the most refractory thermocouples made of tungsten-rhenium alloys. The temperature in the combustion wave is often measured using tungsten-rhenium thermocouples WRe5/WRe20 with a diameter of $d = 100$ μm, the thermal relaxation time of which is $\tau \approx d^2/4a \approx 4 \times 10^{-5}$ s ($a \approx 0.65$ cm^2/s is the thermal diffusivity of tungsten). Thus such thermocouples allow not only accurately measure the temperature profiles of postcombustion and cooling stages but also grab some details of the reaction front. If more accurate measurements of the temperature profile of the combustion front are required, it is necessary to use micro-thermocouples with a thickness of about 10 μm or less [1,2], which should be protected by boron nitride coating. The procedure for the incorporation of the thin thermocouples into the samples is rather complex and time-consuming, because each thermocouple is used only once since usually it is destroyed in the burned sample. The accuracy of high-temperature thermocouple measurements in the SHS wave is about 10–50 degrees.

Contactless methods for temperature measurement, e.g., with a pyrometer, are more productive [3,4], but the accuracy of the obtained data is typically lower than that for the thermocouples. This is primarily because of the uncertainty in the emissivity at different wavelengths. The combination of a photodetector with a microscope enables micro-pyrometric measurements to be taken within a region about 10 μm in size [5]. The most advanced method for measuring the temperature of SHS reactions is apparently thermovision. A thermovision camera registers the thermal radiation of the sample in the far-infrared (IR) range (2–5 μm) and shows the evolution of temperature distribution along the investigated surface. The recording rate for IR images is typically 30 frames per second. However, there are also high-speed IR cameras with a recording rate up to 15,000 frames per second, and they are also equipped with infrared lenses that improve the spatial resolution to tens of micrometers. However, these thermal imagers are still quite expensive and must be used in a reaction chamber with special high-cost windows (fluorite, NaCl, etc.). It is worth noting that all the preceding methods also work well in studies on synthesis in a thermal explosion mode [6,7].

Concise Encyclopedia of Self-Propagating High-Temperature Synthesis
http://dx.doi.org/10.1016/B978-0-12-804173-4.00152-6

REFERENCES

[1] Zenin AA, Merzhanov AG, Nersisyan GA. Structure of heat-wave in some processes of self-propagating high temperature synthesis. Dokl Akad Nauk SSSR 1980;250(4):880–4.
[2] Zenin AA, Merzhanov AG, Nersisian GA. Thermal wave structure in SHS processes (on the example of the boride synthesis). Combust Explos Shock Waves 1981;16(1):63–71.
[3] Andreev VA, Maltsev VM, Seleznev VA. Combustion of Hf–B mixtures as studied by optical pyrometry. Combust Explos Shock Waves 1980;16(4):374–8.
[4] Anselmi-Tamburini U, Maglia F, Spinolo G, Munir ZA. Use of two-color array pyrometry for characterization of combustion synthesis waves. J Mater Res 2000;15(2):572–80.
[5] Garkol DA, Gulyaev PYu, Yevstigneev VV, Mukhachev AB. A new high-speed brightness pyrometry method to investigate self-propagating high-temperature synthesis. Combust Explos Shock Waves 1994;30(1):72–6.
[6] Rogachev AS, Mukasyan AS, Varma A. Volume combustion modes in heterogeneous reaction systems. J Mater Synth Process 2002;10(1):31–6.
[7] Rogachev AS, Mukasyan AS. Combustion for material synthesis. Boca Raton, London, New York: CRC Press Taylor & Francis Group; 2015. 397 p.

Template-Assisted Solution Combustion Synthesis

Khachatur V. Manukyan

Solution combustion synthesis (SCS) reactions take place in homogeneous solutions of metal containing oxidizers (metal nitrates) and fuels, such as water-soluble organic amines, acids, and amino acids [1–3]. These redox reactions are completed within a short time (seconds) with maximum temperatures as high as ~1500°C. High temperatures favor formation of crystalline nanoscale materials, including binary and complex oxides, metals, alloys, and sulfides. Despite many advantages, a limitation of SCS has been a relatively low level of control over the morphological uniformity of the fabricated materials. The preparation of regular, uniform, nonagglomerated nanoparticles or ordered porous nanostructures is a challenging task because of the rapid and complex nature of the process. Recently, several novel approaches were coupled with reactive solutions for synthesizing morphologically tailored advanced nanoscale materials. Incorporation of hard and soft templates is one such approaches.

For example, hydroxylapatite nanotubes were prepared by use of a sol-gel combustion process in an anodized aluminum oxide hard template. In this synthesis, $Ca(NO_3)_2 \cdot 4H_2O$ and $PO(CH_3O)_3$ as the Ca and P sources and ethylene glycol as the polymeric matrix were dissolved in ethanol as the solvent [4]. Then an alumina template was immersed in the sol containing reactants and ethylene glycol. The template was taken out from the sol, dried, and heated in a furnace to initiate combustion. The template was removed by concentrated sodium hydroxide solution. As-prepared hydroxylapatite nanotubes have a length of 10–50 μm and diameter 100 nm (Fig. 1A).

Ultrasmall (2–5 nm) hematite (α-Fe_2O_3) nanoparticles were prepared by template-assisted combustion reaction of iron and ammonium nitrates with glycine (Fig. 1B) [5]. Ordered mesoporous silica (SBA-15) was used as a template. SBA-15 was impregnated with the reactive solution. After drying, the powder was cold pressed to a pellet, and the combustion process was initiated by the local preheating of the sample in the air by a resistively heated tungsten wire. As-prepared materials were then subjected to selective removal of the template using concentrated basic solutions. Because of the ultralow sizes and high crystallinity, the combustion-derived α-Fe_2O_3 nanoparticles exhibited superparamagnetism in temperatures ranging from 70 to 300 K. The high specific surface

Concise Encyclopedia of Self-Propagating High-Temperature Synthesis
http://dx.doi.org/10.1016/B978-0-12-804173-4.00153-8

Fig. 1 SEM/TEM images of hydroxylapatite nanotubes (A), hematite nanoparticles (B), and proros CeO_2 catalyst (C) praperd by hard tamplate-asisted SCS

area (132 m^2/g) of α-Fe_2O_3 indicated the important role of surface magnetic spins resulting in remarkably high magnetization (21 emu/g) at 300 K.

The template-assisted synthesis was also used to produce porous CeO_2 catalysts (Fig. 1C). Colloidal SiO_2 nanoparticles with size ~20 nm were added to cerium nitrate-glycine solutions, which were heated to 150°C on a hot plate for initiation of the combustion reaction. Combustion products were treated with a sodium hydroxide solution to remove the template. A highly porous CeO_2 with uniform spherical pores with size ~20 nm is evident in Fig. 1C. This simple process allowed the production of high surface area catalysts (~80 m^2/g) with high pore volume (0.6 cm^3/g). Owing to the unique mesoporous structure, this catalyst showed excellent potential for soot and CO oxidation.

CeO_2:Sm^{3+} nanoscale luminescent phosphors were also synthesized using polyvinyl alcohol as a soft sacrificial template. In this process, the cerium and samarium nitrates, urea, and polyvinyl alcohol were mixed and the aqueous solutions transferred into a preheated furnace to initiate a combustion reaction at 550°C [6]. The polyvinyl alcohol serves as a soft template limiting the agglomeration and produces fine spherical nanoparticles with a size of ~10 nm. The luminescence intensity of the calcined product of polymer-assisted synthesis was nearly tenfold higher in comparison with that of the solid-state method. The suitable excitation wavelength (370 nm), regular shape, and intense luminescence make this material an excellent candidate for preparation of ultraviolet light-emitting diode applications.

These results indicate that a combination of combustible reactive solutions with hard and soft templates provides new possibilities for the development of novel nanoscale materials with tailored structure and properties.

REFERENCES

[1] Aruna ST, Mukasyan AS. Curr Opin Solid State Mater Sci 2008;12:44–50.
[2] Li F, Ran J, Jaroniec M, Qiao SZ. Nanoscale 2015;7:17590–610.

[3] Cross A, Roslyakov S, Manukyan KV, Rouvimov S, Rogachev AS, Kovalev D, et al. J Phys Chem C 2014;118:26191–8.
[4] Yuan Y, Liu C, Zhang Y, Shan XS. Mater Chem Phys 2008;112:275–80.
[5] Manukyan KV, Chen Y-S, Rouvimov S, Li P, Li X, Dong S, et al. J Phys Chem C 2014;118:16264–71.
[6] Wu J, Shi S, Wang X, Li J, Zong R, Chen W. J Mater Chem C 2014;2:2786–92.

Thermal Explosion Mode of Combustion Synthesis

Alexander S. Mukasyan, Alexander S. Rogachev

Combustion synthesis (CS) can be accomplished in two modes, i.e., thermal explosion (TE) and self-propagating high-temperature synthesis. The general theory of TE was developed by N.N. Semenov, a Nobel Laureate in Chemistry [1]. A significant contribution to understanding TE in condensed systems was made by A.G. Merzhanov and coworkers cf. [2,3]. In the classical case, it is assumed that a homogeneously premixed reactive mixture is placed in a vessel of volume V and the area of the external surface S, that the temperature and concentration are uniform at each point of the vessel, and that heat exchange (q_-) with the environment occurs according to Newton's law

$$q_- = -\alpha(T - T_0)S \tag{1}$$

where α is the heat transfer coefficient and T_0 is the temperature of the environment. The heat that is generated (q_+) throughout the volume of the reaction mixture is a function of reaction kinetics, which assuming an Arrhenius-type dependence of conversion rate vs T, can be presented as

$$q_+ = k_0\, \varphi(C_i)\, VQ\, e^{-\frac{E_a}{RT}} \tag{2}$$

Q is the heat of the reaction and E_a is the activation energy. For zero-order reaction, $\varphi(C_i) = 1$ and

$$q_+ = k_0\, VQ\, e^{-\frac{E}{RT}} \tag{3}$$

Eqs. (1), (3) describe the heat balance in the reaction vessel. It is useful to present this relationship graphically, in the form of a Semenov diagram (Fig. 1A).

Curve *1* shows the rate of heat generation in accordance with Eq. (3), and straight lines *2–6* display the heat transfer to the environment in accordance with Eq. (2). Consider, for example, the dependences *1* and *2*, assuming that the initial reaction mixture has the ambient temperature T_0. Since, at this temperature, the q_+ is higher than q_-, the self-preheating process starts and continues until heat balance is reached at point A_1, where the rates of heat transfer and heat dissipation are equal:

$$\alpha S(T - T_0) = k_0\, VQ\, e^{-\frac{E}{RT}} \tag{4}$$

Concise Encyclopedia of Self-Propagating High-Temperature Synthesis
http://dx.doi.org/10.1016/B978-0-12-804173-4.00154-X

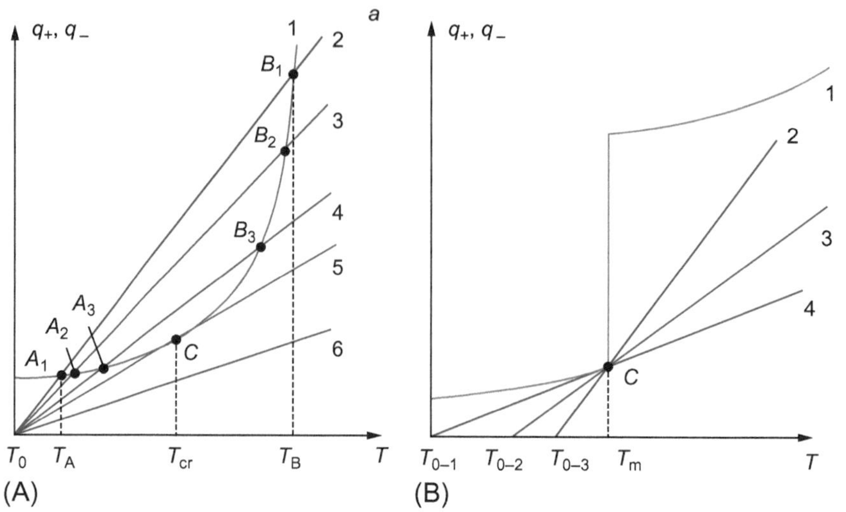

Fig. 1 TE diagrams: (A) classical Semenov; (B) with phase transformation.

It can be seen that the equilibrium at point A_1 is stable: in the case of overheating, the heat losses begin to prevail; in the over cooling regime, the heat generation returns the system to point A_1. This mode corresponds to the steady-state process in a chemical reactor. If you reduce the heat transfer coefficient α (lines *3–5* in Fig. 1A), then the point of intersection (A_i) shifts to higher temperatures until the heat loss line and heat release curve are tangent at the point C.

With further decrease of α, the heat release rate (curve 1) is above the heat loss (e.g., line 6) at any temperature, i.e., the steady state cannot exist, and the system is self-heated with acceleration. This process is called TE. Tangency at point C is the critical condition, which separates the "calm" reaction from the TE:

$$\alpha S = k_0 VQ \frac{E}{RT^2} e^{-\frac{E}{RT}} \tag{5}$$

Dividing Eq. (4) by Eq. (5), Semenov obtained an expression for the critical temperature:

$$T_{\text{cr}} - T_0 = \frac{RT_{\text{cr}}^2}{E} \tag{6}$$

Thus, in general, there is no intrinsic T_{cr} for the reactive system, since it depends on kinetics and boundary conditions [4]. However, the CS systems typically have a nonmonotonic dependence of reaction rate on temperature. For example, melting of the precursor is accompanied by a sharp increase in the reaction rate. Such temperature dependence for q_+ is shown in Fig. 1B. Up to temperature T_{melt}, only solid-type reactions with low rates occur in a mixture of the reagents. At the T_{melt} point, the system jumps to the high-temperature branch of the heat release function, where the reactions occur at a much higher rate. In this case, the critical TE temperature can be attached to the phase transition temperature, e.g., T_{melt}, and thus is weakly dependent on the other parameters, e.g., T_0. This characteristic feature of TE for CS systems was experimentally confirmed in many works cf. [5,6].

REFERENCES

[1] Semenov NN. Z Phys 1928;48(7-8):71–582.
[2] Merzhanov AG, Barzukin VV, Abramov VG. Chem Phys Rep 1996;15(6):793–837.
[3] Barzukin VV. Int J SHS 1993;4:390–405.
[4] Zel'dovich YaB, Barenblatt GI, Librovich VB, Makhviladze GM. Mathematic theory of combustion and explosions. New York: Consultant Bureau; 1985.
[5] Thiers L, Mukasyan AS, Varma A. Combust Flame 2002;131(1-2):198–209.
[6] Knyazik VA, Merzhanov AG, Steinberg AS. Dokl Phys Chem 1988;301(4):689–91.

Thermal Spray Coatings Using Solution Combustion Synthesized Powders

Singanahally T. Aruna

Thermal spraying is an efficient and environmentally friendly method of applying coatings for a wide range of industrial applications [1]. These techniques are coating processes in which a melted material is sprayed onto a surface and in the process marrying two dissimilar materials to improve, in a synergistic way, the performance of the whole. Thermal spraying, especially air plasma spraying (APS), is a low-cost method compared with other techniques such as electron beam physical vapor deposition, sputtering, and so on. In APS the material to be deposited (feedstock) is typically taken as a powder, sometimes as a liquid, suspension, or wire. Then it is introduced into the plasma jet, emanating from a plasma torch. APS process is also rapid and easy for automation, making the process potentially very well suited for mass production [2].

Plasma sprayable powders play a pivotal role in determining the properties of the coatings. Plasma spray technology requires powders with good flowability and a large particle size (20–100 μm), as smaller particles cannot be thermally sprayed because of their low mass and the resultant inability to be carried in a moving gas stream and being deposited onto a substrate [2]. Powders for a variety of industrial applications are being produced by spray drying, fluidized bed sintering, agglomeration, fusion and crushing, plasma spheroidizing, atomizing, and sol-gel processes. Most of the methods employ an additional agglomeration step to get plasma sprayable powders, thus these processes are more expensive and laborious. The solution combustion process employed for the preparation of nanosize oxides has been tailored to prepare plasma sprayable oxide powders with an average agglomerated particle size >50 μm [3,4]. These powders have been used for the fabrication of plasma spray coatings that can find applications in energy, aerospace, engineering, and biomedical sectors [3].

COATINGS FOR BIOMEDICAL IMPLANTS

Biomedical Ti alloys are prone to degradation because of the combined effect of wear and corrosion. To overcome these problems, thick plasma sprayed hydroxyapatite (HAp) coatings are now being routinely applied to the shafts of hip endoprosthetic devices and to dental root implant [2]. By employing calcium nitrate as oxidizer and calcium

Concise Encyclopedia of Self-Propagating High-Temperature Synthesis
http://dx.doi.org/10.1016/B978-0-12-804173-4.00155-1

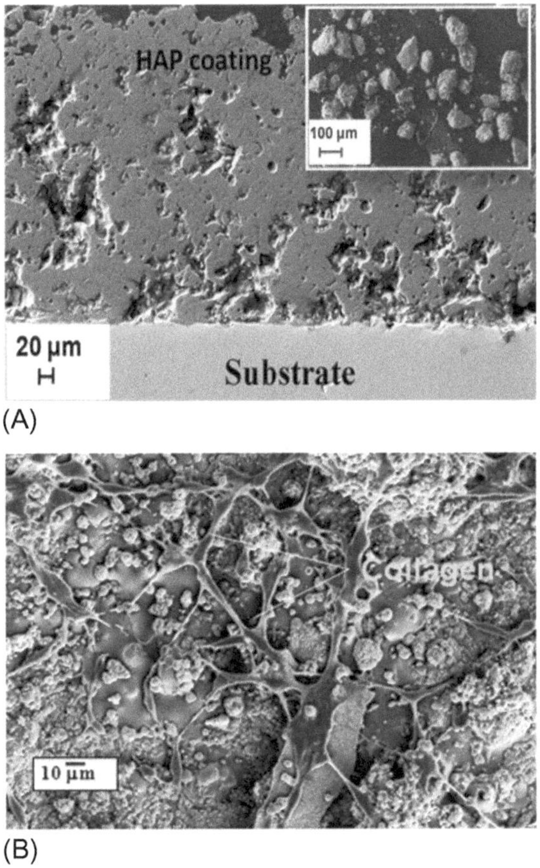

Fig. 1 (A) Plasma sprayed HAp coating fabricated from SCS HAp powder and inset shows plasma sprayable HAp powder; (B) scanning electron image of plasma sprayed HAp coating obtained by SCS method after in vitro studies.

acetate, urea and ammonium acetate as fuels, plasma sprayable-grade HAp powder with a $d_{50} \sim 50$ μm has been prepared and successfully plasma sprayed. The obtained coating is shown in Fig. 1A [5]. The coating showed good biocompatibility as indicated by the collagen synthesis on its surface during in vitro studies (Fig. 1B) indicating good biocompatibility of the HAp coating from SCS powder.

COATINGS FOR AEROSPACE AND ENERGY SECTORS

In aerospace and energy sectors, thermal barrier coatings (TBCs) have been widely used for the protection of gas turbine components [2]. Some of the promising ceramic coatings used in TBCs are YSZ, $La_2Zr_2O_7$, $La_2Ce_2O_7$, and $Gd_2Zr_2O_7$ [6]. All these powders were prepared by using a solution combustion method and by employing fuels such as urea, hexamine, and a mixture of fuels. Plasma sprayable $La_2Zr_2O_7$ powders were synthesized by a single-step solution combustion method without the agglomeration step, and the obtained coating exhibited a thermal conductivity of 1.08 W $m^{-1}K^{-1}$ at 900°C [6]. Plasma sprayable grade $La_2Ce_2O_7$ powder was successfully prepared and plasma sprayed, and the coating exhibited the required properties, such as conductivity and thermal expansion.

In order to commercialize the solid oxide fuel cells, researchers are focusing on the development of metal-supported solid oxide fuel cells (SOFCs) [7]. Plasma spraying is the most suitable method for the fabrication of metal-supported SOFC [7]. Plasma sprayable grade anode powder NiO-YSZ, Sr doped $LaMnO_3$, and Sr doped $LaCrO_3$ have been prepared by a solution combustion process using different fuels, and the powders have been successfully used for the fabrication of metal supported SOFC [8]. Gadolinia doped ceria ($Ce_{0.9}Gd_{0.1}O_{2-\delta}$) powder prepared by using a mixture of glycine and ammonium acetate exhibited good flowability for plasma spraying [9]. The combustion of fuel-deficient a glycine nitrate mixture yielded flowable $Ce_{0.9}Gd_{0.1}O_{2-\delta}$ powder.

The SCS process can be tailored to synthesize all types of flowable ceramic oxide powders for the fabrication of functional coatings. The challenges for commercialization of SCS process for the preparation of thermal spray grade powders are large-scale preparation, production of hard agglomerates, and spherical shaped particles.

REFERENCES

[1] Berndt CC, Bernecki T, editors. Thermal spray coatings: research, design and applications. Materials Park, OH: ASM International; 1993.
[2] Heimann RB. Plasma spray coating principles and applications. Germany: Wiley-VCH; 2008.
[3] Aruna ST, Rajam KS. Ceram Int 2009;35:1353–5.
[4] Aruna ST. CSIR News Lett 2013;63(15-16):179–84.
[5] Shilpa M. Studies on the development of air plasma sprayed bio-medical ceramic coatings on Ti-alloy [MTech Thesis Dissertation]. Karnataka: Visvesvaraya Technological University; 2014.
[6] Aruna ST, Sanjeeviraja C, Balaji N, Manikandanath NT. Surf Coat Technol 2013;219:131–8.
[7] Tucker MC. J Power Sources 2010;195:4570–82.
[8] Manthalkar S. Development of materials and coatings for metal supported SOFC [MTech Thesis Dissertation]. Visvesvaraya Technological University; 2014.
[9] Shri Prakash B, William Grips VK, Aruna ST. J Power Sources 2012;214:358–64.

Thermodynamics of SHS Processes

Anatoly A. Shiriev, Alexander S. Mukasyan

Thermodynamics are typically used to calculate the energy balance of a system, to define possible irreversible reaction routes, and to determine the conditions of chemical equilibrium and their change under the influence of external factors. In the case of combustion processes, a thermodynamic approach is used for calculating the adiabatic combustion temperature (T_{ad}) and equilibrium products. Consider the simple reaction $Ti+2B=TiB_2$. The precursors are elements; therefore their standard enthalpies are equal to zero. The standard enthalpy of formation for titanium diboride (TiB_2) is $\Delta H^0(TiB_2)=-293$ kJ/mol. The difference between the standard enthalpies of the reactants and products is the amount of heat released during the reaction, which is called the heat of reaction (Q). In this example, $Q=-\Delta H^0(TiB_2)=293$ kJ/mol. Under adiabatic conditions, this heat is used to preheat the reaction product to a temperature, which is called the adiabatic combustion temperature and can be determined by solving the equation

$$Q=\int_{T_0}^{T_{ad}} c(T)dT \tag{1}$$

where $c(T)$ is the specific heat of the products. The temperature dependence of the substance's specific heat is typically presented in the form of polynomials:

$$c(T)=A+BT+CT^2 \tag{2}$$

Substituting Eq. (2) into Eq. (1), we obtain an equation for determining the T_{ad}:

$$\frac{C}{3}T_{ad}^3+\frac{B}{2}T_{ad}^2+AT_{ad}-\left[Q+\frac{C}{3}T_0^3+\frac{B}{2}T_0^2+T_0\right]=0 \tag{3}$$

For TiB_2: $c(T)=30.2+0.048\ T$J (mol K)$^{-1}$, and assuming $T_0=300$ K, we obtain an equation:

$$0.024T_{ad}^2+30.2T_{ad}-304,220=0 \tag{4}$$

The solution of this equation gives the value for $T_{ad}=2986$ K. Congruent phase transitions (melting, boiling, etc.) are common for most SHS systems. In this case, the latent heats of these phase transformations should be taken into account in the balance of enthalpy,

Concise Encyclopedia of Self-Propagating High-Temperature Synthesis
http://dx.doi.org/10.1016/B978-0-12-804173-4.00156-3

$$Q-\sum_{i=1}^{n}\Delta H_{\mathrm{ph}}^{i}-\mu_{\mathrm{ph}}\Delta H_{\mathrm{ph}}=\int_{T_0}^{T_{\mathrm{ad}}}c(T)dT \tag{5}$$

where $\Delta H_{\mathrm{ph}}^{i}$ is the enthalpy of phase transitions at $T_{\mathrm{ph}}^{i} < T_{\mathrm{ad}}$ (T_{ph}^{i} is the temperature of the ith phase transition), and ΔH_{ph} and μ_{ph} are the enthalpy of phase transition and the mole fraction of the high-temperature phase (e.g., melt) in the product, respectively. The third term in Eq. (5) appears, when the combustion temperature strictly equals the temperature of the product's phase transition (T_{ph}), i.e. $T_{\mathrm{ph}}=T_{\mathrm{ad}}$.

Two-component compositions are only a small part of CS systems. In general CS reactions can be represented as

$$\sum_{i=1}^{N_X}\nu_i X_i^{(s)}+\sum_{i=1}^{N_Y}\nu_i Y_i^{(g)}+\sum_{i=1}^{N_{\mathrm{MO}_x}}\nu_i(\mathrm{MO}_x)_i^{(s)}+\sum_{i=1}^{N_Z}\nu_i Z_i^{(s)}=\sum_{j=1}^{N_P}\nu_j P_j^{(s)}+\sum_{j=1}^{N_M}\nu_j M_j^{(s)} \tag{6}$$

where $X_i^{(s)}$ is metallic or nonmetallic reagents, $Y_{(g)}^{i}$ is gas-phase reactants, $(\mathrm{MO}_x)_i^{(s)}$ is oxides, $Z_i^{(s)}$ is the reduced metals, $P_i^{(s)}$ is solid-phase products (carbides, nitrides, oxides), and $M_i^{(s)}$ is the reduced metals. The equation for determination of the T_{ad} for a multicomponent system is also more complicated,

$$\sum_{i=1}^{N_0}H_i(T_0)=\sum_{k=1}^{N(g)}n_k H_k\left(T_{\mathrm{c}}^{\mathrm{ad}}\right)+\sum_{l=1}^{N(s)}n_l H_l\left(T_{\mathrm{c}}^{\mathrm{ad}}\right) \tag{7}$$

where the enthalpy of each ith component is

$$H_i(T)=\Delta H_{f,i}^{0}+\int_{T_0}^{T}c_{p,i}dT+\sum\Delta H_{s,i} \tag{8}$$

$\Delta H_{f,i}^{0}$ is the enthalpy of formation of the i-component at a pressure of 1 atm and a temperature of T_0, $c_{p,i}$ is heat capacity, and $\Delta H_{s,i}$ is the heat of the sth phase transition of the ith component. The calculations by formulas (7) and (8) are quite complex, but this is not the only problem. It should be noted that the preceding calculations are made based on the a priori assumption about the equilibrium products, which are formed during the reaction. In general, we do not know the equilibrium products corresponding to the specific system and conditions. Thus the equilibrium product composition should also be identified by thermodynamic calculations simultaneously with the combustion temperature. Such calculations are possible only with the help of computer software packages. At present, many such programs exist for calculation of thermodynamic equilibrium in chemically reacting systems (Table 1).

Prepared specifically as a combustion synthesis processes package, THERMO is particularly noteworthy [1,2]. To determine the equilibrium state, the program searches for

Table 1 Software packages of thermodynamic calculations

Name	Developer	Source
CHEMKIN	Sandia National Laboratory (The United States)	http://www.ReactionDesign.com
CEA	National Agency for Aeronautics and Space Administration (NASA, The United States)	http://www.grc.nasa.gov/WWW/CEAWeb/ceaHome.htm
MTDATA	Materials Thermochemistry Section at National Physical Laboratory (The United Kingdom)	http://www.npl.co.uk/npl/cmmt/mtdata
Thermo-Calc	Dept. Materials Science and Technology Royal Institute of Technology (Sweden)	http://www.thermocalc.com
THERMO	Inst. Structural Macrokinetics and Mat. Sci. Russian Academy of Sciences (Russia)	http://www.ism.ac.ru

the minimum of the thermodynamic potential (F) of the system accounting for the contributions of all compounds contained in the system. The calculation of the equilibrium combustion products can be carried out in two regimes: constant pressure or constant volume. In the former case, the program searches for the minimum of Gibbs free energy and, in the latter case, for the minimum of the Helmholtz free energy.

REFERENCES

[1] Shiryaev AA. Distinctive features of thermodynamic analysis in SHS investigations. J Eng Phys Thermophys 1993;65(4):957–62.
[2] Shiryaev AA. Thermodynamics of SHS processes: advanced approach. Int J Self-Propag High-Temp Synth 1995;4:1995.

Time-Resolved X-Ray and Synchrotron-Ray Diffraction

Dmitry Yu. Kovalev, Vasiliy I. Ponomarev, Alexander S. Rogachev, Alexander S. Mukasyan

Time-resolved X-ray diffraction (TRXRD) analysis allows monitoring the phase composition of the sample in real time directly in combustion, postcombustion, and cooling zones. To apply such methods for the fast processes, such as SHS, two tasks should be accomplished: increasing the intensity of the probing beam (X-ray source) and increasing the registration rate of the diffraction patterns (X-ray detector). For a probing beam, you can utilize the radiation from a conventional X-ray tube, but a maximum time resolution can be achieved only when synchrotron radiation source is used [1]. The intensity of synchrotron radiation, produced by passing electron and positron beams accelerated to relativistic velocities through the rapidly changing magnetic field, is several orders of magnitude higher than the intensity of radiation from X-ray tubes. Rapidly changing diffraction patterns are recorded using linear and matrix gas-discharge detectors. Such detectors consist of several thousands of conventional point detectors, operating simultaneously, thus allowing at once monitoring of the diffracted radiation over a wide range of diffraction angles.

The method's scheme is shown in Fig. 1. The sample is placed in a reaction chamber equipped with a transparent (to X-rays) window made from beryllium foil or plastic Mylar film. The chamber is normally filled with helium, which absorbs much lower amounts of X-rays than other inert gases. The probing beam is directed to the area of interest on the sample surface. The combustion process is initiated, and the sequence of diffraction patterns is recorded (called "diffraction movie") with the highest possible spatial and time resolutions. Frame-by-frame analysis of the obtained "movie" provides valuable information about the dynamics of the changes in the crystal structure of solid phases arising/disappearing in the course of the entire SHS process, i.e., from ignition of green mixture to cooling down of combustion products.

Fig. 2 shows different schemes of focusing the probing beam of synchrotron (X-ray) radiation on the burning sample. The scheme depicted in Fig. 2A is the most commonly used one. This geometry was used in the first TRXRD SHS experiments carried out in Novosibirsk [2,3]. For combustion of the Ni + Al mixture,

Concise Encyclopedia of Self-Propagating High-Temperature Synthesis
http://dx.doi.org/10.1016/B978-0-12-804173-4.00157-5

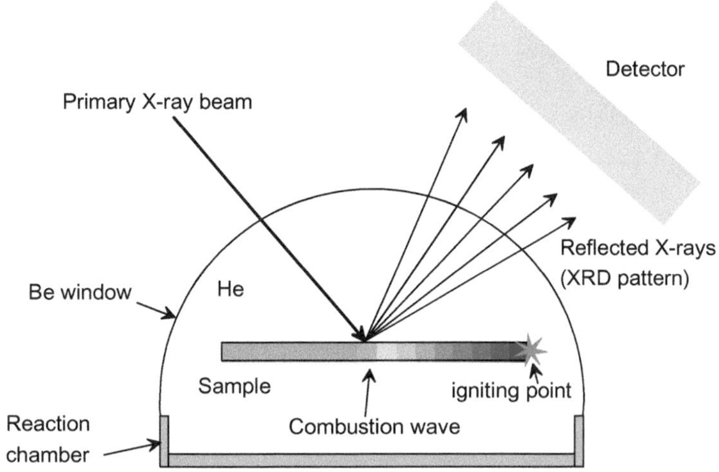

Fig. 1 Diagram of experimental set-up for TRXRD.

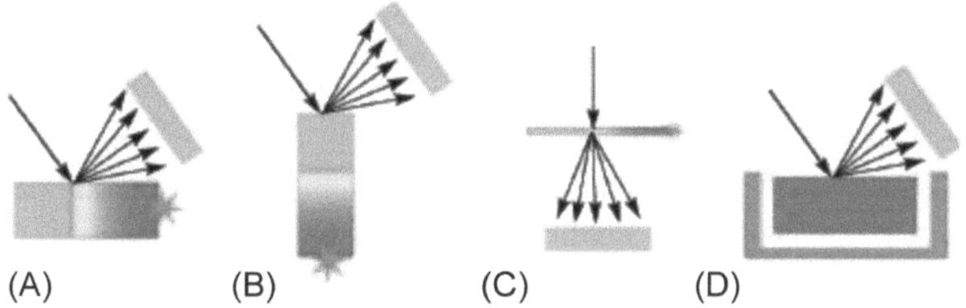

Fig. 2 Different schemes of time-resolved X-ray diffraction: (A) focusing on the lateral surface of a bulk sample; (B) focusing on the end of a bulk sample; (C) recording of a thin sample survey "in transmission"; (D) recording in the thermal explosion mode.

the first sequences of time-resolved XRD patterns of Ni and NiAl were obtained at a time resolution of 0.3 s. Later these experiments were developed in the United States, and the time resolution 0.02 s was achieved [4,5]. In these experiments, two position-sensitive detectors were used in order to simultaneously record two regions of the XRD spectrum.

French researchers used a curved arc-shaped linear detector, which covers a wide diffraction angular range. A specially designed portable reaction chamber that is easy to install on any source of synchrotron radiation was designed and built [6]. The time resolution of 40 ms was obtained, while the probing beam is focused on the side surface of the sample (Fig. 2A). Note that in this case, the time resolution is limited not only by beam intensity and time response of detector, but also by the size of the investigated area on the sample surface. Usually, synchrotron radiation is focused on the side surface of the sample in the form of a strip with a width of about 1 mm, which is oriented parallel to the plane of the combustion front. If the combustion wave travels at a speed of 10 mm/s, the wave front passes the analyzed region in 100 ms. During this time, the analyzed region includes both the initial reagents and reaction products, and their diffraction spectra will be superimposed on each other. Therefore enhancement in analysis time resolution makes sense only with simultaneous decrease of the width of the area under study. Note that the reduction in area leads to a decrease in signal intensity.

This issue can be resolved if the synchrotron radiation is focused on the end of the sample, as shown in Fig. 2B. Penetration depth into the sample is 10–50 μm, and the combustion front travels this distance within 1–5 ms. In this case, the analysis region on the sample surface can be sufficiently larger, for example, 5 × 5 mm, which allows keeping high signal intensity. The only requirement is that the combustion front must be flat and parallel to the end face of the sample. Using a similar approach and the two-axis high-speed detector, time resolution of 0.1–1 ms in the study of combustion of Ni-Al system was obtained [7].

The combustion of thin samples can be studied using "in transmission" detection, as shown in Fig. 2C, which may provide higher spatial and temporal resolutions. A record 55-μs time resolution was obtained when the primary beam was focused to a spot with a diameter of 60 μm on the surface of the multilayer Ni/Al foil 30 μm thick, and the detector was located on the opposite side of the foil [8]. Thus current TRXRD methods can record the transformation of the crystal structure directly in the reaction zone of the combustion wave, as well as in the postcombustion and cooling zones.

In the case of thermal explosion mode, the heating scheme should be changed (Fig. 2D), while a focusing geometry can be any of those previously mentioned. Typically, the reaction in a thermal explosion mode is slower than in the combustion wave, and therefore the time resolution of the order of 0.1–1.0 s is sufficient [9].

Use of synchrotron radiation is quite expensive and therefore, for many tasks that do not require extremely high values of temporal resolution, it is convenient to apply the conventional X-ray tubes [10]. Such equipment allows you to explore the processes of structure formation behind the combustion wave at a time resolution in the range of 0.1–10 s. Fig. 3 shows an example of evolution of the XRD pattern during SHS of the titanium aluminide, obtained using a standard X-ray tube and linear detector.

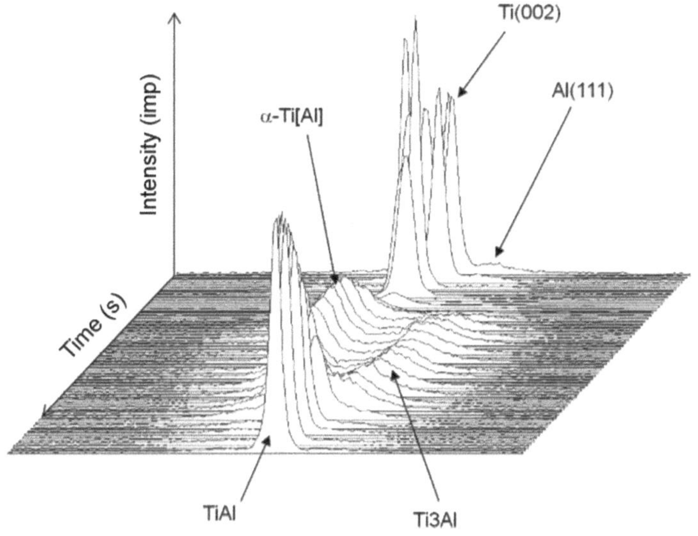

Fig. 3 Phase transformations during the SHS process Ti + Al = TiAl (wave propagation mode).

REFERENCES

[1] Kulipanov GN, editor. Diffractometry using synchrotron radiation. Novosibirsk: Nauka; 1989.

[2] Boldyrev VV, Aleksandrov VV, Korchagin MA, Tolochko BP, Gusenko SN, Sokolov AS, et al. The study of dynamics of phase formation during nickel monoaluminade synthesis in the combustion regime. Dokl Akad Nauk SSSR 1981;259(5):722–4.

[3] Boldyrev VV, et al. Study of SHS processes by X-ray phase analysis using synchrotron radiation, self-propagating high-temperature synthesis by the method of X-ray diffraction analysis using synchrotron radiation. Combust Explos Shock Waves 1983;19(4):430–1.

[4] Holt JB, Wong J, Larson E, Waide P, Rupp B, Frahm RA. A new experimental approach to study solid combustion reactions using synchrotron radiations, In: Proc. 1st US–Japanese workshop on combustion synthesis, Tokyo; 1990. p. 107–13.

[5] Wong J, Larson EM, Holt JB, et al. Time-resolved X-ray diffraction study of solid combustion reactions. Science 1990;249:1406–9.

[6] Held O, Vrel D, Gachon JC. Self-propagating high-temperature synthesis (SHS) of $Ni_{50}Ti_{50}$, $Ni_{45}Ti_{55}$, $Ni_{50-y}Ti_{50-y}Al_{2y}$ $(0 < y < 3)$ observed by time resolved X-ray diffraction (TRXRD). Int J SHS 2000;9:341–56.

[7] Rogachev AS, Tolochko BP, Lyakhov NZ, Sharafutdinov MR, Popkov NA, Pirogov BJ, et al. Characteritisc features of structure formation of nickel monoaluminade in a gasless combustion wave. Crystallogr Rep 2003;48(3):466–8.

[8] Trenkle JC, Koerner LJ, Tate MW, Gruner SM, Weihs TP, Hufnagel TC. Phase transformations during rapid heating of Ni/Al multilayer foils. Appl Phys Lett 2008;93:081903.

[9] Pismenskaya EB, Rogachev AS, Kovalev DYu, Ponomarev VI. The mechanism of formation of copper aluminides in the thermal explosion mode. Russ Chem Bull 2000;49(12):1954–9.

[10] Khomenko IO, Mukasyan AS, Ponomarev VI, Borovinskaya IP, Merzhanov AG. Dynamics of phase-forming processes in metal-gas system during combustion. Combust Flame 1993;92:201–8.

Titanium Aluminides

Victoria V. Kurbatkina

TiAl is a highly promising material for engine elements in aircraft and aerospace engineering. However, TiAl alloys are difficult-to-form materials [1–4], and hard to obtain a homogeneous melt and gravity segregation (liquation). Activities performed for years by GfE Metalle und Materialien (Germany) [1] showed that the most applicable, traditional technology for producing TiAl alloy ingots is vacuum arc remelting. The disadvantages of TiAl alloys are a high impurity level, primarily with interstitial impurities (oxygen, nitrogen, etc.), and difficult formation of a nonporous material, which have a negative impact on mechanical properties.

The potential means of producing TiAl intermetallic compounds are SHS-consolidation [3,4], vacuum SHS rolling [5,6], and centrifugal SHS-molding [7,8]. Force SHS-pressing technology enables the production of TiAl alloys with a dispersed no-liquation structure in a very short cycling process that prevents segregation, which is important for systems with significant differences in the density of their basic components. From the perspective of practical synthesis of TiAl, the most suitable SHS schemes are titanium oxide reduction reaction because the TiAl formation heat is 73.6 kJ/mole and the adiabatic combustion temperature is 1246°C, which is lower than titanium (1668°C) and TiAl (1447°C) melting temperatures. SHS in this system is possible only when the reaction mixture is heated to a temperature higher than the aluminum melting point (660°C). Fig. 1 shows schematically the TiAl phase composition and melt structure formation mechanism. The initial structure formation stage involves aluminum melting and its further spread along capillary-porous-medium channels. Further diffusion of aluminum atoms into the titanium lattice leads to the formation of a $TiAl_3$ intermetallic compound layer (Fig. 1A). Gradual growth of the $TiAl_3$ layer leads to depletion of the aluminum melt and further formation of TiAl titanium monoaluminide. When the process spreads inside the titanium particle (Fig. 1B), the concentration of aluminum decreases leading to formation of a Ti_3Al intermetallic compound (Fig. 1C). The last structure formation stage is adjustment of the composition of intermetallic layers by the recrystallization of Ti_3Al into TiAl (Fig. 1D). Full completion of this stage depends on multiple factors, such as the composition of the initial reaction mixture, modes of cooling under SHS, and so on.

Concise Encyclopedia of Self-Propagating High-Temperature Synthesis
http://dx.doi.org/10.1016/B978-0-12-804173-4.00158-7

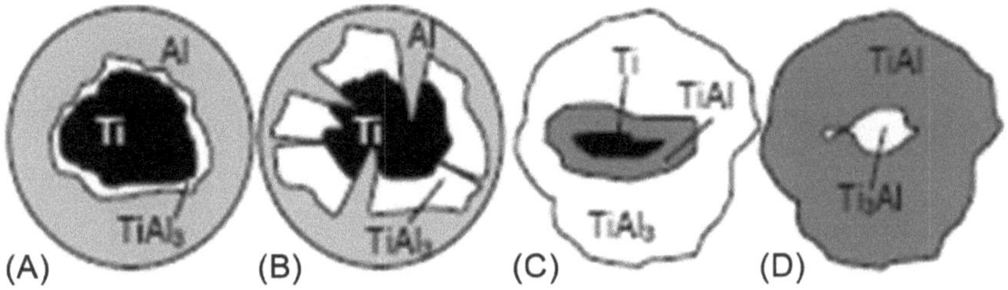

Fig. 1 Main stages of structure formation in Al-Ti system.

To produce molded titanium aluminides using centrifugal molding technology, Ti + Al or TiO_2 + Al hybrid mixtures and CaO_2 + Al high-energy additive are used. The combustion temperature of $3CaO_2 + 2Al$ is 3700 K. The addition of 40% (CaO_2 + Al) into Ti + Al mixtures increases the combustion temperature to 2200–2300 K and produces a melt of Ti_xAl_y and $3CaO_xAl_2O_3$ combustion products. A small difference in the specific gravities of the metal and oxide phases of the combustion products with a high content of Al in Ti_xAl_y should be noted. Therefore, in the centrifugal field, at an overload exceeding 250 g, molded products easily segregate into two layers, a metal layer and an oxide layer. Variation of aluminum content in the basic mixture allows control over the chemical composition of the metal phase and production of Ti_3Al, TiAl, and $TiAl_3$ phases. Oxide combustion products, CaO-Al_2O_3, are inert in relation to the metal melt. Calcium oxide formed as a combustion product causes a decrease in the viscosity of the oxide (slag) phase, which allows almost full recovery of the target components as a metal ingot.

REFERENCES

[1] Guther V, et al. In: Kim YW, Clemens H, Rosenberger AH, editors. Proceedings of gamma titanium aluminides, USA: Minerals Metals and Materials Society; 2003. p. 241–7.
[2] Cao J, Feng JC, Li ZR. J Mater Sci 2006;41:4720–4.
[3] Agote I, Coleto a J, Gutiérrez M, Sargsyan A, García de Cortazar M, Lagos MA, et al. Intermetallics 2008;16:1310–6.
[4] Bertolino N, Monagheddu M, Tacca A, Giuliani P, Zanotti C. Intermetallics 2003;11:41–9.
[5] Osipov EE, Levashov EA, Chernyshov VN, Merzhanov AG, Borovinskaya IP. Int J SHS 1992; 2(1):314–7.
[6] Chernyshov VN, Osipov EE, Levashov EA, Merzhanov AG, Biyachi L. Int J SHS 1993;3(2):315–21.
[7] Andreev DE, Sanin VN, Yukhvid VI. Inorg Mater 2009;45(8):867–72.
[8] Sanin V, Andreev D, Ikornikov D, Yukhvid V. Cast intermetallic alloys by SHS under high gravity. Acta Phys Pol A 2011;(2):331–5.

Titanium Carbide

Valentina K. Prokudina

Titanium carbide possesses a variety of valuable properties. Thus this compound inevitably attracted close attention of SHS processes researchers. It was one of the first compounds obtained by the SHS route. TiC is a refractory compound with a high melting point (3260 ± 150 K). It also possesses high chemical endurance as well as very high microhardness (30.4 GPa) that remains unchanged even when a compound is heated to high temperatures.

In the first SHS-related articles [1,2], the possibility of the production of titanium carbide from elements on laboratory equipment was demonstrated. Combustion front propagation velocity as a function of argon pressure and titanium particle size was investigated. It was shown that combustion velocity does not depend on argon pressure. An increase of particle size from 40 to 200 μm resulted in a decrease in combustion velocity from 3.2 to 0.5 cm/s.

Development of the SHS technology for TiC production started on gasless combustion installation, created by ISMAN. The installation included four sealed reactors, each with a swept volume of 8 L. Productivity of 2 kg/h was achieved for the reactor [3]. A stoichiometric mixture of titanium powder and soot was loaded into reactor either in the form of filling or of pressed briquettes (tablets). The final product was in the form of a TiC-sintered powder mass that eventually was milled to a powder state using standard powder metallurgy equipment (crushers, ball mills, etc.).

With the development of new types of SHS reactors—multipurpose reactors—the technology was upgraded. Productivity of 30 kg/h was obtained for multipurpose 30-L SHS reactors [4]. Multipurpose SHS reactors in quantities up to six are usually united in a single installation with one common remote control that includes gas supply, water cooling, electro ignition, and recording device systems. In the case of TiC synthesis, multipurpose reactors with inner walls of porous graphite are used. This particular use allows compacting the powder mixture within the reactor to fill all of its swept volume. Such types of loading preclude lamination of a reaction mixture that can occur in a gasless reactor and ensures good contact between the solid components in combustion. Controlled dropping of gases is performed through the porous inner walls during the synthesis process. This dropping promotes refining gas impurities (including adsorbed oxygen) from a synthesized product. Gasifiable impurities escape powder mixtures during combustion. This self-refining is widely known as an SHS phenomenon [5].

Concise Encyclopedia of Self-Propagating High-Temperature Synthesis
http://dx.doi.org/10.1016/B978-0-12-804173-4.00159-9

Oxygen is an extremely deteriorating and undesirable impurity in carbide, because when carbide is converted into hard alloys, oxygen will make the alloys very brittle. Additional refinement of a synthesized product occurs while a reactor is vacuumed after synthesis. The goal of the designed technology was to obtain TiC with chemical composition that would be the most approximate to stoichiometric one (80.0 at% Ti and 20.0 at% C). This composition is equal to the maximum content of bonded carbon ($C_{bonded} = C_{general} - C_{free}$) and the highest degree of carbonization ($m = C_{bonded}/C_{stoich}$). Minimal content of main impurities (oxygen and free carbon in the form of graphite) was also desired. To find the optimal conditions of synthesis, an investigation of the influence of the main technological parameters on the quality of synthesized TiC was performed, including morphology of surface, particles diameter and specific surface area of Ti and C powders (with C in the form of soot or graphite), conditions of drying of initial powders, powder mixture composition (degree of dilution by final product, excess of C), homogeneity of powder mixture, density and amount of mixture load in reactor, and synthesis atmosphere (protective atmosphere, vacuuming of reactor before and after the synthesis, gas dropping during the synthesis). Some of the results of this investigation are showed on Figs. 1 and 2.

The highest degree of carbonization $m = 0.99$ with a content of C_{free} <0.1% was obtained in the synthesis of the initially pressed powder mixture in a gasless reactor. Ti powder mark PT, produced by a reduction of titanium dioxide by calcium hydride, was used in the powder mixture. When the synthesis was performed in a multipurpose SHS reactor with the same initial Ti powder, the highest degree of carbonization ranged from 0.97 to 0.98, with the C_{free} content <0.2%, and the oxygen content in a range of

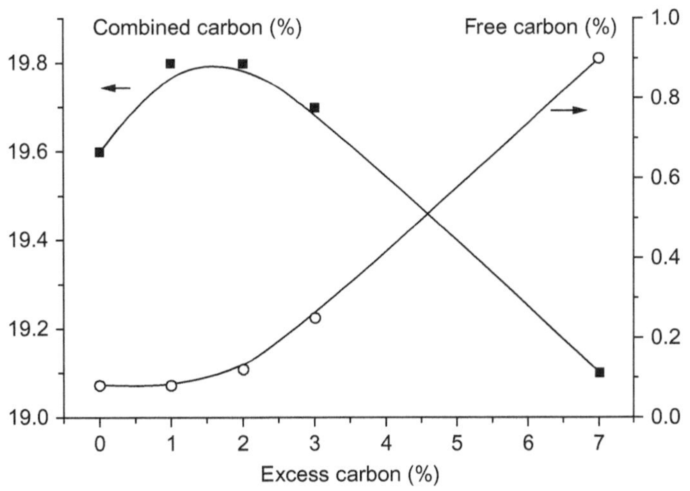

Fig. 1 Dependence of amount of bonded and free carbon in TiC on excess of soot in initial powder mixture.

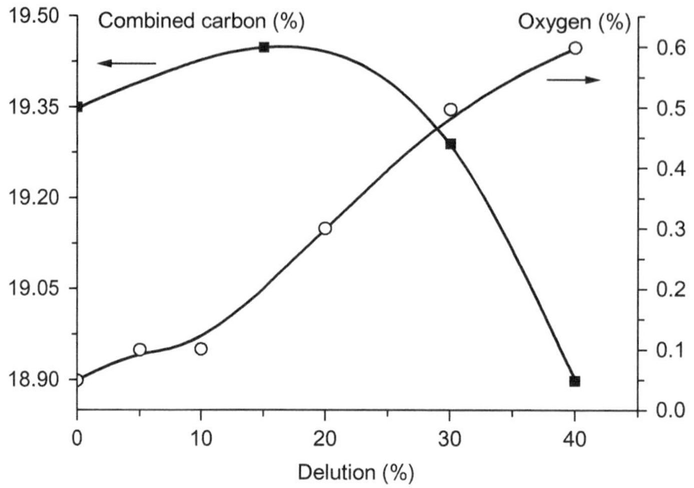

Fig. 2 Influence of degree of dilution on chemical composition of TiC.

0.1%–0.3% (while the initial titanium contained 0.3%–0.6% oxygen) [6]. TiC mark SHS (obtained by element synthesis) was produced in a trial-production facility under the following conditions: in a multipurpose SHS-30 reactor, from titanium powders, with the particles' diameter no more than 200 μm, and with soot with a specific surface area approximately equal to 15 m^3/g. The product obtained was characterized by a degree of carbonization *m* of 0.90–0.96, C_{free} content <0.5, and oxygen content <1%.

The technology for the production of TiC with the preceding composition was standardized and implemented in commercial production in a pilot plant of powder metallurgy and in other facilities, with more than 200 tones of TiC mark SHS being produced. The widest application of TiC mark SHS-E is as an abrasive material (in grinding powders, micropolishing powders, abrasive pastes, and pencils) as well as filler in carbide steels.

Grinding pastes (abrasives) mark KT and KTIOL were produced commercially for more than 10 years and were used in more than 3000 facilities with considerable economic results. So-called "carbide-steel" is a composite of TiC and tool steel as a binder that are consolidated by solid-phase sintering and consequent deformation (about 85%–90%). The strength of carbide steels is similar to the strength of highly doped high-speed steel, and its hardness is similar to that of hard alloys mark VK-8 to VK-10 (hard alloys consisting of WC and 8%–10% Co). Unfortunately, TiC mark SHS did not achieve use as a component of hard alloys, because TiC mark SHS is hard to mill as compared with conventionally used TiC, which is reduced from oxide in furnaces. It is a consequence of increased hardness, peculiar shape of grains, and high abrasive properties of TiC [7].

It is possible to produce "soft" TiC by various routes in order to use it as a component. One of these routes is to decrease the sintering ability of the reaction mixture by inserting gasifying additives in the mixture. When haloid-containing organic salts were used as such additives, TiC with particles of a size no more than 50 μm was obtained. However, this route is unwarranted because of its ecological dangers, including the risk of spoilage of toxic fluor-containing compounds in the final product.

Another route is to perform SHS with a stage of magnesium-thermal reduction of titanium dioxide, with an after-synthesis acid washing treatment. The TiC produced by a magnesium-thermal route has $m = 0.90$ and C_{free} content <1.0%; however, it contains impurities of residual magnesium up to 0.5%. The particle's size of this TiC is <45 μm.

Complex compound based on TiC and TiCN were produced by doping an initial mixture with molybdenum. These compounds were used to produce hard alloys of commercial marks (TH-20, KTH-16) as well as novel marks (CKH, CKC) with strength properties exceeding those of analogous furnace-reduced titanium dioxide-based alloys.

REFERENCES

[1] Merzhanov AG, Borovinskaya IP. Self-spreading high-temperature synthesis of refractory compounds. Dokl Chem 1972;204(2):429–31.
[2] Shkiro VM, Borovinskaya IP. Combustion processes in chemical technology. Chernogolovka: BIChPh AN SSSR; 1975. p.253–8 [in Russian].
[3] Prokudina VK, Ratnikov VI, Maslov VM, Borovinskaya IP, Merzhanov AG, Dubovitskii FI. Combustion processes in chemical technology and metallurgy. Chernogolovka: BIChPh AN SSSR; 1975. p.136–41.
[4] Ratnikov VI, Borovinskaya IP, Prokudina VK. Izvestia Vyzov, Poroshkovaya metallyrgia I funkcionalnye pokrytia. Russ J Non-Ferrous Met 2013;(1):34–41.
[5] Martirosyan NA, Dolukhanyan SK, Mkrtychan GM, Borovinskaya IP, Merzhanov AG. Poroshk Metall 1977;7(175):36–40 [in Russian].
[6] Borovinskaya IP, Prokudina VK, Ratnikov VI. Izvestia Vyzov, Poroshkovaya metallyrgia I funkcionalnye pokrytia. Russ J Non-Ferrous Met 2010;(4):26–33 [in Russian].
[7] Merzhanov AG, Borovinskaya IP, Prokudina VK, Pesotskaya NS, Nasonova MA. Nauka-Proizv 1998;8 (10):5–12 [in Russian].

Titanium Nitride

Valentina K. Prokudina

Titanium nitride is a refractory compound that possesses a number of valuable properties, such as high microhardness and chemical and thermal stability. TiN has a variety of applications: as a component in special refractories and cermets, as a material for crucibles for anoxic casting of metals, and as a precursor for wear-resistant and decorative "gold-like" coating.

Investigations of the combustion of compacted Ti powder samples in nitrogen showed that the filtration velocity of nitrogen within the titanium is a principal factor that affects the combustion process. Filtration velocity depends mainly on the porosity of titanium compact and ambient nitrogen pressure. In addition, it was found that the maximum thermodynamic temperature of combustion of titanium in nitrogen ($T_c = 4470$ K) far exceeds the melting point of titanium. Thus two types of nitridation are peculiar to the system: solid-phase and liquid-phase nitridation. Solid-phase nitridation occurs when nitrogen pressures are low. Consequently, the depth of nitridation in combustion front is small, and the real temperature of combustion T_c is less than the melting point of titanium T_m. Inserting additives that decrease combustion temperature within the initial titanium can also keep the nitridation process in solid-phase [1]. Development in technology of SHS production of TiN powders was conducted using multipurpose SHS reactors with 8 L of swept volume (SHS-8). The initial powder mixture (up to 3 kg) was loaded (with bulk density) on the graphite substrate, and then introduced into a reactor. The final product was obtained in the form of a sintered powder mass that doubled the form of substrate and was fragile enough to be milled into a powder of any required dispersity.

The quality of the obtained product was estimated by two parameters: the degree of nitridation and oxygen impurities content. The degree of nitridation m was N_{exp}/N_{calc}, where N_{exp} was the experimentally determined content of nitrogen in synthesized product and N_{calc} the calculated content of nitrogen in stoichiometric product, that is, 22.6%. To find the optimal condition for synthesis, the influence of the main technologic parameters on the quality of the final product was investigated, which included chemical purity and particle size of initial titanium and diluent, purity of nitrogen, degree of dilution of initial powder mixture by final product, nitrogen pressure in reactor, mass and type of powder mixture loading, and so on. Conditions that were found optimal for synthesis of TiN with a maximum degree of nitridation are as follows. Initial titanium (in the state

Concise Encyclopedia of Self-Propagating High-Temperature Synthesis
http://dx.doi.org/10.1016/B978-0-12-804173-4.00160-5

of supply) must be with a purity >98.9% and with an average particle size of 0.08–0.63 μm, ergo it must not contain excessively large particles that cannot burn completely as well as very tiny particles that can impede filtration of nitrogen. Nitrogen must have a purity not less than ultrapure, with a humidity not more than 0.005 g/m^3, and an oxygen content not higher than 0.001%. Initial nitrogen pressure in the reactor should not be lower than 10 MPa.

Another important characteristic of this process is the dispersity of diluent, the final product. To ensure good filtration of nitrogen to the reaction zone, particles size of the diluent must be 0.6–0.8 μm. The degree of dilution of the initial mixture by the final product must be 40%–50%. Optimal conditions for production of TiN with an average degree of nitridation $m=0.90$–0.95 differs from the one mentioned before in one aspect—titanium powders with wider particle-size distribution (0.04–5.0 mm) can be used. All other synthesis parameters, except for the degree of dilution by the final product, are identical to those that are used for the production of TiN with a maximum degree of nitridation.

As is shown in Fig. 1, for the titanium with maximum particle size (from 0.63 to 5 mm), the highest degree of nitridation is relatively low ($m=0.92$), and the degree of dilution (20 wt.%) is relevantly moderate. For the titanium with minimal particle sizes (<0.04 mm), the highest degree of nitridation is also relatively low ($m=0.90$) with the degree of dilution 50%. The highest degree of nitridation ($m=0.97$) with a degree

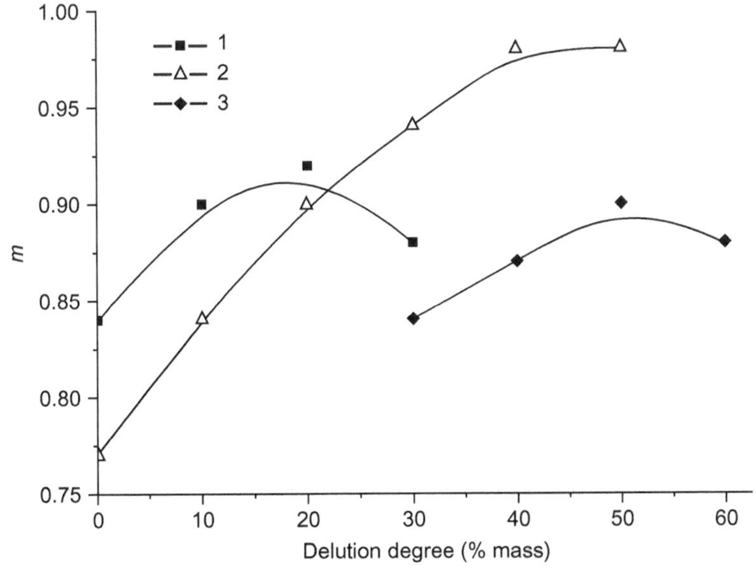

Fig. 1 Dependence of degree of nitridation of TiN on dilution of initial titanium powders with various particle size: (1) 0.63–5 mm, (2) 0.18–0.63 mm, (3) 0.04 mm.

of dilution 40%–50% was produced using titanium powders with medium dispersity (particle size from 0.18 to 0.63 mm).

An increase in humidity and oxygen in nitrogen leads to spoilage of the final product by oxygen impurities. Optimal conditions for TiN synthesis do not depend on the swept volume of the reactor. The productivity of one SHS-30 reactor is 7–9 kg/h of final product [2].

It is known that titanium sponge is a purer, cheaper, and more conventional source of titanium than titanium powders. Therefore investigation of the synthesis of TiN from titanium sponge with a particles diameter between 2 and 12 mm were made. The process of nitridation goes through two stages: during the first stage, pure titanium sponge is reacted and then milled again. The maximum degree of nitridation in sponge after the first stage does not exceed $m=0.7$ (Fig. 2); however, this is enough to make the sponge brittle and its consequent milling a medium dispersity powder. In the case of long-time milling (which continues more than 10 h), spoilage of TiN by unwanted impurities of iron of oxygen occurs. When milling is conducted for 5–10 h, its effect on nitrogen and impurities content in the final product obtained after the second stage of synthesis is negligible. The maximum degree of nitridation after the second stage of the process is $m=0.80$–0.95. The use of titanium sponge considerably widens the raw material base for SHS technology, not only for the production of TiN but also for the production of other titanium compounds [3].

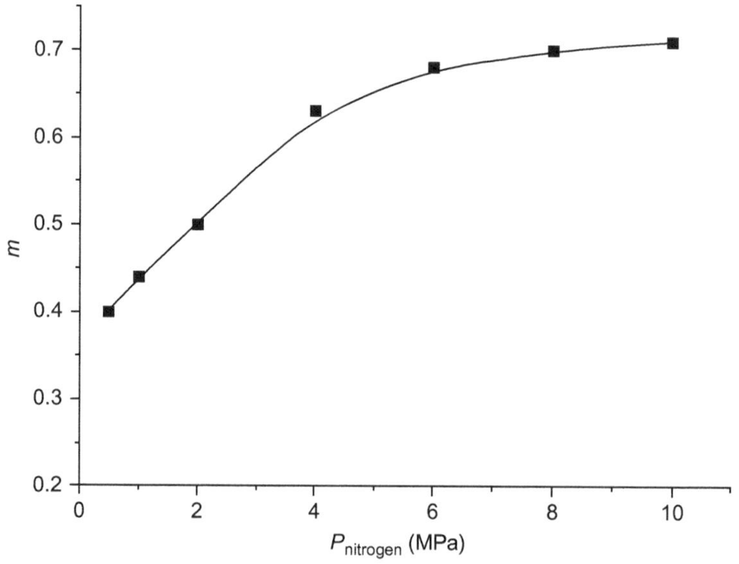

Fig. 2 Dependence of degree of nitridation on nitrogen pressure on the first stage of synthesis of TiN with titanium sponge as a raw material.

TiN could be produced by SHS technology not only in form of powder but also in form of coatings, produced by SHS gas-phase transport. Coatings with a thickness of 7–12 μm are applied to the cutting tool made from hard alloys mark VK-6 (WC +6%Co), VK-8 (WC + 8%Co), VK-30 (WC + 30%Co), with an aim to increase their wear resistancy [4]. In this method, a special unit for SHS gas-transport coating deposition was used. It was characterized by the production of 0.17 kg of treated cutting blades per hour. The coatings increased the wear resistance of the cutting blades during the process of mechanical treatment of the cast iron by 2.5–3 times, being measured until the wear of rear edge grew to 0.5 mm.

In regard to TiN, it is necessary to mention another compound with a similar synthesis, $TiC_{0.5}N_{0.5}$, and that is in high demand as a material for tungsten-free hard alloys. For the production of this compound, a mixture of titanium and soot in relation to Ti +0.5C was used. Combustion took place in a nitrogen atmosphere and went through two stages: carbonization and nitridation. Optimal conditions for SHS production of $TiC_{0.5}N_{0.5}$ differ from those of TiN in one aspect: to ensure more complete carbonization, a finer fraction of titanium powder was needed (<0.04 mm), so the fraction of raw titanium powder with a particle size up to 0.063 mm was used. Excess of soot over the stoichiometry was 8%–10%. Considering the amount of free carbon in the diluent, the degree of dilution of the initial powder mixture in the final product was 30%. Developed technology was implicated in a powder metallurgy for production of TiN and $TiC_{0.5}N_{0.5}$ SHS powders, with overall productivity up to 5 tons per year.

REFERENCES

[1] Merzhanov AG, Borovinskaya IP. Self-spreading high-temperature synthesis of refractory compounds. Dokl Chem 1972;204(2):429–31.
[2] Merzhanov AG, Borovinskaya IP, Ratnikov VI, Prokudina VK. Production of nitride and carbonitride of titanium on the pilot SHS-equipment; Preprint of BIChPh AN SSSR, 31 p. [in Russian].
[3] Borovinskaya IP, Prokudina VK, Ratnikov VI. Izvestia Vyzov, Poroshkovaya metallyrgia I funkcionalnye pokrytia. Russ J Non-Ferrous Met 2010;(4):26–33 [in Russian].
[4] Kogan YaD, Sereda VP, Shtessel EA, Kostogorov EP. Metallovedenie I termicheskaya obrabotka metallov. Met Sci Heat Treat 1991;(6):39–40 [in Russian].

Tubular Structures Including Nanotubes

Inna P. Borovinskaya

One of the promising directions of SHS is the combustion in gas phase resulting in the formation of condensed products as particles or films (coatings). The combustion of gas (or vapor) results in solid products or combustion of a gas suspension containing a solid phase and forming a solid phase of another composition. These processes allow obtaining nanosize powders, nanotubes, and tubular structures of various modifications. Combustion of gases and gas suspensions gave rise to obtaining nanosize particles of various refractory inorganic compounds (nanopowders, nanofibers, nanorods, whiskers, etc.) by SHS, and demonstrated new possibilities of SHS in synthesizing carbon and noncarbon inorganic nanotubes and tubular structures of various morphological types. Nanopowders of inorganic carbides (WC, TiC, TiC-WC, SiC), nitrides (Si_3N_4, AlN, BN, etc.), borides (TiB_2), sialons, and so on have been obtained. The values of the specific surface area of the powders reach $\geq 70\ m^2/g$, and their particle sizes range from ≤ 100 to 200 nm. The particle morphology varies: uniaxial particles, rods, threads, strips, whiskers, agglomerations like nanostraps, nanoflowers ("roses," "chrysanthemums"), and so on [1].

Structures such as nanotubes and tubules are of particular interest though the mechanism of their formation is not sufficiently studied. Alekseev et al. reported on obtaining carbon nanotubes during SHS using initial reagents, one of which was a carbon carrier and the other its efficient reductant [2]. The authors reported on reactions with the following products: $(C_2F_4)_n + 4nLi + C \rightarrow 4nLiF + 2nC$; $Na_2CO_3 + 3Mg \rightarrow 3MgO + C + 2Na$; $CaCO_3 + 2Mg \rightarrow MgO + CaO + C$; and $(C_2F_4)n + 2nMg \rightarrow 2nMgF_2 + 2nC$. The scheme of the experiment was conventional for solid-phase SHS. The primary heat release was provided by igniting exothermic mixtures, for example, fluoroplastic lithium. The experiments were carried out in argon or in the air. Carbon nanotubes were formed from the catalyst of NiO (~10%) or iron. They had a varied morphology similar to that of nanotubes in CVD processes: either straight multiwall or curved filled with the catalyst. The output of the nanotubes was ~2–4%.

The Chinese scientists Wang et al. carried out the SHS process on boron nitride nanotubes [3]. The synthesis was done by annealing a porous precursor in flowing NH_3 gas at $T = 1150°C$. The porous precursor of $B_{18}Ca_2(MgO)_9$ composition was obtained by SHS from a mixture of Mg, B_2O_3, and CaB_6 as the starting material, which played an important role in the synthesis of BN nanotubes in large quantities. BN nanotubes of 10–150 nm in diameter and 5 μm in length were obtained. Chemically active particles

Concise Encyclopedia of Self-Propagating High-Temperature Synthesis
http://dx.doi.org/10.1016/B978-0-12-804173-4.00161-7

of B (vapor), nitrogen, and hydrogen took part in the process of nanotube formation. MgO acted as a catalyst for the reaction through a VLS mechanism.

Tubules of boron-carbon compounds of various compositions were synthesized by Russian scientists (Borovinskaya et al.). They studied the possibility of SHS of boron nitride nanotubes [4]. Tubules of boron-carbon compounds within homogeneous boron carbide were synthesized at combustion of elemental boron in CO_2 in a closed reactor at P_{gas} up to 9 MPa. Depending on the SHS conditions, the main reaction products were boron oxide (B_2O_3), carbon, boron carbide B_4C (including single crystals of up to 3 μm), as well as boron-carbon nonstoichiometric compounds of various morphologies, representing practically all known configurations of noncarbon tubules: continuous straight tubes (cylindrical), "bamboo" tubes with parts of the intermittent growth having closed ends, cones, funnels, and horns. In some experiments, X- and Y-shaped tubes and peapod nanotubes with solid particles inside were observed (Fig. 1A and B). The tubule diameter ranged from 20 to ≥100 nm, which is typical of noncarbon nanostructures, particularly boron-carbon, because of the high instability of the tubular shape at a significant incurvature of the walls. The tubule wall thickness was several nanometers.

The mechanism for the formation of BC_x tubules was shown to be closely connected with the possibility of the reaction $2B + 2CO_2 \rightarrow B_2O_2 + CO + C$. Carbon was released as a fine and very active powder. When it reacted with boron oxide, gaseous products such as dimeric $B_2O_{2(g)}$, $BO_{(g)}$, $BO_{2(g)}$, and so on were formed. In addition, high combustion temperatures provided the evaporation of solid reagents with the generation of $B_2O_{3(v)}$ and $B_{(v)}$ vapors and an accumulation of the gases $CO_{(g)}$ and $CO_{2(g)}$. So the original solid-gas (VS) mechanism of the reaction $B + CO_2$ partially changes the gas-gas or vapor-vapor mechanism, which is one of the necessary conditions of nanostructure formation. The tubular structures were separated from other solid combustion products by chemical and thermal treatment at 600–1000°C. The thermal treatment at 1000°C allowed synthesizing decomposition nanostructures: nanofibers of up to dozens

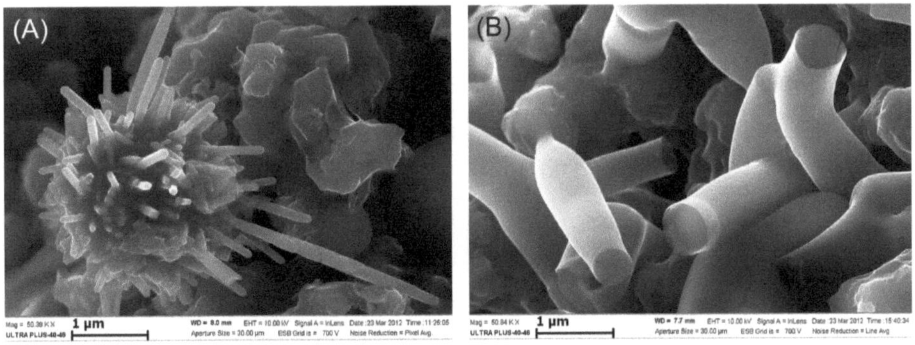

Fig. 1 (A, B) Morphological variety of boron-carbon tubules.

of microns in length, nanofiber nets, and nanosize particles of boron carbide (Fig. 2). Fig. 3A and B shows a TEM image of a boron-carbon tubule.

Upon combustion of elemental boron in gaseous nitrogen at high pressure (up to 300 MPa), temperatures can reach and even exceed 3000°C because of the high

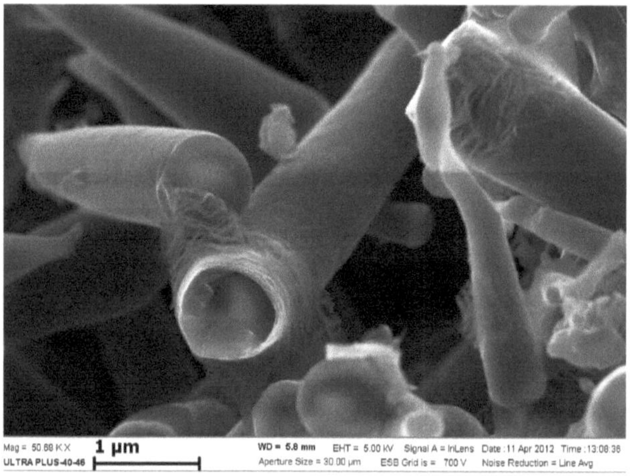

Fig. 2 SEM image of boron-carbon nanotube decomposition resulting in fiber formation.

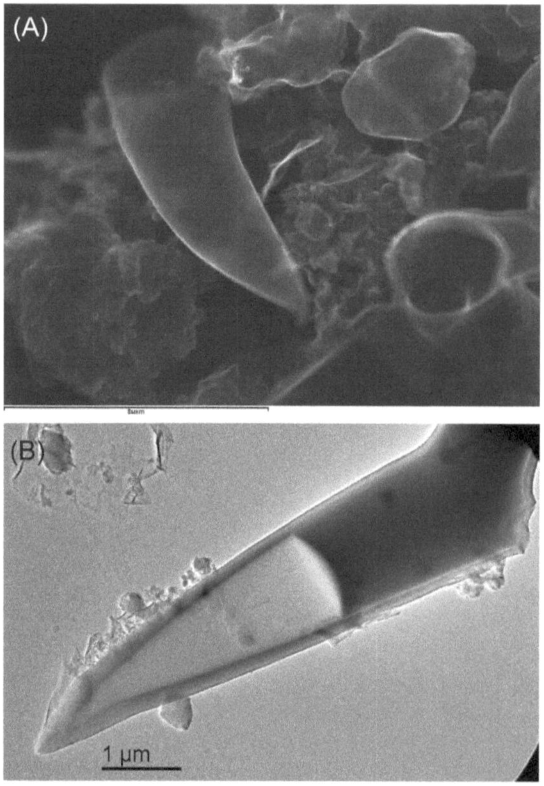

Fig. 3 (A, B) TEM image of B_xC nanotubule ("horn").

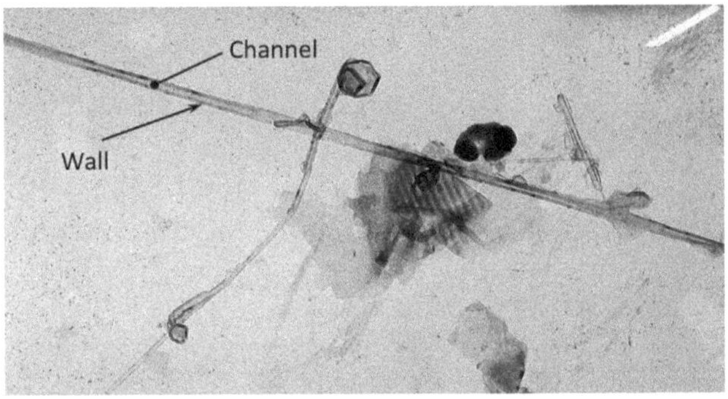

Fig. 4 TEM image of cylindrical boron nitride nanotubes (×40,000).

exothermic effect of the boron nitride formation. This temperature is much higher than that of boron nitride dissociation at atmospheric pressure. These peculiarities provided extreme conditions for nitride synthesis that lead to the evaporation of reagents, gaseous intermediate product formation and to a combustion mechanism that caused transition from a solid-gas to a gas-gas or vapor-vapor and the formation of nonstandard nitride structures. Gas-phase forms of BN were observed in the combustion products: fibers, "cotton," and nanotubes similar to carbon ones. Such BN particles were formed because oxygen-containing, highly volatile reactive boron oxides were present in the boron particles or because of oxygen, which is always present even in highly pure gaseous nitrogen. It is known that at 1500–1600 K, $B_2O_{2(g)}$, $BO_{(g)}$, and $BO_{2(g)}$ vapors already exist in $B + B_2O_3$ mixture. When burned in nitrogen, these products appeared to be sources for BN nanostructure formation, including nanotubes. According to TEM analysis, boron nitride nanotubes were cylindrical in shape. Sometimes "faceted" nanotubes were formed; they may also be products of restructuring (Fig. 4). The diameter of BN nanotubes did not exceed 20–50 nm as a rule.

REFERENCES

[1] Borovinskaya IP, Barinova TV, Vershinnirov VI, Ignatieva TI. SHS of ultrafine and nanosized refractory powders: an autoreview. Int J SHS 2010;19(2):114–9.
[2] Alekseev NI, Izotova SG, Osipov YuG, Polovtsev SV, Semenov KN, Sirotkin AK, et al. Obtaining of carbon nanotubes in reactions of self-propagating high-temperature synthesis. Zhurnal tekhnicheskoi fiziki 2006;76(2):84–9.
[3] Wang J, Gu Y, Zhang L, Zhao G, Zhang Zh. Synthesis of boron nitride nanotubes by self-propagation high-temperature synthesis and annealing method. J Nanomater 2010;2010:6.
[4] Borovinskaya IP, Akopdzhanyan TG, Chemagina EA. Formation of nanostructured particles of boron, aluminum and silicon nitrides at thermo-coupled nitriding reactions at self-propagating high-temperature synthesis. Int J SHS 2016;25(2):119–24.

Tungsten Carbide

Inna P. Borovinskaya, Vladimir I. Vershinnikov, Tatyana I. Ignatieva

Tungsten carbide is characterized by a high melting point, essential hardness, low friction coefficient, chemical stability, oxidation resistance, and excellent electric conductivity.

There are different methods for obtaining tungsten carbide powders. They can be obtained by direct carbonization of tungsten powder [1], synthesis at low temperatures [2], and thermochemical synthesis of nanophased WC powders [3].

The scientists and researchers of the Institute of Structural Macrokinetics and Materials Science of RAS developed the method for producing micron, submicron, and nanosize WC powders by SHS. The synthesis is carried out at the combustion mode with the green mixture corresponding to Eqs. (1)–(3)

$$WO_3 + Mg + C + R \rightarrow WC \cdot MgO \cdot Mg + R + Q, \tag{1}$$

$$W + C = WC \text{ does not occur at the SHS mode} \tag{2}$$

$$Mg(ClO_4)_2 + 8Mg = 8MgO + MgCl_2 + Q \;\; T_{ad} = 3771 \text{ K} \tag{3}$$

where R is a regulating additive.

Different regulating additives—sodium chloride, magnesium oxide, and a $WC \cdot MgO$ semiproduct—are used to suppress the WC powder recrystallization.

The final product is separated by chemical dispergation of the milled cake in solutions of sulfuric (1:4) and hydrochloric (1:1) acids:

$$WC \cdot MgO + H_2SO_4 \rightarrow WC + MgSO_4 + H_2O \tag{4}$$

$$WC \cdot MgO + 2HCl \rightarrow WC + MgCl_2 + H_2O. \tag{5}$$

Treatment of WC powder with sulfochromic acid at $T \leq 100°C$ allows extracting free carbon without dissolving the main product. Tungsten carbides are easily dissolved in dilute alkali solutions. After refinement, tungsten carbide contained free carbon $C_{free} \leq 0.02\%$ and oxygen $O \leq 0.03\%$.

Fig. 1 demonstrates a microphoto and particle size distribution of tungsten carbide synthesized by reaction 1.

It is well known that tungsten carbide can't be obtained from elements by reaction (2) under the SHS mode because the reaction heat is insufficient for the process. That's why the green mixture of reaction (3) and (1) was added to the green mixture of reaction (2).

Concise Encyclopedia of Self-Propagating High-Temperature Synthesis
http://dx.doi.org/10.1016/B978-0-12-804173-4.00162-9

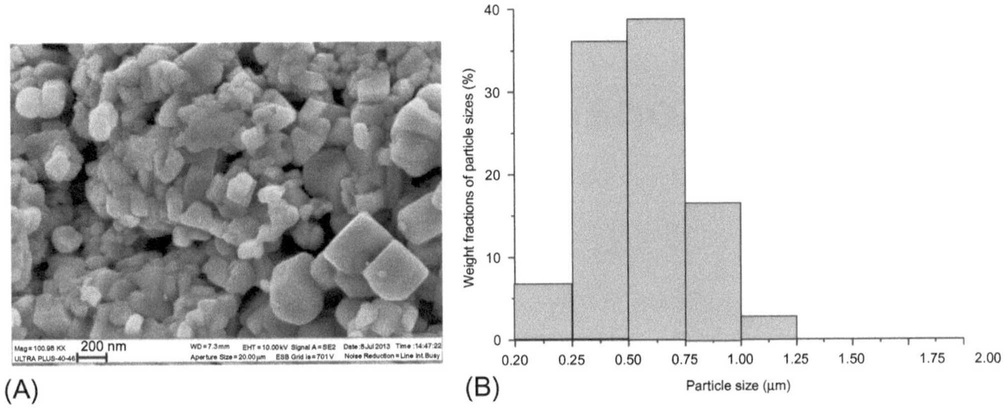

Fig. 1 SEM image (A) and particle size distribution (B) of refined WC.

When the green mixture of Eq. (2) was used with the 10% addition of that of Eq. (3), tungsten carbide with an insignificant amount of W_2C was obtained after combustion, acid enrichment, and refinement. Single-phase tungsten carbide by reaction (2) was obtained in the case of 50% addition of the green mixture of Eq. (1) to that of Eq. (2). Depending on the particle size of tungsten powder, the particle size of tungsten carbide ranged from 200 nm to 30 μm.

REFERENCES

[1] Brookes KJ. World directory and handbook of hard metals and hard materials. Hertfordshire, UK: Int. Carbide Data; 1992. p. 88.

[2] Medeiros F, De Oliveira S, De Souza C, Da Silva A, Gomes U, De Souza J. Synthesis of WC through gas-solid reaction at low temperature. Mater Sci Eng 2001;315(1–2):58–62.

[3] Gao L, Kear B. Low temperature carburization of high surface area tungsten powders. Nanostruct Mater 1995;5:555–69.

Two-Dimensional Materials

Khachatur V. Manukyan

Two-dimensional (2D) materials are layered crystalline solids that have strong bonding in one crystal plane while the neighboring atomic planes are held together by much weaker van-der-Waals forces [1]. Graphene (single sheets of graphite) and its structural derivatives (bi- or few-layered graphene, graphene oxide (GO)) are typical 2D materials. These crystalline solids have attracted great attention because of their electronic, thermal, optical, and mechanical properties. Other 2D-layered materials include hexagonal boron nitride (h-BN), transition metal dichalcogenides (e.g., MoS_2), metal oxides, and hydroxides. Being structurally related to graphene, but having their own distinctive properties, these materials offer the possibility of tuning material characteristics and devices to better fit a particular technology or to be used in combination with graphene.

The combustion-based methods for synthesis of thin 2D crystals do not require an external energy source, as they occur in a self-sustained manner [2–5]. Combustion synthesis of graphene materials involves highly exothermic reactions that form elemental carbon. For example, the self-propagating reaction in SiC + polytetrafluoroethylene was proposed for preparation of graphene sheets [3]. In this process, the polymer decomposes into C_2F_4 gas, which then exothermically reacts with silicon carbide resulting in SiF_4 gas and solid crystalline carbon sheets. High reaction temperatures (above 1000°C) facilitate the formation of crystalline carbon nanostructures. Similar processes were proposed to utilize an exothermic reaction in Si + polytetrafluoroethylene [4] and CuO + Zn + polyvinyl chloride [5] mixtures. These combustion processes enable the production of graphene nanodots or graphene-based porous materials. The exothermic reaction of magnesium with carbon dioxide was also considered for production of few-layer graphene in large quantities [6]. This synthetic process is cost-effective and can be used to produce few-layer graphene in large quantities using nontoxic chemicals and recyclable materials. Self-propagating deoxygenation of GO for preparation of reduced graphene was also reported [7]. Energetic and thermally unstable GO can undergo exothermic disproportionation reactions to produce reduced graphene oxide (r-GO) and carbon dioxide. GO reduction to r-GO does not require toxic reducers (e.g., hydrazine), which are currently used for liquid-phase processing of GO.

Synthesis of thin h-BN crystals by combustion was reported by combustion of $B_2O_3 + Mg + NH_4Cl$ mixtures in a nitrogen atmosphere [8]. Prepared h-BN consists

Concise Encyclopedia of Self-Propagating High-Temperature Synthesis
http://dx.doi.org/10.1016/B978-0-12-804173-4.00164-2

Fig. 1 Transmission electron microscopy image of WO_3 crystals.

of nanoplates with an edge length ranging from 50 nm to several micrometers and a thickness from 10 to 30 nm.

Pure and doped WO_3 nanosheets were prepared using a two-stage solution combustion synthesis method [2]. At the first stage, calcium tungstate ($CaWO_4$) was synthesized by combustion of a solution that contained tungsten peroxo complexes, calcium and ammonium nitrates, and citric acid. At the second stage, $CaWO_4$ powders were subjected to sonication in a diluted HCl solution in the presence of surfactants (sodium dodecylsulfate) followed by dehydrogenation of the resulting tungstic acid at $\sim$250°C. This process enables the preparation of WO_3 nanosheets with a thickness of $\sim$5 nm (Fig. 1).

Ball-milling–induced synthesis of MoS_2 nanosheets was also proposed using the exothermic reaction in a $MoCl_5 + Na_2S$ mixture in an inert gas atmosphere [2]. The characterization of MoS_2 revealed that the thickness of the sheets changes in the 3–5 nm range, which corresponds to 4–7 S-Mo-S layers, and high-resolution transmission electron microscopy indicated the crystalline nature of these sheets.

Thus the combustion of processes offers environmentally friendly routes for the preparation of 2D materials. Those materials show great potential to be integrated into electrochemical energy storage devices (batteries and supercapacitors) and solar water splitting technologies and to be used as hydrogen storage materials.

REFERENCES

[1] Xu M, Liang T, Shi M, Chen H. Chem Rev 2013;113(5):3766–98.
[2] Mukasyan AS, Manukyan KV. Curr Opin Chem Eng 2015;7:16–22.
[3] Manukyan KV, Rouvimov S, Wolf EE, Mukasyan AS. Carbon 2013;62:302–11.
[4] Liu Y, Gao B, Qiao Z, Hu Y, Zheng W, Zhang L, et al. Chem Mater 2015;27:4319–27.
[5] Nersisyan HH, Kim SH, Lee TH, Lee JH. Mater Lett 2015;142:320–3.

[6] Chakrabarti A, Lu J, Skrabutenas JF, Xu T, Xiao Z, Maguire JA, et al. J Mater Chem 2011;21:9491–3.
[7] Kim F, Luo J, Cruz-Silva R, Cote LJ, Sohn K, Huang J. Adv Funct Mater 2010;20(17):2867–73.
[8] Nersisyan HH, Lee T-H, Lee K-H, Jeong S-U, Kang K-S, Bae K-K, et al. J Solid State Chem 2015;225:13–8.

Ultrasonic Field Impact on SHS

Boris B. Khina, Mikhail M. Kulak, Evgeny A. Levashov, Yury M. Maksimov

The application of high-power ultrasonic field (USF) in self-propagating high-temperature synthesis (SHS) is motivated by the idea of exerting an influence on heterogeneous reactions and mass transfer in the SHS wave in order to modify the product structure and composition in situ. Most traditional methods are aimed at controlling only the initial reactive mixture, for example, by changing its composition, particle size, preheating, and changing the reaction mixture microstructure by mechanical activation; that is, they cannot affect the SHS wave after it is initiated.

An experimental setup [1] for SHS in the USF consists of (i) a reaction chamber equipped with a gas supply system and windows for video recording and measuring the combustion temperature, (ii) a USF generator, (iii) a system for feeding USF to a specimen, which includes a transducer, booster, and waveguide, (iv) sensors, (v) a loading system to ensure tight contact of the specimen to the waveguide, and (vi) a control system (Fig. 1).

The impact of USF on SHS depends on the nature of reactants, relative density of the specimen, particle size, and experimental conditions. On the macrokinetic level, this impact is ascribed to the enhancement of capillary spreading of a transient melt by ultrasonic oscillations and also to a change of nucleation conditions during crystallization of the product phase.

In one of the first experimental works [2], the combustion velocity in the Ti-2B system at relative density $\rho_{rel} = 0.6$ was found to substantially increase under the action of USF for coarse Ti particles (about 1100 μm), where capillary spreading is essential, whereas for smaller particles (100 μm), it remained almost unchanged. The latter is connected with the diffusion regime of interaction which is weakly affected by USF. At a constant particle size and relative density, increasing the oscillation amplitude causes the SHS velocity either to decrease monotonously until a constant value is attained (systems Ti-0.6C, Ti-1.0C, (Ti-1.0C) + 20% TiC in Ref. [3], and (Ti-1.0C) + 15% (Ni + Mo) in Ref. [1]) or to change nonmonotonously: first decrease and then increase somewhat (systems (Ti-1.0C) + 20% Ni or 20% (Ni + Mo), (Ti-1.0C) + 50% Ni) [4]. Application of USF also changes the thermal structure of SHS waves [3] and decreases the combustion temperature [1,3]. Evaluation of the heat transfer coefficient, α, on the surface of a pellet permitted segregation of the thermal effect of US on SHS from

Concise Encyclopedia of Self-Propagating High-Temperature Synthesis
http://dx.doi.org/10.1016/B978-0-12-804173-4.00165-4

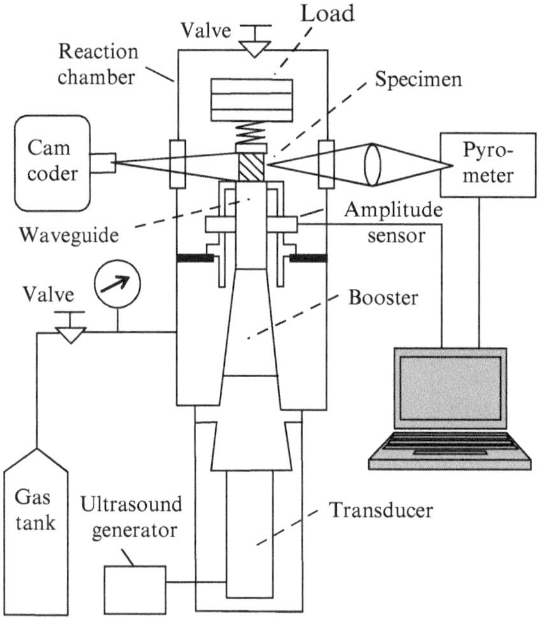

Fig. 1 Layout of an experimental setup.

the nonthermal, or physical, effect [1]. The former is connected with an increase of α because of forced convection of ambient gas around oscillating specimen, which is to a certain extent responsible for the decrease in the combustion temperature and velocity.

Theoretical description of the physical effects of USF on SHS in reactive systems diluted with an inert metal (e.g., Ni) is based on the theory of capillary spreading and accounts for the specimen porosity, particle sizes, and a characteristic spreading distance of two melting metals [5]. It has been shown that the effect of USF on the SHS wave velocity and completeness of conversion is connected with competition between the spreading rates of the reactive and inert metals. In particular, intensification of capillary flow for the inert decreases the combustion velocity.

Metallographic study [1] has revealed the influence of USF on microstructure of SHS products in the system (Ti-1.0C) + 15% (Ni + Mo). Increasing the amplitude not only decreases the average size of TiC particles but also changes their size distribution, which confirms the effect of USF on melt spreading and crystallization in the SHS wave. An optimal amplitude has been found to exist at which the uniformity of grain size is substantially improved. Thus optimal regimes of USF-affected SHS permit the production of cermets with uniform fine-grained structures that are favorable, for example, for grinding tools and for materials with poor machinability.

REFERENCES

[1] Khina BB, Kulak MM. J Alloys Compd 2013;578:595–601.
[2] Kirdyashkin AI, Maksimov YuM, Merzhanov AG. Combust Explos Shock Waves 1981;17(6):591–5.
[3] Piskovskiy SV, Seleznev VA, Mal'tsev VM. Him Fiz 1990;3(6):830–4.
[4] Andreev VA, Levashov EA, Mal'tsev VM, Khavskii NN. Combust Explos Shock Waves 1987;23(6):725–9.
[5] Andreev VA, Levashov EA, Mal'tsev VM, Khavskii NN. Combust Explos Shock Waves 1988,;24(2):73–7.

Unstable Combustion Mode

Yury M. Maksimov, Oleg V. Lapshin

A very interesting phenomena was discovered and described in the studies of SHS processes during unstable propagation of the exothermic reaction front (self-oscillating and spin combustion modes of condensed systems) [1]. Pulsating combustion modes were observed mainly in gas-free systems, and spin combustion modes were observed in hybrid systems [2,3]. At the same time, theoretical [4] and experimental [5,6] studies have shown that both of these modes can be observed during combustion of the same system. These modes can be realized through significant dilution of the initial gas-free mixture by a melting additive.

Several modes of unstable combustion are observed during dilution and can be divided into four basic types.

1. Self-oscillating combustion. In this case, there are two characteristic types of front perturbations: (1) The combustion front executes vibrations in the vertical direction without visible lateral perturbations. The samples are cracked and form the cakes with an equal diameter and thickness; (2) There is pulsation in some of the front sites not in the entire front. In this case, only partial discontinuity of the product is observed on the side surface of burned samples.
2. Transition from self-oscillating combustion to spin combustion. For most systems diluted with a final hard-melting product, the self-oscillating mode is limited, and the combustion process stops with further dilution of the system. For titanium-boron, titanium-carbon, and hafnium-boron systems, there is a change in the mode of combustion caused by dilution with a low-melting product. Here the development of two-dimensional instability on the surface of the sample causes the occurrence of a luminous point that can disappear and then reappear moving freely in a circle and changing the direction of motion to the opposite motion.
3. Stable spin combustion. A visual picture of spin gas-free system combustion does not differ from spin combustion of the metal-nitrogen system. As is the case in Ref. [2], a luminous spot occurs on the lateral surface and retains the bright trace moving along a spiral. However, in contrast to Ref. [2], a helix line is observed on the burned samples with a constant step of the helix (trace of chemical reactions) (Fig. 1).

   In addition, in this case, the reaction takes place, in contrast to Ref. [2], not only on the surface but also in the sample volume. The light and more melted lines are the traces of passing the chemical reaction; the dark lines are the boundaries between them. X-ray phase analysis of light and dark areas did not find any differences in the phase composition.

Concise Encyclopedia of Self-Propagating High-Temperature Synthesis
http://dx.doi.org/10.1016/B978-0-12-804173-4.00143-5

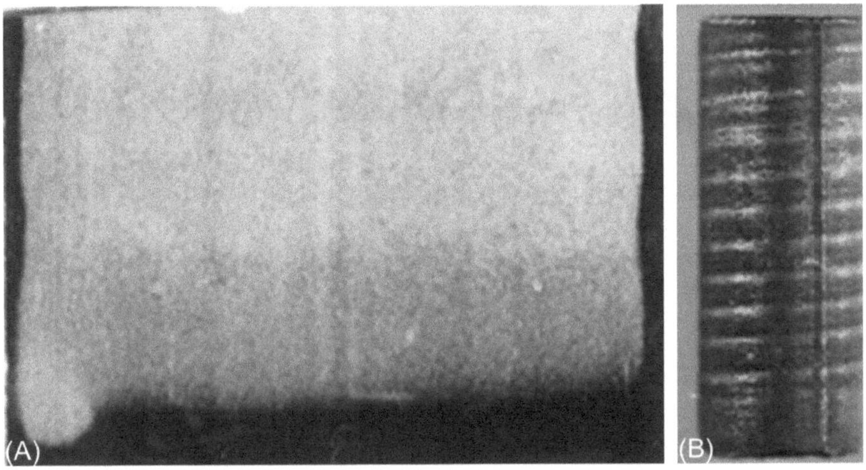

Fig. 1 (A) Image of spin combustion for the Ti-B-αCu system and (B) the external view of the burned cylinder.

4. The maximum combustion mode. The spin mode is not the limit of unstable combustion. A qualitatively different mode takes place after further dilution of the mixture with a low-melting point component. The phenomenon is characterized by the fact that the luminous spot appears on the lateral surface of the sample, spreads over a circle, and forms two points moving toward each other. The points join together to form a bright ring. Then there is the heating of the next layer, the appearance of a light spot, and the picture is repeated. The observed spin and self-oscillating combustion can be explained by the concepts developed in Refs. [4,7] and can be considered as a consequence of one-dimensional instability of the combustion front for longitudinal perturbations.

In this case, the oscillating mode characterized by pulsations or alternations of flashes and depressions is the result of two main factors: the presence of heat excess in the combustion front and a heat flow of the reaction products to the conversion zone.

REFERENCES

[1] Barzykin VV, Merzhanov AG. Unstable combustion in heterogeneous systems with condensed reaction products—a review. Int J SHS 1997;6(4):377–98.
[2] Merzhanov AG, Filonenko AK, Borovinskaya IP. New phenomena during combustion in condensed systems. Proc USSR Acad Sci 1973;208(4):892–4.
[3] Filonenko AK, Vershinnikov VI. Regularities for spin combustion of titanium in nitrogen. Combust Explos Shock Waves 1975;11(3):353–62.
[4] Ivleva TP, Merzhanov AG, Shkadinsky KG. Regularities of the spin mode during propagation of the combustion front. Combust Explos Shock Waves 1980;16(2):3–10.
[5] Maximov YuM, Pak AT, et al. Spin combustion of gas-free systems. Combust Explos Shock Waves 1979;15(3):156–9.
[6] Merzhanov AG, Pak AT, Kuchkin MN. Unstable combustion modes of gas-free systems. Combust Explos Shock Waves 1981;17(4):51–8.
[7] Shkadinsky KG, Chaikin BI, Merzhanov AG. Pulsing front propagation of the exothermic reaction in the condensed phase. Combust Explos Shock Waves 1971;7(1):19–28.

Unsteady Regimes of Combustion

Konstantin G. Shkadinsky

On a macroscopic scale, SHS can be carried out in the form of a stationary autowave of exothermic chemical reaction [1]. The temperature of the medium T depends on the coordinate (x) and time (t) according to the law $T(x,t) = T(x - ut)$, where u is the constant velocity of the combustion front propagation. Thus the structure of the wave front in the coordinate system moving with the front becomes stationary. If the front is stable, small disturbances smooth out, and the front "forgets" their impact, which can be considered to be in a steady-state regime (intermediate asymptotic: relatively long period of stationary propagation between ignition and degradation of the reaction wave [2]).

In other cases, the front of the exothermic chemical reaction spreads, if it exists, in the transient mode. Nonstationary combustion modes can be conventionally divided into two groups. The nonstationary combustion relaxation group includes nonstationary processes, where unsteadiness may be due to heterogeneity in the original environment including nonuniform initial temperature, the process of initiating combustion (ignition), time-dependent influences from the environment, and the geometry of the sample. In the absence of these factors, the process relaxes to a stable stationary regime of burning.

Another group of nonstationary combustion processes is due to instability in the stationary structure of the combustion front. In this case, extinction of the front or switching to another mode of combustion may occur, for example, switching to nonstationary periodic regimes. In this case, the spatial structure of the front changes periodically in time. These regimes of combustion include pulsed regimes, spin regimes, and gasless burning. Pulsed cellular combustion is observed in filtration combustion of porous compositions. Periodic modes can also lose their stability and transform into new stable periodic regimes with a substantially different structure of the front.

Stable modes of combustion may also lose stability and transition to unsteady chaotic regimes, such as the focal mode for gasless combustion [3] or the fingering mode [4] for filtration combustion. The chaotic regimes also include the cellular combustion mode. Traces of the unstable regimes usually remain in the solid products of SHS.

The classification of combustion modes is conditional. Even a stationary, stable combustion regime is formulated under conditions that permit us to neglect processes of chemical interactions in the initial medium ahead of the front when compared with the rate of chemical interaction in the front. In the filtration combustion front, structure

Concise Encyclopedia of Self-Propagating High-Temperature Synthesis
http://dx.doi.org/10.1016/B978-0-12-804173-4.00167-8

of the front and velocity of combustion strongly depend on the distance from the outer surface of the sample, which varies during combustion. However, this fact does not prevent us from analyzing the front in a quasistationary approximation.

REFERENCES

[1] Merzhanov AG. Self-propagating high-temperature synthesis: twenty years of search and findings. In: Munir ZA, Holt JB, editors. Combustion and plasma synthesis of high-temperature materials. New York, NY: VCH Publishers; 1990. p. 1–53.
[2] Barenblatt GI, Zeldovich YaB. Self-similar solutions as intermediate asymptotics. Ann Rev Fluid Mech 1972;4:285–312.
[3] Strunina AG, Dvoryankin AV, Merzhanov AG. Non-stable regimes of thermite systems combustion. Fiz Goren Vzryva 1983;19(2):30–6.
[4] Kagan L, Sivashinsky G. Pattern formation in flame spread over thin solid fuels. Combust Theory Model 2008;12(2):269–81.

Vapor-Liquid-Solid Mechanism of Structure Formation

Alexander S. Mukasyan

There are two primary mechanisms of crystal growth that produce whisker or filament morphology from partially or completely gasified solid reagents [1]. The first mechanism, vapor-crystal (VC), is based on the phenomenon of gas-phase reactions on top of a screw dislocation that, while continuously relocating to the growing solid-phase surface, forms a self-perpetuating step [2]. The second mechanism, vapor-liquid-solid (VLS), which was proposed in 1964 [3], has been used widely for explaining the growth of columnar crystals of silicon in the process of chemical vapor deposition in the presence of liquid gold drops on the substrate. The scheme of the VLS mechanism is shown in Fig. 1.

At a temperature below the melting point of silicon but above the melting point of Au-Si eutectics, crystal growth is catalyzed by liquid droplets of the gold-silicon eutectic. The reduction reaction of $SiCl_4$ by hydrogen takes place on the surface of the molten cap after which the silicon atoms diffuse through the liquid and crystallize on the solid-phase surface lifting the cap from this surface and thereby facilitating continuous whisker growth by the VLS mechanism.

Generally speaking, a drop of the catalyzed melt should wet the solid-phase surface, and therefore the contact angle (β_0) must be less than 90°C (Fig. 1). The shape of the cap is determined by the balance of surface forces, surface tension at the liquid-gas (VL) and liquid-solid (LS) and gas-solid (VS) interfaces [4].

The kinetics of growth for these crystals is discussed in detail in the classic work of Givargizov [5]. Generally, the driving force behind growth is the supersaturation of the solution. In other words, there is a gradient chemical potential along the liquid drop associated with different equilibrium solubility values of the element at the convex surface at the gas-solid interface and the flat surface at the liquid-solid interface. According to the Gibbs-Thompson law, this difference in the chemical potentials of silicon in the gas phase and the crystal can be summarized as

Concise Encyclopedia of Self-Propagating High-Temperature Synthesis
http://dx.doi.org/10.1016/B978-0-12-804173-4.00168-X

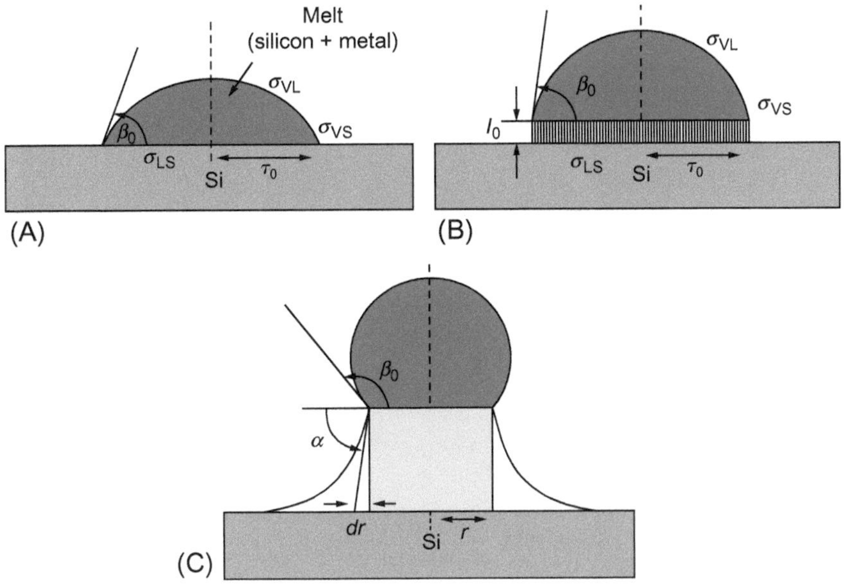

Fig. 1 Scheme of the initial stage of crystal growth of a columnar crystal via the VLS mechanism.

$$\Delta\mu = \Delta\mu_0 - \frac{4\Omega\alpha}{d} \quad (1)$$

where Ω is the atomic volume of silicon, α is the surface tension energy of the drop, and d is the drop diameter. This relation shows that the smaller the drop diameter, the smaller the driving force of the process. Apparently, there is a critical diameter at which crystal growth under equilibrium conditions is impossible. In many cases, the crystal growth rate (V) obeys the power law

$$V = b\left(\frac{\Delta\mu}{kT}\right)^n \quad (2)$$

where b is the coefficient of proportionality (independent of supersaturation), which can be determined experimentally.

Many SHS powders can form through the VC and VLC mechanisms. Moreover, the mechanisms can be implemented in the systems both with and without the addition of gasifying agents [6–10]. Microstructures of SHS whiskers, i.e., SiAlON, Si_3N_4, and SiC, produced by VLC mechanisms are presented in Fig. 2.

Fig. 2 Microstructure of nanopowders produced by different SHS routes: (A) SiAlON: salt-assisted SHS [8]; (B) Si_3N_4: "catalytic" SHS [9]; (C) SiC: PTEF SHS [10].

REFERENCES

[1] Andrievskii RA, Spivak II. Silicon nitride and related materials. Moscow: Metallurgiya; 1984 [Russia].
[2] Vainshtein VK. Modern crystallography. Moscow: Nauka; 1980 p. 271–8 [Russia].
[3] Wagner RS, Ellis WC. Appl Phys Lett 1964;4(5):89–90.
[4] Rogachev AS, Mukasyan AS. Combustion for materials synthesis. New York, NY: CRC Press/Taylor & Francis; 2015 p. 44–204.
[5] Givargizov EI. Fundamental aspects of VLS growth. J Cryst Growth 1975;31:20–30.
[6] Mukasyan AS. Proc Combust Inst 2005;30:2529–35.
[7] Kharatyan SL, Nersisyan HH. Int J SHS 1993;3(1):17–25.
[8] Yeh CL, Wu FS, Chen YL. J Alloys Comp 2014;604:260–5.
[9] Xia M, Ge C, Guo H. Adv Eng Mat 2012;14:166–9.
[10] Dabrowska A, Husczko A, Soszynski M, et al. Phys Status Solid B 2012;248:2704–7.

X-Ray Generation by SHS

Vitaliy G. Salamatov, Alexander I. Kirdyashkin, Yury M. Maksimov, Viktor Tarasenko

High-energy X-rays can be generated during SHS (self-propagating high-temperature synthesis) in the Ti-B system [1,2]. It should be noted that information on the generation of X-rays during any combustion process is still lacking. The presence of high-energy X-ray emission was detected using an ionizing X-ray dosimeter (Arrow Tech, Inc., Model 138) according to the accumulated exposure dose and the data of photographic photometry on a Kodak RAR 2497 film. To eliminate the influence of the optical radiation spectrum and thermal exposure of combustion products, the detectors were placed behind the protective screens such as metal foils and black paper. For the experiments, we used two types of reaction samples:

One type cf. [1] (Fig. 1): Layer of nanopowder mixture of Ti + 31.5 wt.% B supported on the surface of the metal foil (15 μm thick aluminum). The thickness of the Ti-B layer was 70 μm. The particle size was 500 nm for B and 100 nm for Ti. The reaction was initiated using an electrical coil in the form of a combustion wave propagating through the layer (Fig. 1A).

Second type cf. [2]: Cylindrical briquette of the powder mixture Ti + 31.5 wt.% B. The particle size was 500 nm for B and 30 μm for Ti. The weight of the briquette was 0.5 g. In this type, SHS was conducted in the thermal explosion mode. The process started after slow heating of the sample to a temperature between 1100 and 1200 K.

Taking into account the known thermodynamic data, the following schemes of chemical transformations can be proposed:

$$\mathrm{Ti} + 2\mathrm{B} = \mathrm{TiB_2} + 293\,\mathrm{kJ/mol}; \quad \mathrm{Ti} + \mathrm{B} = \mathrm{TiB} + \geq 122\,\mathrm{kJ/mol}.$$

According to the video data, the combustion front (1) propagated in the 70 μm Ti-B layer (3) with an average linear velocity $V_c \approx 20$ cm/s. The observation showed that the local reaction spots were emerging stochastically in the combustion wave (Fig. 1A). The size of the reaction spots (2) was up to $d_c \approx 1.2$ mm. The maximal temperature in the center of the reaction spots was reached during an incubation period that lasted from 2 to 5 ms. The process was accompanied by dispersion of the condensed reaction products because of gas generation in the combustion wave, which produced gas dynamic flows. This was the result of partial evaporation of the main components and volatile impurities in the system. During SHS, the exposure spots (4) (size up to $d_r \approx 1.5$ mm) appeared on the film (Fig. 1B) placed behind a supporting aluminum foil. The spots indicate the existence of

Concise Encyclopedia of Self-Propagating High-Temperature Synthesis
http://dx.doi.org/10.1016/B978-0-12-804173-4.00169-1

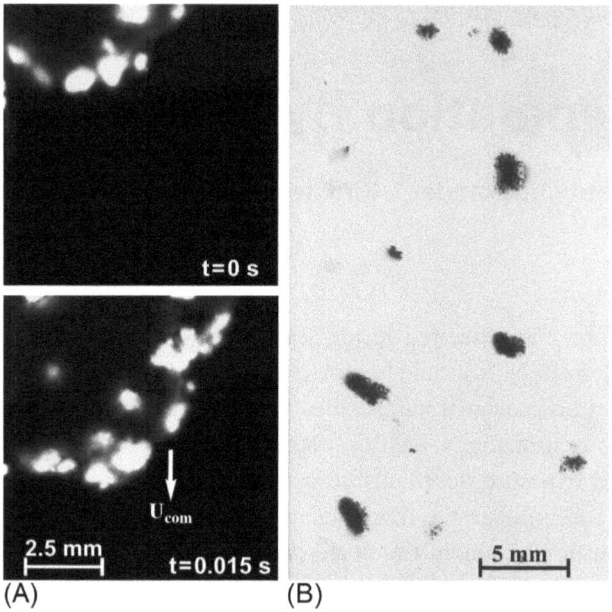

Fig. 1 (A) Video recording frames of the reaction wave in type 1 samples; (B) light spots on the photographic film Kodak RAR-2497 wrapped with the protective shields (titanium foils were 20 μm, and the black paper was 120 μm).

X-ray emission. Considering the closeness of the quantities d_s and d_r and their locations, the main sources of X-ray emission, apparently, were the reaction centers. The radiographic density of exposition in the light spots (4) was highly heterogeneous, which can be explained by the heterogeneity of the reaction rate in the reaction centers.

During thermal explosion of the powder mixture Ti + 31.5 wt.%, the sample dispersion that occurred led to the formation of an expanding cloud of evaporated material. The cloud contained the reacting particles of the mixture with sizes of about 50–300 μm and the ionized gas. In the process of particle cloud expansion, the particles came into contact with the outer surface of the shield.

The X-ray film was exposed in the form of points (diameter ~100–800 μm; Fig. 2A) and a finger (the transverse dimension of ~200–400 μm; Fig. 2B) during thermal explosion of the sample. Exposure of the Kodak film gave evidence that favored the existence of X-rays. Taking into account that the particle size in the cloud is close to the diameter of spots and fingers, we can assume that the main sources of the X-rays were the reacting particles. The images of fingers that appeared on the X-ray film were probably the result of particle propagation along the surface of the shield.

The total exposure dose of radiation measured by the dosimeter reached 10 mR in a series of 24 explosions. The radiation doses in individual explosions differed substantially, which means that maximal radiation doses can reach 2.5 mR during a single explosion.

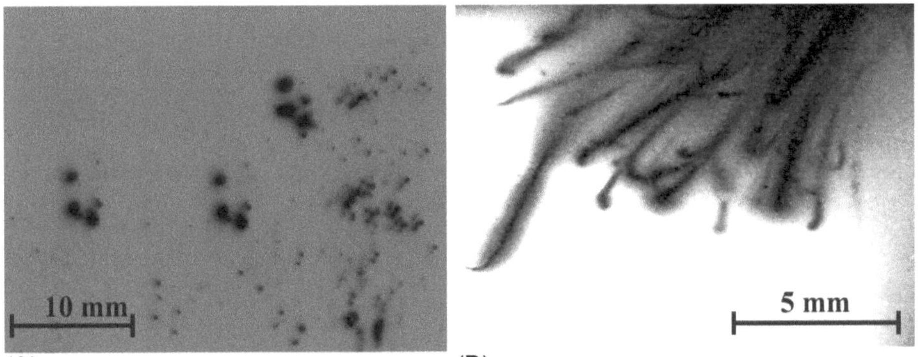

(A) (B)

Fig. 2 Exposure spots (A) and fingers (B) on the Kodak film that appeared during thermal explosion powder mixture Ti + 31.5 wt.% B. The film was wrapped with protective shields (titanium foils 20 μm + black paper 120 μm).

Analysis of the known characteristics of X-ray passage [3] through the applied protective screens (metal foils, black paper) shows that the energy of X-ray photons emitted during the combustion synthesis and forced reaction in the Ti-B, Ti-Si systems was not <2000–3000 eV. The origin of the X-ray emission may be as follows:

1. The formation of charges in the reaction zone because of mechanical micro-separation of the charges can lead to electrical breakdown into gas gaps. This can occur during cracking of the solid dielectric, or, for example, generation of X-ray radiation occurs during the tearing of a sticky tape cf. [4,5]. The generation of X-ray radiation can also occur as a result of charge separation in the gas flow with particles (e.g., X-ray radiation is observed in the moments preceding an electrical discharge in thunderclouds [6]).
2. X-ray radiation can be the result of collision of intensive microwaves that are generated during phase transition of the medium. The generation of X-ray radiation occurs, for example, during cavitation of liquid media [7].

The first mechanism is consistent with the existence of short electric pulses of 0.1 μs during thermal explosion of powder mixture Ti + 31.5 wt.% [8]. These pulses can be the result of micro-discharges within the cloud suspension. The second mechanism is consistent with the existence of a powerful acoustic wave that was detected during combustion synthesis [9].

Further studies will shed light on the origin of the X-ray emission mechanism.

REFERENCES

[1] Tarasenko VF, Maksimov YuM, Kirdyashkin AI, Salamatov VG, Sosnin EA, Gabbasov RM. Int J Eng Sci Invent 2014;10(3):6–12.
[2] Kirdyashkin AI, Salamatov VG, Maksimov YuM, Sosnin EA, Tarasenko VF, Gabbasov RM. Combust Explos Shock Waves 2008;44(6):729–31.

[3] Henke BL, Gullikson EM, Davis JC. At Data Nucl Data Tables 1993;54:181–342.
[4] Karasev VV, Krotova NA, Deryagin BW. Dokl Akad Nauk 1953;88:777–80 [in Russian].
[5] Camara G, Escobar V, Jonathan R, Putterman J. Nature 2008;455:1089–92.
[6] Rakov VA, Uman MA, Rambo KJ. Atmos Res 2005;76:503–17.
[7] Kornilova AA, Vysotskii VI, Sysoev NN, Litvin NK, Tomak VI, Barzov AA. Mosc Univ Phys Bull 2010;65(1):46–50.
[8] Kirdyashkin AI, Salamatov VG, Maksimov YuM, Sosnin EA, Tarasenko VF, Gabbasov RM. J Chem Chem Eng 2012;6:292–8.
[9] Kirdyashkin AI, Gabbasov RM, Maksimov YuM, Salamatov VG. Combust Explos Shock Waves 2013;49:676–81.

INDEX

Note: Page numbers followed by *f* indicate figures, *t* indicate tables, and *ge* indicate glossary terms.

D

N

Kolibri 360 GmbH
Ferdinand-Jühlke-Straße 7
99095 Erfurt, Deutschland
produktsicherheit@kolibri360.de